CIGARETTES, INC.

CIGARETTES, INC.

An Intimate History of Corporate Imperialism

NAN ENSTAD

The University of Chicago Press
Chicago and London

The University of Chicago Press, Chicago 60637
The University of Chicago Press, Ltd., London
© 2018 by The University of Chicago
All rights reserved. No part of this book may be used or reproduced in any manner whatsoever without written permission, except in the case of brief quotations in critical articles and reviews. For more information, contact the University of Chicago Press, 1427 E. 60th St., Chicago, IL 60637.
Published 2018
Printed and bound by CPI Group (UK) Ltd, Croydon, CR0 4YY

27 26 25 24 23 22 21 20 19 18   1 2 3 4 5

ISBN-13: 978-0-226-53328-5 (cloth)
ISBN-13: 978-0-226-53331-5 (paper)
ISBN-13: 978-0-226-53345-2 (e-book)
DOI: https://doi.org/10.7208/chicago/9780226533452.001.0001

Library of Congress Cataloging-in-Publication Data
Names: Enstad, Nan, author.
Title: Cigarettes, inc. : an intimate history of corporate imperialism / Nan Enstad.
Description: Chicago ; London : The University of Chicago Press, 2018. | Includes bibliographical references and index.
Identifiers: LCCN 2018008652 | ISBN 9780226533285 (cloth : alk. paper) | ISBN 9780226533315 (pbk. : alk. paper) | ISBN 9780226533452 (e-book)
Subjects: LCSH: Cigarette industry — United States — History. | Cigarette industry — China — History. | British American Tobacco Company — History. | British-American Tobacco Co. (China) — History.
Classification: LCC HD9149.C5 U5535 2018 | DDC 338.7/679730973 — dc23
LC record available at https://lccn.loc.gov/2018008652

♾ This paper meets the requirements of ANSI/NISO Z39.48-1992 (Permanence of Paper).

For Finn, Always

CONTENTS

Preface: Who Counts in the Corporation?   *ix*

	Introduction	*1*
1	The Bright Leaf Cigarette in the Age of Empire	*16*
2	Corporate Enchantment	*52*
3	The Bright Leaf Tobacco Network	*86*
4	Making a Transnational Cigarette Factory Labor Force	*120*
5	Of Camels and Ruby Queens	*154*
6	The Intimate Dance of Jazz and Cigarettes	*187*
7	Where the Races Meet	*221*
	Conclusion: Called to Account	*260*

Acknowledgments   *269*
Notes   *273*
Index   *325*

PREFACE

Who Counts in the Corporation?

I wrote this book out of my conviction that our collective conversation about the business corporation is strangely impoverished and that this impoverishment flows from the structure of the corporation itself. The problem can be summed up as, "who counts?"[1] For at least a hundred years, commentators have pointed out that the organizational structure of the business corporation creates borders of belonging that confine corporate membership to stockholders and the board of directors. This structure means that corporations' sole, immutable charge is to make money for its stockholders, leaving the concerns of workers, consumers, and the environment off the corporate balance sheet—except as "costs." Stockholders count; stakeholders, as they have come to be called, do not count unless some kind of reform measure is taken to make them count. The organization of unions and various kinds of regulatory bodies make "other" interests matter, but these remain external to the corporation. In fact, such interests are commonly called "externals." Purist ideas of shareholder profit maximization repeatedly wax, as they have in recent years, driving down regulatory and union power. No matter how indispensable workers, including managers, are to the corporation, no matter that they are the ones that actually create goods and profits, they are literally not members unless they buy stock.[2]

Who counts in the corporation has also shaped who counts in corporate history. While there is no shortage of critical perspectives on corporate power, a wide divide opened decades ago between labor and business history that reflected labor's position as an outsider. This divide became the template for more of the same. Social history boomed as the baby boomers came of age, but their magnificent histories of people of color, women, immigrants, LGBT people, workers, and consumers are not typically categorized with histories of corporations, even when economics plays a central role in the story. These myriad new histories prompted historians to investigate the shifting borders of national belonging, literal and ideological, but the borders of the corporation seemed static. Historians now agree that the nation is more than presidents and politicians, but it remains difficult to see the corporation as more than businessmen and bureaucrats.

This book tells a new story of corporate empowerment rooted in the simple premise that the multinational business corporation, specifically in the cigarette industry, is made by people acting within culture. I began my research by investigating a group of several hundred white men from the segregated US south who traveled to China to build the cigarette industry for the British American Tobacco Company (BAT)-China between 1905 and 1937. Some of these men became executives in this massive endeavor, but most were entry-level workers from rural backgrounds, away from home for the first time. From their experiences, I followed the outgoing and incoming streams of connection and influence in the US and China to build a new picture of the rise of the cigarette in both countries. Readers will encounter herein the lives—not just business decisions—of entrepreneurs, managers, distributors, factory workers, farmers, servants, wives, sex workers, advertisers, jazz musicians, and consumers—a multinational group from the US, China, and Britain.[3] It is here that the corporation was made.

As I formed this project, I took inspiration from historical debates about corporate power that approached the boundaries of corporate belonging in ways not seen today. In the 1950s, many commentators attempted to come up with a way to arrange corporate membership and governance that would enfranchise more stakeholders. Lawyer Abram Chayes suggested that "a more spacious conception of [corporate] 'membership,' and one closer to the facts of corporate life, would include all those having a relation of sufficient intimacy with the corporation or subject to its power in a sufficiently specialized way. [They rightfully would have a] share in decisions on the exercise of corporate power."[4] Chayes reminds us that the corporation is not just a legal and financial but also a social organization that employs and affects a great many people. It is beyond my power to revise corporate governance procedures, but a comparatively simple matter to change the boundaries of history books. Chayes's conception of corporate membership generates a narrative frame that hews closer to "the facts of corporate life," based on the emerging webs of social and economic connection rather than formal corporate boundaries. It also denaturalizes the corporation and makes visible the ways the more narrow legal definition of the corporation affected the unfolding of history as well as our collective imagination about economic possibility. Chayes's intriguing phrase, "sufficient intimacy," helped me see the corporation as an intimate form, teeming with life, even if its product was an endangerment to health, as in the cigarette industry.

Which brings me to another defining point: nearly everyone in this book—American, Chinese, and British—smoked cigarettes. Readers should imagine almost every space that appears in this book as smoke filled, whether in the US, China, or Britain. Corporate employees smoked in corporate boardrooms, divisional offices, factory bathrooms, gentlemen's clubs, shops, restaurants, and cabarets, and at home. As individual items, cigarettes were so cheap and portable that they became a part of the culture of the corporation at every level. Free cigarettes flowed to executives and workers in patterns that created privilege and hierarchy. Ritual smokes opened conversations and sealed deals. When I realized the prevalence of smoking, I had two insights that transformed my approach. First, the unmistakably intimate, ritual nature of cigarette exchange and smoking helped me grasp the fact that every space and transaction in the corporation was one of cultural formation. There was no "economy" separate from "culture"—they were intertwined at every level. I determined to treat them that way.

Second, everyone who smoked, from the captains of industry to factory workers, was subject to cigarettes' dramatic risks to health, and they sickened and died from smoking much as other consumers. This rather obvious point was nonetheless jarring to me because it revealed that the agents of corporate power and those that are subject to that power do not so easily divide into two distinct groups. Even highly placed executives found themselves incorporated by the corporation, subject to its hierarchies, pleasures, and risks. In asking the question "who counts?" in the corporation, my goal was to humanize all corporate actors, even the very privileged ones, while also attending to the ways the corporation generated inequality and suffering both among its employees and in the wider society. While I do not refrain from documenting treachery, and there is plenty of it in the global cigarette industry, readers will not find here a morality tale of villains and victims. Indeed, I explain how, in the case of tobacco, this narrative reflex has sometimes constrained rather than furthered ethical critique of corporate power.

Finally, the reach of this story's significance is manifest in the devastating health crisis that followed close behind the skyrocketing smoking rates of the 1920s, as incidence of heart disease, emphysema, and lung cancer all spiked. In the United States, after decades of corporate obfuscation and court battles, smoking and disease rates have recently declined, but rates continue to increase in places like China, where aggressive cigarette marketing remains unregulated. Currently, 350 million people in China alone smoke cigarettes. This history, originating in a small area of North Carolina and Virginia, came to be a global story of the powers, pleasures, and perils of corporate capitalism.

Though this book charts the ascent of cigarettes, it does not pin the story to the rising scientific evidence of cigarettes' dangers as a medical history might do. Tracking corporate empowerment, this study instead finds a natural ending in the late 1930s, when cigarettes containing bright leaf had captured a mass consumer base in both the US and China as well as across much of the rest of the globe. This is also when the Japanese invasion of China made BAT's work more difficult and just before World War II changed the equation for corporate power in the US. The most dramatic events in the medical history of cigarettes occurred after World War II, when scientific evidence of cigarettes' impact on health became definitive and US cigarette corporations colluded to obfuscate this truth. Fortunately, there are several recent expert books on this post–World War II story.[5] In this book, I focus instead on the cigarette industry's generative role in creating new forms of global corporate power in the Jim Crow South and treaty port China.

One of the great joys of researching and writing this book has been delving into Chinese history. As someone trained as a US historian, I could not have anticipated that my interest in corporate power would take me to China in pursuit of the white Southerners of the British American Tobacco Company. There I found a story far more fascinating than I could have imagined. Though the full acknowledgments for my book can be found near the end, I owe a special debt to three Chinese translators who enabled me to do research at the Shanghai Academy of Social Sciences, which holds the records of BAT-China, and to utilize a range of Chinese sources throughout this book: Guo Jue, Wang Haochen, and Xu Zhanqi. While the story of BAT-China and its Southern connection is virtually unknown in US history, Chinese historians are well familiar with BAT-China because it was one of the largest and most influential foreign companies into the interwar period. Chinese historians have explored the business, labor, agricultural, and advertising history of BAT in considerable detail focusing, of course, on Chinese people primarily; foreigners are critical to their histories but they rarely come with much backstory. I see this book, therefore, as exploring the growth of the cigarette corporation in both the US and China, with a focus on people from the US, but including Chinese and, to a lesser extent, British corporate workers. In this way, the book fits well into the category of "the US in the world" or "transnational US history."

Each chapter of this book takes as its subject a different aspect of corporate empowerment and the rise of the cigarette, and the chapters follow a roughly chronological order. I cannot come close to covering the full lives of the myriad "members" of the corporation in the space I have here, so I have used the tools of cultural history to investigate salient sites and moments when the corporation is made, examining the people who came to the fore at key moments as well as the ways new categories, cultural/economic systems, and values—including market values—became established. The book thus takes the reader from the development of the first bright leaf cigarette market in England, to the legal and financial enhancement of corporate power in the US, to the formation of BAT's agricultural and factory system in China, to the establishment of BAT-China's foreign business culture, to the rise of big brands in both the US and China, to the circulation of jazz and cigarettes, and to the revisions in BAT-China's corporate structure in the interwar period.

Though it all, I attend to the key paradox between value and power in corporate life: creativity, innovation, productivity, and profit emanated from all corners of the transnational corporation, but the emerging relationship between racial segregation, new corporate entitlements and structures, and imperial ambitions repressed this fact, gave undue credit to CEOs and entrepreneurs, and naturalized—even celebrated—undemocratic corporate governing structures. Through this intimate history of corporate empowerment, I invite readers to reconsider who counts in the corporation.

INTRODUCTION

In 1916, Lee Parker left his father's tobacco farm in Ahoskie, North Carolina, for Shanghai, China. He wrote sixty years later in his memoir, "I was fresh from the United States, sent by BAT, the British American Tobacco Company to 'put a cigarette between the lips of every man and woman in China.'" Parker's father had sent him to Wake Forest College in hopes he would, upon graduation, join the small white professional class. But even with a college degree, "jobs was hard to come by for a country fellow," Parker recalled. He had heard that a buyer at the tobacco market in Wilson, North Carolina, hired young men for jobs in China, so he borrowed five dollars from his brother and made the journey to Wilson. After an interview on the tobacco warehouse floor that lasted "between thirty seconds and two minutes,"[1] Parker's life path veered sharply east, and he headed to China to work as a cigarette salesman for one of the world's first multinational corporations.

Parker was one of hundreds of young white men who journeyed from the bright leaf tobacco–growing states of Virginia and North Carolina to work for BAT-China from 1905 to 1937, the very years that cigarette consumption skyrocketed worldwide. Southerners filled positions in every department. Richard Henry Gregory from Granville County, North Carolina, ran the agricultural department, where US Southerners introduced bright leaf tobacco and the flue curing system to Chinese farmers. Ivy Riddick, from Raleigh, North Carolina, managed Shanghai's massive cigarette factories during the 1920s, a decade of dramatic strikes and anti-imperialist protest. James N. Joyner, born in Goldsboro, North Carolina, worked in sales in China from 1912 to 1935, including heading two large sales divisions. And James A. Thomas of Reidsville, North Carolina, was at the helm, steering the China branch of BAT during its years of rapid expansion.[2] For every career man, there were dozens of other, mostly rural white Southerners who worked for BAT-China for one or more four-year terms; hundreds of Southerners went to China over the course of BAT-China's tenure there.

These men were part of a network that selected and sorted white men from the Upper South's bright leaf tobacco region into emerging corporate opportunities in the United States and China. The men who went to China left behind fathers, brothers, and cousins who performed similar work in the cigarette corporations that were emerging simultaneously in the United States. To be part of the transnational network, one had to be white and male and to have some connection to farming, curing, auctioning, or manufacturing bright leaf tobacco. What mattered was not so much *what* one knew, but *who* one knew, as those who had a father or brother in the tobacco trade found entrance even without any particular training or experience. This network, then, functioned as a managerial system, coordinating the hire and placement of white-collar managers in both the US and China; it was as foundational to the transnational tobacco corporate structure as the board of directors, and more significant for shaping daily decision-making and corporate culture.

We have not heard the stories of the men who went to China on the bright leaf network before now partly because, as rural Southerners, they seem unlikely global capitalists. Most had never left their home states before they embarked on the long voyage to China. The majority came from farm backgrounds; Shanghai was by far the most modern and cosmopolitan place they had ever seen. They were not, as a group, highly educated. While some, like Parker, had college degrees from Wake Forest College in Winston-Salem or Trinity College in Durham, others traded schoolbooks for full-time employment while still children, as James A. Thomas did at age ten. They certainly were not cigarette smokers. Especially before the 1920s, cigarettes were for city slickers; Southern men chewed their tobacco. A pipe was relaxing in the evening; a cigar was crucial for business meetings; but cigarettes—not so much. In China, men of the bright leaf network became their own first converts to cigarettes, an easy "sell" because they got the cigarettes for free from the company. For contemporary readers who may be accustomed to seeing rural life, especially in the Jim Crow South, as isolated and parochial, the globe-trotting of the men on the bright leaf network might seem incongruous. And yet despite their geographic and cultural distance from the financial metropole of New York City, these men became the domestic and foreign representatives of one of the world's first multinational corporations, pushing a commodity that became associated with modernity itself.

Only when considering the operations of this network, including its dealings with a host of employees at all levels, can we understand the simultaneous and meteoric rise of cigarette brands containing bright leaf tobacco in the US and China. In the 1920s, Chinese consumers catapulted Ruby Queen cigarettes, made of 100 percent bright leaf, to unprecedented sales; in the US, Camel cigarettes, made of a tobacco blend including bright leaf, soared far beyond its competitors. By capturing these two huge markets, Camels and Ruby Queens became the two most popular cigarette brands in the world, and it is not too much to say that they also changed the world. With their success, cigarette corporations moved to the cutting edge of the science of branding, experimenting with new ways to capture consumers' imaginations and cash. Cigarettes became especially visible parts of urban, "modern" life around the globe while their capacity for personal and shared associations made them potent symbols in diverse disputes. Corporations did not achieve this feat simply by exporting a US cigarette to China. Rather, this story of the cigarette is one of cross-cultural encounter in both countries, as innovation and production occurred on the ground through the daily life of business.

The bright leaf cigarette is a big story in itself but it is also the flash accompanying the even more profound ignition of corporate empowerment. The US cigarette industry began as business partnerships, not corporations, but it incorporated and expanded just as a larger transformation of corporate power made new kinds of business practices possible. Lax new incorporation laws, the Fourteenth Amendment, the rising importance of the stock market, and avenues for expansion paved by imperial ventures all created fresh potentials; cigarette companies jostled to turn those potentials into profits and prestige.[3] The industry had reached international markets from its inception, long before incorporation. Lewis Ginter of Richmond, Virginia, achieved the first big success with bright leaf cigarettes in London. The owners of the five major native-born cigarette companies, all partnerships, understood that the future of the industry depended on continued access to foreign markets. They incorporated as the

American Tobacco Company (ATC) in New Jersey in order to maintain overseas control of the most efficient cigarette-making machine, the Bonsack.

After incorporation, however, James B. Duke, the owner of one of the ATC's original companies, wrested control of the ATC from Ginter and implemented an aggressive plan for expansion. In a dramatic and highly contested process, the ATC soon took over hundreds of chewing and pipe tobacco companies, giving the corporation nearly complete control over the supply of bright leaf tobacco.[4] Though its business practices faced court challenges, the ATC evaded government regulation by drawing on new entitlements granted by New Jersey's incorporation laws and the Fourteenth Amendment's protections of private property and due process.

The ATC also immediately pursued an ambitious plan for foreign expansion. Partly using capital raised in the stock market, it took over cigarette companies in Germany, Australia, and Japan, and in 1902, it merged with the Imperial Tobacco Company of Britain to form the British American Tobacco Company (BAT). The ATC and Imperial agreed to stay out of each other's domestic markets but merged their foreign holdings, casting BAT as a multinational company wholly dedicated to foreign expansion. The ATC owned 60 percent of BAT's stock, and James B. Duke served as chairman of the board of both corporations. The ATC's, and later BAT's, foreign expansion became part of the story of British and US imperialism, both by drawing on the privileges that imperial powers won for foreign companies and by becoming mechanisms for the extraction of profits and the imposition of foreign control. BAT soon spread over much of the world, and China became its largest outpost.[5]

The bright leaf network was integral to corporate expansion because it directed and managed the actual creation of large and geographically disparate farming and factory systems on the ground in both China and the United States. Along with selecting for the hire of certain white-collar corporate employees by race, gender, and region, this transnational network served as a conduit for bright leaf tobacco, cigarettes, bright leaf tobacco seeds, and managerial knowledge—knowledge that was entangled with racial understandings forged in the Jim Crow South. In addition, even as the network functioned as part of the corporate structure, it also played a critical role in materializing US and British empire overseas. Empires are best understood not solely as occupying or governing forces but as "hierarchical networks of migration, information, force and rule making that carried and connected laborers, settlers and administrators moving across global space."⁶ By this definition, we could say that the bright leaf network was a manifestation of corporate imperialism, one with dramatic effects on both China and "home."

Corporate empowerment occurred through daily innovations and transactions among myriad people; cigarettes' transnational success, in turn, emerged from those efforts. In the process of daily business practice, corporate power waxed in inextricable relationship with the emergence of Jim Crow segregation and imperialism—a relationship that would define the United States' place in the twentieth-century global order. The bright leaf network serves as the ballast for this book's inquiry as it tacks between the US and China. As members of a managerial class, the men of this network depended utterly on a host of people who also built the corporation, including Chinese entrepreneurs, managers, salesmen, factory workers, farmers, and servants, as well as African American and white farmers and factory workers and African American servants. As social organizations, the cigarette corporations in the US and China were profoundly diverse entities involving daily contacts across racial differences, contacts without which no cigarettes would be made or sold.

This history has remained buried until now because we are mired in fables of capitalism. There are two linked stories about the cigarette industry that have been repeated for so many decades and across so many platforms that they have come to seem like common sense. The first arises from the cult of the entrepreneur. This story holds that James B. Duke gained control over the industry from the outset because of his entrepreneurial innovation. While his competitors turned out the hand-rolled cigarette, the story goes, Duke forged ahead by introducing the more efficiently produced machine-made variety, driving down costs, capturing profits, and forcing his competitors to merge with him in the American Tobacco Company. The story attributes much of this success to Duke's personality: brash and risk-taking, he refused to play by the established rules of the game and followed instead his exceptional business foresight.

The second story is the related mythos of modernity that posits that superior technology, like the cigarette machine, led to a uniform pattern of West-to-East spread of modern business forms and commodities, such as the corporation and the cigarette. The value placed on this process varied: it was the benevolent spread of progress; it was the violent spread of empire. Whether celebratory or critical, proponents of this story agree that Western cigarette corporate representatives developed new technologies, business forms, and commodities at home and then scaled them up and exported them whole cloth around the world, where they transformed more "passive" and "primitive" (or "undeveloped") societies.

I call these two stories fables because they are heroic tales of mastery that are not borne out by the historical evidence. They are, instead, theories of capitalism that have been laid over cigarette history by patently disregarding the facts. James B. Duke and the cigarette corporations were indeed powerful, but how we render that power makes a world of difference. Telling a new story about cigarette corporations necessarily also means developing new ways of understanding capitalist innovation and expansion.

Rethinking Innovation

In collective legend, James B. Duke has become a revered archetype: the brilliant, innovative entrepreneur. The tale of Duke's restructuring of the cigarette industry around the cigarette machine serves as a textbook case of economist Joseph Schumpeter's theory of creative destruction, a theory enjoying a heady revival in the early twenty-first century. Working in the decades surrounding World War II, Schumpeter identified a pattern of innovation in which an entrepreneur disrupted established business practices by using a new technology to create a cheaper, lower-quality product. The innovator then lowered prices, catching sluggish competitors unaware, and restructured the industry around the newly victorious model.[7]

Schumpeter himself called for scholars to seek historical examples for this model and, in the 1960s, researchers newly began to present Duke as the quintessential creative destructor, with cursory attention to the actual historical record.[8] The famed business historian Alfred D. Chandler, Jr., picked up this thread and elaborated it in his classic book, *The Visible Hand*.[9] Historians never looked back. The tidy story has appeared in virtually every history of tobacco or cigarettes written since, whether critical or celebratory.[10] The fable has circulated from business journals to popular magazine and website profiles and from business schools to high schools, where it is currently part of the Advanced Placement US History curriculum.[11] In other words, the story of Duke's innovation with the cigarette machine has shaped widely held beliefs about how capitalism operates.

The problem is that factually and substantively this story is simply wrong.

Duke was not especially innovative with the cigarette or the machine. Indeed, some of the innovations attributed to Duke occurred before his family's company even began making cigarettes. The cigarette machine was significant, but all major producers had access to it. The issue, rather, was control of the machine in overseas markets. By merging, the major cigarette manufacturers strengthened their hand with the Bonsack Company, which was threatening to award exclusive overseas patents to only one US company or perhaps to a foreign company. By erroneously claiming for Duke dominance over the early cigarette market, historians missed a more nuanced story about how power — including incorporation — operated in global capitalist expansion. Duke's eventual very considerable power came not from innovation with technology but from his ability to wrangle managerial and financial control over the ATC after its formation. Nevertheless, the fable of Duke's innovation has been so powerfully persuasive that no one has reassessed the early industry for over a half century.

The two entrepreneurs who made the most significant innovations with the bright leaf cigarette globally were Lewis Ginter and Zheng Bozhao. Ginter was the first to develop the mass-produced bright leaf cigarette, and the first to market it overseas, casting it as an exotic product from the US. From that base, he expanded sales through the US, Europe, Australia, and New Zealand and into Asia. Zheng Bozhao worked with BAT-China from day one and was the single most significant entrepreneur in the branding and marketing of Ruby Queen cigarettes. Zheng gave Ruby Queen cigarettes their very consequential Chinese name — Da Ying, or Great Britain. Zheng also created and had full control over the sales system that made the cigarette brand famous. Both Ginter and Zheng achieved their innovative vision not by being newcomers, upstarts, or rogues but by virtue of their particular business and personal experience. Ginter's experience as an importer and as a man who partnered with men was critical to his insight; for his part, Zheng drew on his experience and connections related to foreign trade as a merchant from Guangdong Province.[12]

Understanding the process of innovation, however, requires looking beyond brilliant individuals to the role played by other entrepreneurs, cultural intermediaries, significant geopolitical events, and the social circulation of goods themselves. Egyptian, Greek, and Jewish entrepreneurs so profoundly affected the conditions faced by Ginter and the ATC that they must be considered. In addition, a range of cultural intermediaries, including London clubmen, Chinese courtesans, African American jazz musicians, and Chinese anti-imperialist protesters shaped bright leaf cigarette markets and the cigarette itself. Specific geopolitical events, such as the British occupation of Egypt and transnational protests against the US Chinese Exclusion Act, left permanent marks on cigarettes and their brands. Even cigarettes themselves were rarely blank slates. As malleable as they were to branding, once they accrued associations they did not readily release them. The story told here, then, does not simply substitute Ginter or Zheng in for Duke, but pursues a new way of exploring innovation.

Rethinking Expansion

When Lee Parker said that his job was to "put a cigarette between the lips of every man and woman in China," he used a formulation consistent with the mythos of modernity. He painted himself as the active Western agent bringing the modern commodity to passive China; Chinese consumers seemingly needed only to part their lips for the cigarettes. The mythos of modernity is a story with several core features: it attributes catalytic power to technology; it features West-to-East movement of fully developed industries, capitalist forms, and commodities; it assumes a time lag between development in the West and the rest; and it ascribes action and ability to transform to the West and receptive passivity to the East.[13] US historians have told the history of cigarettes in just this way. Highlighting the role of cigarette machine technology, they credited Duke with development of a successful industry and national market in the US, which he then spread across the world.[14] We are so accustomed to the mythos of modernity that this story might sound like common sense, but it is highly distorted. Yet often repeated: Parker's memoir echoed this story of Western capitalist expansion even though his own personal knowledge and experiences contradicted it in key ways.

One thing Parker surely knew, but did not mention, is that the Turkish tobacco cigarette was the dominant cigarette in the US when he departed for China. In fact, the cigarette industry had a bidirectional flow: the Turkish tobacco cigarette, made famous and mobile by the Egyptian industry, flowed from East to West, while the bright leaf cigarette flowed from West to East. This was common knowledge in the US at the time. For Ginter's and Duke's entire cigarette careers, they battled the primacy of the Turkish tobacco cigarette in the United States.[15]

Parker did not tell the whole story about his own experience in BAT either.

The discrepancy came out in the 1970s, in an interview with historians of China who wished to know how BAT opened up Chinese markets to cigarettes. They asked Parker, but he had no idea. "This question always embarrasses me," he said. "The Chinese knew where the markets were and how to make sales. ... Window dressing is what I was."[16] Now BAT's Chinese employees seem the active agents and Parker the passive "window dressing." Chinese consumers, it should be noted, were likewise far from passive. Indeed, they had a pesky tendency to boycott rather than smoke BAT cigarettes, and they did so in an organized movement in 1905 and again in 1925.[17] The mythos of modernity, it seems, requires skipping over a large number of interesting facts.

It is not possible to dispense with the idea of modernity entirely, however, precisely because it shaped both the way that Parker and his counterparts thought about what they were doing and also, therefore, what they actually did. Operating with a mental map that sharply distinguished between modern and primitive, they accepted imperial privileges in the treaty ports, including access to Chinese servants and sex workers, as the perquisites of their own putative modernity. Their assumptions of Chinese primitivism shaped all of their relationships with Chinese people and became foundational to BAT's foreign business culture. So accustomed to seeing themselves as modern in contrast to Chinese primitivism, men of the bright leaf network never noted in their letters, memoirs, or interviews that their home states were experiencing the "modernizing" development of the cigarette industry at the same time, though they were certainly well aware of it.

BAT foreigners also attributed to the cigarette a powerful ability to transform Chinese people into modern subjects. By presenting and marketing cigarettes as Western, modern goods, they helped shape the avenues through which cigarettes entered Chinese culture. In both the US and China, cigarettes became tightly linked with another globally circulating commodity associated with the West and modernity: jazz music.[18] Chinese people forged cultural and economic practices and political movements using, and thereby changing, this same notion of modernity. As a powerful imaginary of the globe and economic and cultural transformation, the notion of modernity affected this story at every turn.

New cigarette machine technology, the empowered corporation, and the rise of US empire were indeed significant developments, but attachment to the cult of the entrepreneur and the mythos of modernity has slowed reassessment of the nature of that power. As a transnational imperial formation, the bright leaf network holds one key to a new story. To discover it, we need to know more about Parker and his counterparts than they were inclined to reveal.

The Origins of the Bright Leaf Network

White men of the bright leaf network commonly said that they "knew tobacco." By this, they meant quite a lot. They meant that they had grown up in the bright leaf tobacco region of North Carolina or Virginia and were familiar with the requirements of growing and curing bright leaf. They meant that they understood something about grading bright leaf tobacco, selling it at auction, and manufacturing it into pipe or chewing tobacco or cigarettes. With the phrase, they also indicated that they shared a cultural background with others who knew tobacco, though not all viewpoints or opinions. They understood that there were plenty of African American men who knew tobacco in the sense of having skills with the agricultural commodity, but they also understood that white-collar jobs in the growing bright leaf tobacco industry were reserved for whites, and this exclusion, too, became a part of bright leaf tobacco culture.

The bright leaf tobacco network was a corporate network of white—but not black—men who knew tobacco, and had its origins in racial struggles. The network hired personnel but also managed expansion: it was a selective conduit for the people, knowledge, seeds, tobacco, cigarettes, and other things that moved when the corporation expanded in the US and internationally. And bright leaf was expanding, both before and after incorporation. Bright leaf tobacco emerged as a new and especially lucrative agricultural commodity just before the Civil War disrupted the Southern economy. Bright leaf's resurgence after the war meant that the development of bright leaf agriculture and manufacturing took place in the context of Reconstruction and its aftermath.[19] Later, the industry reorganized and incorporated in the context of the rise of Jim Crow segregation. The bright leaf network developed as a technique for infusing racial hierarchy into the new social and economic structures that came with capitalist expansion.

From African Americans' perspective just after the Civil War, bright leaf tobacco held tremendous potential for their upward mobility. There are three things about bright leaf itself that affect this story. First, bright leaf developed shortly before the Civil War in just three counties on the North Carolina–Virginia border: Halifax and Pittsylvania in Virginia and Caswell in North Carolina. Second, it was enormously lucrative when manufactured into pipe tobacco because it produced a very mild smoke and was an attractive golden color. Because it grew on soil poorly suited for other crops, its spread promised a dramatic rise in land prices and untold future profits. Third, bright leaf was fussy. Just having the seed was not enough; bright tobacco required sandy ridge soil and skillful tending to produce a suitable leaf, after which it needed to be heat-cured by a particular method that was so persnickety that some referred to it as an art form. African Americans had every reason to believe the rapid spread of bright leaf after the Civil War would benefit them because under slavery, they had performed all of the skilled and unskilled labor.[20] They knew bright leaf better than anyone.

Instead, whites captured control over the budding postwar industry and ensured that new white-collar opportunities in manufacturing, seed development, leaf grading and sales, and consulting work went to whites only. This was a lengthy and violent process that began with struggle over land and labor systems. Instances of violence had a seasonal nature matching the moments in the bright leaf process when black labor was most required. More dramatic events also connected to the bright leaf industry. The 1870 Ku Klux Klan assassination of Republican state senator and bright leaf tobacco buyer John Stephens, in Caswell County, took place in the context of pitched battle over blacks' right to sell the tobacco they grew. The 1883 Danville, Virginia, riot put down a political interracial coalition that had asserted its power by threatening to withhold black labor in the bright leaf industry. In the end, African Americans and poor whites, as elsewhere in the South, became ensnared in a sharecropping system that meant few would ever gain enough money to purchase even a scrap of land.[21]

And a scrap would have been enough in the early years. Consider the white-collar career ladders of James B. Duke, R. J. Reynolds, and so many other tobacco tycoons: their families' businesses began after the Civil War with very little capital on small farms. They newly grew bright leaf, manufactured their own chewing and smoking tobacco on site, and sold the product by horse and wagon through the South. By seizing this moment of opportunity, some soon built larger manufacturing facilities and bought their tobacco from area farmers and sharecroppers. By the 1880s, hundreds of small and midsize manufacturers spread in a latticework across the bright belt, all white-owned, and profits from bright leaf continued to grow.[22] There was nothing taken for granted or incidental about the whiteness of this industry.

Also in the 1880s, a deluge of stories spread through Southern newspapers and other media about primitive blacks who could not understand technology, could not be trusted to work without supervision, and lacked the judgment needed for management. In other words, this discourse denied what everyone in the locale knew about black people's skill with bright leaf and cast African Americans as racially unfit for white-collar positions. In 1866, for example, the *Pittsylvania Tribune* published a story about well-known early promoters of the bright leaf curing technique, Abisha Slade and his brothers, from Caswell County. This story, however, focused on one of their slaves, Stephen, who was supposedly interviewed at the Danville, Virginia, tobacco auction. Now an old man, Stephen explained that he had invented the curing method, not by tinkering with heat and airflow, but by accident when he fell asleep on the job, allowing the fires he was tending to go out and then fanning them to life again. He also expressed admiration for the Democratic Party and longing for the simpler days of slavery, saying, "I wish he [Abisha] was alive today and I was his slave."[23] Southern newspapers were full of such fabrications that repressed the fact of black skill in tobacco while casting blacks as primitives in relationship to modern, technically astute whites. In doing so, such stories also linked Jim Crow, emerging from local conflicts, to internationally circulating imperial discourses of civilization and primitiveness.

The Bright Leaf Network and Corporate Imperialism

After the ATC incorporated, it aggressively took over the bright leaf chewing and pipe tobacco companies and transformed the bright leaf industry's white-collar class into the bright leaf network. As a product and expression of Jim Crow, the bright leaf network served as a mechanism for replicating segregation's hierarchies as the corporation expanded.[24] The ATC continued its domestic expansion for two decades. Former factory owners and managers funneled into emerging corporate positions, both at New York City headquarters and in its rapidly growing overseas holdings.

The ATC first bought companies in Australia and Canada, building on its foundational overseas market in the British Commonwealth, but its ambitions kept pace with waxing US imperial might, especially looking to East and Southeast Asia.[25] In 1887, the US gained control over Pearl Harbor in Hawaiʻi and the harbor at Pago Pago in Samoa. From the War of 1898, the US gained possession of the Philippines, Puerto Rico, and Guam and temporary control of Cuba. Also in 1898, the US annexed Hawaiʻi.[26] The corporation was one key institution that deployed imperial power; conversely, corporations like the ATC gained advantages from imperial privileges won through war and leveraged diplomacy. When the US occupied the Philippines, the ATC sent James A. Thomas to Manila to sell tobacco products to the US military market.[27] In 1899, the ATC purchased the Murai Brothers Tobacco Company of Kyoto, Japan, which it used as its production center for expanding bright leaf cigarette sales through ports in East Asia.[28]

In 1902, the ATC expanded in a new way. Over the previous decade, it had expanded overseas by purchasing established cigarette companies and using them as a base in a new market. The merger of ATC with the British Imperial Tobacco Company to form BAT joined the two largest bright leaf companies in the world and set BAT on a mission of foreign expansion exclusively. The merger also provided access to the considerable infrastructural resources and reach of the British Empire. Duke celebrated the formation of BAT in an interview with a British tobacco trade journal: "Is it not a grand thing in every way that England and America should join hands in a vast enterprise rather than be in competition? Come along with me and together we will conquer the rest of the world."[29] BAT had no exclusive relationship to either the US or Britain but was self-consciously imperialist in its own right.

Corporations' connection to imperialism was by no means new at this time, but the relationship was transforming. Joint-stock companies arose in the 1500s as arms of the empires that chartered them. The English East India Company, the Dutch East India Company, the Hudson Bay Company, and many others served to extract resources and develop markets in colonial outposts. As the primary means of colonization, many of these companies had both economic and governance functions. Only in 1857, after the tyranny of the English East India Company prompted an uprising, did the British government revoke the company's charter and institute crown rule in India.[30] Corporations could be and were chartered for a plethora of other reasons on home soil, including schools, churches, utilities, and other kinds of projects. Nevertheless, the imperial business corporation's three-hundred-year history did not so much come to a close as set the stage for a new relationship that would develop between "private" business corporations and imperial states. Multinational corporations like the ATC and BAT were not chartered by single governing bodies expressly as colonizing forces but gained their entitlements and generated uneven economic and political power in relationship to multiple imperial entities.[31]

BAT's massive expansion in China was another new departure for the bright leaf corporations. BAT's predecessor companies had sold cigarettes in China for over a decade, but the 1905 decision to build factories was a significant departure for the company. As long as the BAT-owned Murai Brothers produced BAT's East Asian exports, the company simply coordinated sales in China through Chinese commission agents. In 1904, however, Japan nationalized the Murai Brothers Tobacco Company. Having lost its outpost in Japan, BAT turned to China. Unlike its ventures elsewhere, there was no successful cigarette company to take over, and BAT soon determined to create a full production center in China, from growing bright leaf tobacco to manufacturing cigarettes and packaging. Rather than sending a handful of foreign managers, it sent dozens. Eventually, hundreds of foreign representatives made the journey.

In China, BAT claimed the benefits won by a half-century of imperial wars and arm-twisting diplomacy by Britain, Germany, the US, France, and Japan. The British Opium Wars ended with a series of treaties that forced China to create treaty ports and give British companies enormous advantages. European countries, the US, and eventually Japan claimed the same privileges.[32] Critically, these privileges included extraterritoriality, which gave foreigners the right to be governed by their own police force and court system in the international settlements, rather than by Chinese authorities. Extraterritoriality denied the Chinese government jurisdiction over people on its own soil, a jurisdiction that was taken as a matter of course among European nations. In 1895, Japan defeated China in the Sino-Japanese War and won new rights for foreign companies, including the right to own property and build factories within treaty ports.[33] By the turn of the century, with the right to build factories secured, foreign companies, including BAT, expanded their operations in Shanghai.

By 1918, more than seven thousand foreign companies had located their headquarters in Shanghai, where an elaborate imperial leisure culture emerged to cater to foreigners.[34] BAT's foreign employees enjoyed a plethora of privileges and services created for them, from cheap and abundant servants to lavish entertainment venues, including fine clubs and restaurants, a racetrack and, eventually, cabarets. Shanghai's foreign settlements and the business district, called the Bund, exploded with growth, as European-style office buildings and houses displaced a large number of Chinese people out to the "Chinese city."[35] For foreign BAT representatives, experience with Jim Crow made conditions in Shanghai feel at once exotic and familiar. As Lee Parker later recalled, "It was wonderful from a young American's point of view because it was during the time of extraterritoriality and we lived in a little small community, segregated you might say."[36]

That Jim Crow, the corporation, and imperialism were coming into a new relationship was clear, but how the matter would unfold in both the US and China was anything but predetermined. Corporate imperial power was considerable but not total. Value had to come from the soil and the hands, backs, and minds of thousands of people. Many constituents coproduced corporate structures, leaving their mark upon them, while others discovered opportunities for sabotage and co-optation. Employees and consumers put cigarettes and brands to unexpected uses. Zheng Bozhao slowly wrested enough power from BAT that he could call some significant shots, forcing BAT foreigners to adapt to forms of Chinese capitalism that they had helped to induce. In other words, the cigarette corporations were vast and messy organizations and things had a way of not staying put.

Our story begins in the 1870s, when Lewis Ginter of Richmond, Virginia, first made cigarettes from bright leaf tobacco and the Egyptian cigarette industry was on a rapid ascent. Ginter was responsible for many of the innovations in cigarette manufacturing and international marketing that shaped the industry and the cigarette.

1

The Bright Leaf Cigarette in the Age of Empire

Lewis Ginter of Richmond, Virginia, saw a potential that no other Southern tobacco manufacturer perceived when he married bright leaf tobacco, locally produced in Virginia and North Carolina, with the foreign novelty of the cigarette. Ginter took this cigarette, named Richmond Gem, to London, where it found its first big success in the elite gentlemen's clubs of the West End. With Richmond Gems as its cash cow, the Allen Tobacco Company — soon to be renamed the Allen & Ginter Tobacco Company — sent the bright leaf cigarette around the world, placing brands like Dubec, Richmond Gem, Richmond Straight Cut, Opera Puff, and Little Beauties via commission agents in France, Belgium, Australia, India, China, and many other places. Only after Ginter had made a large and widely publicized success with the bright leaf cigarette did his neighbors to the south, W. Duke and Sons, venture into them. In 1889, Ginter, James B. Duke, and three other manufacturers merged the companies to form the American Tobacco Company (ATC). To this new venture, Ginter brought more years of experience, more financial savvy, and more extensive distribution in foreign markets than did Duke, though Ginter has since been largely forgotten. The story of entrepreneurial innovation in the US cigarette industry properly begins with Ginter rather than Duke.

Telling the story of the inauguration of the bright leaf cigarette's global career requires dethroning Duke, but the point is not simply to hoist Ginter to Duke's vacated perch. Rather, Ginter's entrepreneurship reveals a new story about innovation within the world of goods.[1] Ginter played two discernable functions as innovator: product development, in combining bright leaf tobacco with the cigarette; and marketing, in his decision to take the cigarette abroad at first opportunity and to market it as a foreign novelty. These were indispensible preconditions to success, and Ginter's vision came from his unique biography. When the first bright leaf cigarettes arrived in the London depot, though, they were inert objects, signifying nothing—at least, not to Londoners. They awaited social circulation within a particular context or group that would give them what for commodities is the breath of life: brand identity and resonant appeal. Tracing that process takes us into public scenes where bright leaf cigarettes became famous and introduces us to additional innovators, like Miltiades Melachrinos, Nestor Gianaclis, and Zheng Bozhao, as well as cultural intermediaries, such as London clubmen and Shanghai courtesans. The bright leaf cigarette's early career faced challenges when the British invasion of Egypt, the US Chinese Exclusion Act, and the rise of sexology shaped cigarette markets in unpredictable ways. In other words, to understand the rise of bright leaf cigarettes, we have to follow them into the world.

One thing neither Ginter nor Duke could escape through the whole of their careers was the innovative powers of the Egyptian cigarette industry. The bright leaf cigarette had to do battle with the Turkish and Egyptian cigarette for decades. From the 1880s through the 1910s—approximately forty years—most people in the US saw the cigarette as originally or most authentically from the Ottoman Empire. Egyptian producers not only exported to Britain, the US, and scores of other countries, but also set up branch factories in the US and around the world. For decades, the Egyptian industry defined cigarette tastes in the US and Britain and led in global trade. Indeed, competition with Egyptian producers only increased at the dawn of the twentieth century. "Turkish" tobaccos had the reputation of being the best for cigarettes and they carried the highest cost.[2] What would the Sultan smoke? Likely not a bright leaf cigarette, which, to put it mildly, lacked the cachet of Turkish tobacco.

US producers did not shake the cigarette loose from its foreign reputation until after World War I. They did successfully make and market 100 percent bright leaf cigarettes in the US with such brand names as Little Beauties and Pinhead but, to be competitive, they also made cigarettes with all or a portion of imported Eastern tobaccos. US producers copied Egyptian branding and marketing techniques, acquired some Egyptian companies, and sold brands like Dubec, Fatima, and Murad to compete with brands like Nestor, Egyptian Deities, and Crocodile. They also improved their access to costly Eastern tobaccos.³ Cheap and proximate bright leaf tobacco was a great advantage for US companies; one might reasonably surmise that they would have shifted more of their domestic offerings to all–bright leaf cigarettes if they could have done so profitably.

In China, however, the cigarette became known as a Western product. Though other countries such as the Philippines, Japan, Russia, India, and Egypt also sold cigarettes in China, the cigarette took on a reputation there variously as "British," "American," or simply "Western." There, British and American producers sold the 100 percent bright leaf cigarette almost exclusively and, unlike in the US, this cigarette became the standard of value. While US and British companies sold brands with Eastern names in the US, in China they sold brands named Pinhead, Pirate, and Ruby Queen; the latter also acquired a Chinese name, Da Ying, which means Great Britain.

These global flows of tobacco and cigarettes created an odd dynamic: the cigarette entered the US as an Eastern product but China as a Western product. People in the US saw the cigarette as Egyptian and favored "Turkish" tobaccos, while people in China saw the cigarette as British, American, or "Western" and favored bright leaf tobacco. Cigarettes first found urban markets in the US and China decades before the masses of people adopted the habit, and the legacy of these early years lived on in material ways. When cigarettes did truly boom in the late 1910s and '20s, the first blockbuster brand in the US was Camel cigarettes, made of a blend of Turkish, bright leaf, and burley tobaccos, and in China, Ruby Queen/Da Ying cigarettes, made of 100 percent bright leaf tobacco. The bidirectional flow of cigarettes, brand imagery, and tobaccos contradicts the capitalist story that globalization and modernity flowed from West to East and that Western companies set the codes for consumption and taste that the rest of the world adopted. In other words, one of the most quintessential modern commodities does not fit our stubbornly persistent models of globalization and modernity.

Key to understanding virtually all cigarettes in the nineteenth century is that their national identification was not simply an objective geographical description of their origin but was inextricable from how people—especially in the US and Europe—were learning to see the world in terms of nations, races, and essences. This was the era of rising nationalism and imperialism, when Western thinkers rushed to chart the world as a hierarchy of peoples at different stages of civilization. The global flow of goods played an important role in the popularization of these ideas. In particular, cigarettes, as a new, transnationally circulating commodity, accrued and delivered value within this framework of nations and races. It became common to believe that the essence of a people inhered in the objects they made as well as in their taste in goods; one could experience something of Egypt, then, by smoking an Egyptian cigarette.[4] Race as an irreducible essence, not the literal location of production, was the point. For native-born New Yorkers and Londoners, for example, immigrant cigarette producers in their city produced "foreign" cigarettes as surely as did producers in France or Egypt, no matter how long they had lived in the city or their citizenship status. In this way, cigarettes were world making; that is, they helped form visions of distant lands and places, however inaccurate. Conversely, imperial fantasies and events repeatedly shaped cigarettes' branding stories and markets. For the bright leaf cigarette, this global story begins in the South in the wake of the Civil War.

Developing the Bright Leaf Cigarette

In 1872, Lewis Ginter was at a turning point in his life. After seven years in New York City, he was returning to Richmond in order to become a partner in the John F. Allen Tobacco Company, maker of chewing and pipe tobaccos and a small line of cigars. Ginter had already worked as the Allen Company's principal salesman in New York for two years; as a partner, he took complete charge of the company's sales and marketing, while Allen managed the manufacturing process. Ginter brought with him the young John Pope, who had worked with him at the Allen Company's New York City depot and now took up a position in the Richmond factory. Two years later, when well established as partner, Ginter convinced Allen to add cigarettes to their line.[5] At fifty-one, Ginter began what would become his most significant work. Within ten years his cigarettes would reach into the far corners of the world and he would be among the wealthiest men of the South. Pope would become Ginter's partner in business and in life, taking Allen's place in the company upon that man's retirement and living with Ginter for over twenty years.[6]

Ginter's life had uniquely prepared him to discern the potential of a cigarette made from bright leaf tobacco and to envision possible markets for it overseas. In Richmond in the 1850s, Ginter certainly heard about the new tobacco type from three counties in the Virginia and North Carolina piedmont that led to such large profits. Though the Civil War disrupted development of bright leaf markets and infrastructure, the industry grew rapidly afterwards. Bright leaf was the rare economic good news that circulated through the cash-poor Upper South. In 1869, one grower in Halifax County urged others in nearby counties with similar soil types to grow "that quality [of tobacco] which is peculiar to this portion of Virginia and North Carolina.... This for the present, at least, furnishes the only rainbow of hope to us."[7] Richmond was not in the bright belt but was only a few hours away by train.

Ginter started out his business life in imports, not tobacco, which gave him a different perspective than men in the bright leaf belt. Born in New York City, Ginter migrated to Richmond at age seventeen with John C. Shafer, a tailor who was just a few years his senior. At age twenty-one, after working in a hardware store and briefly owning a toyshop, Ginter opened the Variety Store on Main Street, a dry goods store specializing in imports and "Fancy Goods." Fancy goods were ornamental goods or novelties that included housewares, artistic pieces, fine fabrics, and accessories. Ginter later opened a second store that became an import wholesaler supplying stores further inland and South.[8] In order to supply this store, Ginter took nearly annual buying trips to European cities, learning commodities, trends, and markets and enjoying European cultures. His import business specialized in Irish linens and Saxony woolens, among other products, which brought him especially often to London.[9]

Ginter's import business built upon Richmond's unique connections to both the surrounding tobacco-growing region and to Atlantic trade. Though Richmond was a small city in the 1840s, steamships carried tobacco from the port city to Eastern seaboard and European markets. Tobacco had been the foundation of Virginia since its inception as a colony, and Virginia tobacco had enjoyed an international reputation for its high quality and mild flavor for over a century. As tobacco went out, other commodities flowed in. Ginter's import wholesale business tapped that flow, tying European and Atlantic markets to Richmond shops as well as to shops in smaller towns that served the rural countryside.[10] Furthermore, some of these towns supplied Richmond warehouses with the new, lucrative commodity of bright leaf tobacco. Ginter's business perspective developed from Richmond's unique status as both a tobacco town and an Atlantic port that connected the city and environs to the North and to Europe.

The massive destruction of the Civil War eventually pushed Ginter to New York City and, eventually, to the tobacco trade. War disrupted shipping through Richmond, cutting the city off from northern and overseas markets. Ginter shuttered his shop and joined the Confederate Army as a commissary.[11] After the war, Richmond's infrastructure was badly damaged and cash was in short supply. Ginter relocated to New York City and opened a short-lived bank. When the bank folded, Ginter lost what remained of his prewar savings. He traveled back to Richmond and convinced his longtime friend John F. Allen that they should open a tobacco factory.[12] Since Ginter had no capital to contribute, the company was owned by Allen and bore his name. Ginter returned to New York City as the company's salesman.[13]

By a twist of fate, then, Ginter was in New York City selling chewing and pipe tobaccos during the years when cigarettes were just appearing on the market. After the close of the Civil War, French, Russian, and Greek cigarettes, made with Eastern "Turkish" tobaccos, predominated in the tiny cigarette import market. If the cigarette market was tiny and dominated by imports, US production was even smaller and was dominated by immigrants.[14] Greek manufacturers, such as the Bedrossian Brothers and Nicholas Coundouris, probably initiated local production in New York City, selling both to immigrants and native-born consumers.[15] The Kinney Brothers Tobacco Company began making cigarettes in 1869, possibly the first native-born producers to set up shop in the city.[16]

It might seem strange that the US was a latecomer to cigarette production, given its later prominence in the global market, but the Civil War disrupted US trade at just the moment that cigarettes circulated in significant numbers. The development of paper-wrapped cigarettes required the industrialization of paper, a feat only accomplished in the 1830s.[17] The French industry took an early lead in cigarette production because the French government had established a tobacco monopoly that socialized the risk of investment in an untried product. French cigarettes circulated through Europe and the Middle East by the 1850s, and French cigarette paper remained known as the world's best for decades. The Ottoman Empire quickly outpaced France in cigarette consumption, and Greek producers, especially, entered the business, as did producers in Russia and other countries.[18] New York City likely saw very small numbers of cigarette imports in the 1850s, a flow that surely would have increased in the early 1860s had not the Civil War disrupted transatlantic trade.[19] After the war, cigarettes again flowed into the US market, and domestic producers entered the competition. The vast majority of cigarettes on the market, however, were clearly marked and marketed as foreign; among them, Turkish tobaccos brought the highest praise.

Though cigarettes were far from a hot item, Ginter's work in tobacco sales and his background as an importer of foreign novelties gave him a vantage point that enabled him to see the promise of cigarettes well before any other Southern producer. He was nothing like the Schumpeterian ideal of the innovator as rogue interloper who gained insight from his *lack* of experience with established practices. Rather, Ginter drew on the depth of his diverse experience to envision a new combination: using a pipe and chewing tobacco in the cigarette, which was strongly identified by the flavor and cultural cachet of Turkish tobacco. In fact, Ginter's innovation matches Schumpeter's less-cited theory in which he defined innovation precisely as the carrying out of new combinations, such as joining new raw material to an established commodity.[20] Ginter's experiences as both Northerner and Southerner, as both an importer and a tobacco salesman, all informed his ability to innovate with product development.

In hindsight, a bright leaf cigarette might seem obvious, but there were a few reasons it was not intuitive to tobacco manufacturers in the bright leaf belt. First, virtually no one in North Carolina or Virginia smoked cigarettes of any kind. Samuel Schooler of Carolina County, Virginia, was aware of cigarettes but saw their consumption as marking someone as alien to the region. In 1868, he described someone to his wife by writing, "Knew he did not belong about here—and moreover he was smoking a cigarette which is unheard of in these parts."[21] In fact, some people had never heard of cigarettes. James A. Thomas grew up farming and manufacturing smoking and chewing tobacco and later made his career promoting the cigarette as the head of BAT-China, but he knew nothing of cigarettes until 1876 when, as a teenager, he attended the Centennial Exposition in Philadelphia. "Although my means were extremely limited," he recalled, "I finally succeeded in getting sufficient money for a trip to Philadelphia. . . . Here I saw my first cigarette. It was made at an exhibit by some people from Egypt."[22] Thomas first experienced the cigarette as a foreign good on display at a world exposition, not in the course of daily life in the South.

In contrast to the insignificance of cigarettes in their lives, Southern tobacco men knew that virtually every man and boy in the South chewed tobacco, and most men smoked or were taking up the pipe. In addition, these markets were not confined to the South. In the wake of the Civil War, chewing tobacco maintained its place as the most popular tobacco product in the country. The *New*

York Times complained, "The national mastication and expectoration are known over the world, and do ample service in all conceptions and caricatures of [the American]."[23] Pipe smoking waxed in popularity through the late nineteenth century, which was good news for bright leaf, as it made an exceptionally good pipe tobacco. The decision of hundreds of Southern entrepreneurs to go into bright leaf chewing and pipe tobacco manufacturing and to ignore cigarettes made perfect sense.

Second, the Turkish tobacco cigarette had such a tight grip on people's conception of what a cigarette should be that the bright leaf cigarette, which offered a very different taste and smoking experience, could disappoint or even offend consumers. Eastern tobaccos had a pungent smell and a very strong flavor and were quite acidic. Because the smoke was harsh, people smoked Turkish tobacco cigarettes like cigars, by holding the smoke in the mouth, where sensitive tissue absorbed the nicotine, rather than by inhaling deeply into the lungs. Bright leaf cigarettes, in contrast, were very mild in flavor—so mild that some people declared them flavorless. In addition, bright leaf's distinctive curing process lowered the pH of the leaf, making it a less acidic smoke and therefore easier on throat and lungs when inhaled.[24] Indeed, bright leaf changed the gestures of smoking to include a deep inhale and delayed exhale. Both mild flavor and ease of inhalation could—and did—become selling points, but it took marketing to change established smokers' conception of the cigarette and to lure new smokers to the bright leaf version.

Ginter responded to this challenge in two ways. First, he balanced the Allen offerings, beginning his venture with the Dubec cigarette, made of imported Turkish tobacco, as well as several 100 percent bright leaf brands. Ginter took the Dubec cigarette to the Philadelphia Centennial Exposition the next year, where they competed with the Egyptian cigarettes noticed by James A. Thomas. Also at the exposition was the FS Kinney Tobacco Company with a cigarette made of a blend of bright leaf, Turkish, and New Orleans perique tobaccos.[25] (Thomas did not later remember either domestic producer.) Almost immediately, Ginter also sought an international market, specifically in London and Paris, the European cities he knew best, where he marketed bright leaf cigarettes as a foreign novelty.

At this moment, Ginter's innovations shifted from product development to marketing. His decision to pursue foreign markets paid off when London became the site of the first public trend of bright leaf cigarettes. As with product development, Ginter drew on his distinctive business and personal experience to discern marketing possibilities in London. In this case, he succeeded in substituting bright leaf cigarettes for pipes and cigars in the cultural scene of London's gentlemen's clubs.

The British Cigarette

Ginter and his commission agent in London, John M. Richards, faced a familiar marketing challenge: in London, as in New York City, cigarette smoking was primarily an immigrant practice. Native-born men smoked pipes and cigars. Richards noted that "the smoking of cigarettes of any sort or kind was extremely limited; in a few shops in localities frequented by foreigners [that is, immigrant neighborhoods] one could buy a packet of 'Caporal' [French] or 'La Ferme' [Russian], also Egyptian and Turkish brands of a few sorts." All of these brands were made with Eastern (Turkish) tobaccos. Richards added, "Englishmen did not then care for cigarettes."[26] There was a significant population of cigarette smokers in Paris, so one might assume Ginter would find his first success there, but it was difficult to convince established smokers to shift from Eastern tobaccos to bright leaf. Despite the fact that Ginter faced essentially the same problem in London as in the US, he found his first big market in London by presenting the bright leaf cigarette as a foreign novelty.

Ginter had a number of advantages from prior experience that enabled him to understand London's market conditions. As an importer, he began traveling to Europe at age nineteen. One friend estimated he crossed the ocean thirty times in his lifetime and noted that Ginter himself "said he knew London as well as he knew Richmond."[27] Of course, Ginter socialized while in London and came to know some of the culture of the city. From his import business, Ginter also well understood the cachet that foreign goods carried in the US and Europe. Still, it is important not to exaggerate Ginter's control. Indeed, Richards explicitly stated that early, extensive advertising of Richmond Gems had virtually no impact.[28] Richmond Gems did not sell in significant numbers until London clubmen became cultural intermediaries, promoting the cigarettes in their social scene and endowing them with character and charisma.

The first major obstacle that Ginter and Richards faced in London also led to the breakthrough for Richmond Gem cigarettes. When Richards attempted to place his product in tobacco retail shops across London, only those in the working-class East End would accept Allen Company cigarettes; retailers in the West End theater, club, and commercial district shut them out. Ginter and Richards then took two actions: first, they purchased tobacco dealer licenses for pharmacists and displayed Allen Company cigarettes on drugstore counters. Second and, according to Richards, more importantly, the Allen Company opened up its own tobacco shop at Piccadilly Circus called the Ole Virginny Cigarette and Tobacco Stores. Richards explained, "The windows were dressed in the most attractive form, and in a very short time the brand became the fashion. Then the cigarettes crept into the hotels, clubs, and public-houses."[29] Located in the heart of the West End, the popular store proved to be the ticket for getting their cigarettes into the hands of elite London men.

The location of the Ole Virginny Cigarette and Tobacco Stores proved crucial, for it was near the West End gentlemen's clubs. Like New York City's Broadway and Times Square, the West End was a vibrant, heterogeneous space with restaurants, theaters and music halls, high-end shops, and gentlemen's clubs. By 1882, the Richmond Chamber of Commerce bragged that "the bulk of the cigarettes used in the London clubs are made in Richmond."[30] It is not that West End clubmen were the only consumers of Richmond Gems — the cigarettes sold in the East End as well — but West End men became *known* as smokers of the cigarettes. That is, the cigarettes became part of clubmen's public image, and the London clubs imbued cigarettes with a particular character that came to be known outside the clubs as well. In other words, the clubs became the first public scene of the bright leaf cigarette.[31]

The gentlemen's clubs, by their numbers alone, represented an enormous market of tobacco consumers, quite capable of making or breaking a new brand. There were nearly two hundred clubs in London when Ginter marketed his bright leaf cigarettes, all of them in the West End, and their membership rolls ranged from a few dozen to several thousand. The term "gentlemen's club" in the US today is a euphemism for strip club, but the British gentlemen's club was an elite and highly respected institution attended by politicians, businessmen, professional men, and men of leisure. Though the gentlemen's club has a long history in Britain, developing out of the eighteenth-century coffeehouses, the heyday of the club trend was from the 1870s to World War I. The best clubs had long waiting lists of wannabe members — as long as sixteen years — which drove the founding of new, less exclusive clubs. When Ginter open his tobacco shop on Piccadilly at Circus, he occupied the heart of clubland: the highest density of elite clubs lined Piccadilly, Pall Mall, and St. James Street, with several large palace-like club buildings on each block.[32]

The store's nostalgic name, "Ole Virginny," gave an exotic spin to bright leaf cigarettes and, harking to the days when Virginia was still a British colony, played to the British imperial imagination. By the late nineteenth century, whites on both sides of the Atlantic read fiction that denied the violence of slavery and presented instead a romantic view of Virginia tobacco plantations as a place of contented slave labor that enabled white colonial refinement and repose.[33] The foreign nature of the cigarettes, then, promised to lend an exotic air to the British smoking experience. An advertisement for Richmond Gem cigarettes in the *London Times* read, "From the Royal palaces to swell Clubs, and under the modest vine and fig tree of the suburban villas, you now sniff the breeze and Ole Virginny tobacco, odours fragrant."[34] The smell of Virginia tobacco blew in from the British colonial past and promised to unify British men—middle class, elite, and royal—in a common luxurious experience.

The Ole Virginny Cigarette and Tobacco Stores also offered London men a rare foreign novelty. At the shop and in advertisements, the company explained the origins and fine qualities of bright leaf tobacco. In contrast to today, when cigarette advertisements rarely draw attention to tobacco type, the Allen Company specified the importance of soil and curing method to bright leaf. One ad explained that "the light gray soil on which they are grown (a small area in Virginia and North Carolina) and the manner of curing, largely reduce the percentage of nicotine; hence they can be smoked without fear of heartburn, dizziness in the head, or blistering of the tongue." Richmond's "proximity to [these] districts . . . enables [the company] to make selections of leaf . . . that are not obtainable by manufacturers elsewhere."[35] Richmond Gem cigarettes promised British clubmen a foreign novelty with superior and not easily attainable tobacco.

Richmond Gems became a new, fashionable item to offer to companions in the clubs, substituting for the pipes and cigars that were already integral to the social scene. At the clubs, men shared meals, conversation, billiards, relaxation, and drink and tobacco with other men—no women allowed. Clubs restricted smoking to particular times and locations. For example, they encouraged men to retire to lavish smoking rooms after meals. The ritualized time and space for smoking, of course, only heightened tobacco's value in the culture and community of the clubs. Offering one's pipe tobacco or cigar case to others became an obligatory part of the male intimacy of the clubs. The relationships forged were significant: men expressed great emotion about their clubs and the sense of community that they found there.[36] Perhaps the West End tobacco retailers were less willing to carry Ginter's cigarettes than their East End counterparts simply because they were not about to share the goose that laid the golden egg with a foreigner pushing a new product.

As cigarettes entered the clubs, they became an auxiliary to elite men's exuberant and intentional loafing. Club culture hewed not to middle-class ethics of work and thrift but to an aristocratic embrace of leisure, even as more middle-class people joined clubs. Written rules proscribed talking business at the club, for example. On offer at the clubs was enjoyment: extravagant meals cooked by esteemed chefs, moments of peace away from the missus to read the papers or answer letters, and especially, conversations in the smoking rooms about politics, art, sports, and women. For both the single and married man, time at the club was a perk of maleness and went hand in hand with a more general male mobility and license, including sexual license within certain constraints of discretion, in the late nineteenth-century metropolis.[37]

The cigarette did not entirely displace pipes and cigars in the clubs—the Allen Company could not even produce that many cigarettes yet—but the embrace of it by many was enough to anchor Ginter's business and build a reputation for the cigarette. As early as 1881, London's *Tobacco Review* associated the cigarette with all the familiar elements of club culture, personified: "The cigarette has become a favourite accessory of luxury, indulged in by *flâneur* and statesmen, epicure and aesthete alike."[38] Oscar Wilde, a member of the Ablemarle Club, famously embraced and epitomized club culture, including its rejection of the work ethic. His character Lady Bracknell, in *The Importance of Being Earnest*, upon hearing that Jack became a cigarette smoker, commented wryly, "I am glad to hear it. A man should always have an occupation of some kind. There are far too many idle men in London as it is."[39] By 1880, Ginter's marketing success in London made him a full partner in the company, now renamed Allen & Ginter.[40]

Ginter had a feeling for this elite culture because he was himself a cosmopolitan man who socialized in a world of men. Ginter had no doubt visited the clubs on business trips. In addition, his circle of friends in Richmond, a port city, participated in the transatlantic flow of cosmopolitan goods, ideas, and cultural values. There, Ginter was at the center of a male social circle whose members lived similarly internationally connected lives, including John C. Shafer, who owned a tailor shop offering men Richmond-made "Paris and European Fashions," Dr. William P. Palmer, a physician with a love for literature, and John R. Thompson, the editor of the *Southern Literary Messenger*.[41] (During the Civil War, Thompson relocated to London and published internationally circulating Confederate propaganda.) Ginter's "rare good taste and fine eye for the artistic," according to the *Richmond State*, made him a central figure in this group and "men of literary taste courted his companionship."[42] Ginter hosted regular dinners for his circle of elite male friends; after dinner, the group would retire to the smoking room for leisured conversation. Ginter and his circle cultivated a cosmopolitan culture in Richmond that, while certainly not identical to the one found in London's clubs, bore a family resemblance to it. Unlike more rural tobacco entrepreneurs, Ginter was well prepared to envision marketing cigarettes in this context.

Ginter and Richards adeptly followed the early success of Richmond Gems with advertising that appealed to the club ethos of cosmopolitan irreverence. For example, they made "mammoth posters" using a Joseph Ducreux self-portrait that was well known in Britain.[43] Ducreux was a French painter famous for breaking with portraiture conventions. "Self portrait of the Artist in the Guise of a Mockingbird" features the angled Ducreux figure looking and pointing directly at the viewer with a smug grin on his face. Ginter's artist made a drawing based on this portrait, leaving out the pointing finger but adding a cigarette to the smirking mouth. It was a highly irreverent move with a classic but rebellious painting; the caption simply read "Smoke Richmond Gems." Richards posted several hundred posters overnight through London. By its expressive sarcasm and its irreverence, the ad appealed to Ginter's target audience of clubmen who rebelled against bourgeois reserve.[44] The scheme was so successful that the image, copyrighted for both Britain and the US, became the brand image for Richmond Gems as well as for the company more generally.

Ginter's cigarette card series, "World's Champions," augmented the homosocial conversations of the clubs that so often centered on sports (fig. 1.1). Ginter was the first to include trade cards in his soft cigarette packs as stiffeners; the baseball-themed cigarette cards were the very first baseball cards and are now rare collectables. The "World's Champions" series featured baseball players, wrestlers, and "pugilists," as well as players of elite sports discussed in the clubs, such as billiard players and rifle shooters. Later, cigarette cards would become

Fig. 1.1 Allen & Ginter's "World Champions" cigarette cards celebrated the male body. Courtesy the Library of Congress.

known for their risqué pictures of women. Washington Duke would complain to his son that the ATC's use of "lascivious photographs with cigarettes" has "pernicious effects" and "has not jingled with my religious impulses."[45] Ginter's pioneering cards, however, displayed the athletic male body, some naked from the waist up, for the male gaze. The exposed male body would not have been read as lascivious but as aesthetic to British contemporaries familiar with the male physical culture movement, a fitness trend based in gymnastics and other sports, and the aesthetic movement. Walter Pater of Oxford University, mentor to Oscar Wilde and the art critic most associated with aestheticism, based a celebration of the male body in scholarly venerations of ancient Greek sculptures.[46] Ginter's athletic cards, noted for their artistic quality, popularized this ideal male body and reflected it back into the elite male culture of the clubs.

Ginter's Richmond Gem cigarettes "held the monopoly of American cigarettes [in Britain] for perhaps two years," according to Richards, "but at the end of that time the demand for cigarettes became so enormous" that both US and British manufacturers produced bright leaf cigarettes for the British market.[47] Not Duke—he was not yet making cigarettes—but the Kinney Tobacco Company likely exported bright leaf cigarettes to Britain. In addition, the British company W. D. and H .O. Wills began producing the Three Castles brand in 1878, made from 100 percent imported bright leaf tobacco, and other British firms soon followed its lead. Because of this popularity, the 100 percent bright leaf tobacco cigarette became known in Britain and the Commonwealth as the "English" or "British" cigarette.[48]

Ginter's success with cigarettes as a foreign novelty encouraged him to continue to focus on overseas markets. The *Richmond Times* recounted that "many orders came from the English agency, and Major Ginter determined to cater to a foreign trade."[49] Richards, who worked for the company for a decade, extended its reach into the Commonwealth, especially to Australia. From this base, Ginter's US sales anchored in a similar demographic of elite, cosmopolitan men in the urban North. The men's club scene was tiny in the US compared to London, but most American clubs had direct ties to London clubs, providing a cultural conduit for British styles. In the US, tellingly, Ginter advertised that Richmond Gems were popular in London.[50] By the mid-1880s, Allen & Ginter had branch factories in New York and London and nineteen salesmen on the road domestically and overseas, had established export depots in Australia, France, Germany, Switzerland, Belgium, and other locations, boasted about established markets in India and South Africa, and delivered its cigarettes to commission agents in "all parts of the world."[51]

Imperial developments threw a wrench into Ginter's plans, however, and the bright leaf cigarette faced sharply elevated competition from the Egyptian cigarette. In 1882, Britain occupied Egypt; steamships, railroads, and telegraph lines quickly formed a coordinated channel for the efficient flow of Egyptian cotton and cigarettes as well as official and newspaper reports, postcards, and photographs to Britain.[52] Before long, fascination with all things Egyptian gained a name: Egyptomania. The US, a rising imperial power in its own right, embraced Egyptomania with nearly the verve of the British. As people in Britain and the US wove imperial fantasies, an Egyptian cigarette proved to be a popular prop. Britain soon became the second greatest importer of Egyptian cigarettes, just behind Germany and edging out India.[53] The Egyptian cigarette transformed cigarette markets in both Britain and the US and shaped the ways Ginter, Duke, and the ATC would develop and market products for decades.

The Egyptian Cigarette

The new Anglo-American markets for Egyptian cigarettes was just one factor in the industry's boom in the 1880s. The Egyptian industry had been growing steadily since the 1860s, when the US Civil War disrupted the sale of Southern US cotton in the international market. Egyptian cotton rushed into the huge void, and the Egyptian economy surged, making new ventures in cigarettes possible. In 1883, the Ottoman Empire established a state-run tobacco monopoly, which forced the closure of approximately three hundred Greek-owned factories. Many of these proprietors relocated to Egypt just in time to serve the new British market.[54] The first globally circulating cigarette was the Egyptian cigarette: Egyptian manufacturers produced far more cigarettes and reached thousands more customers than the newcomer, bright leaf.

Egyptian entrepreneurs' cigarette branding methods played to British and US orientalist fascination by depicting on their packages and in advertisements iconic images of pyramids, obelisks, the Sphinx, Islamic architectural domes, stereotypically sensual women of the Nile, and feluccas (ancient Egyptian rowing boats). Nominally part of the Ottoman Empire, but not subject to its state-run tobacco monopoly, Egypt could also claim the legacy of the Turkish tobacco cigarette, with the appeal of the Sultan and the harem.[55] As one commentator

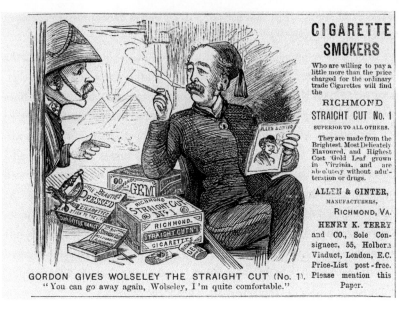

Fig. 1.2 Allen & Ginter advertisement, *The Illustrated London News*, October 18, 1884.

put it, "The Egyptian cigarette . . . is merely a Turkish cigarette made in Egypt."[56] Still, the reputation of Egyptian cigarettes for both quality and mystique rose exponentially and the Egyptian industry became the standard bearer in many parts of the world, including Britain and the US.

Ginter responded to the rise of the Egyptian cigarette with an irreverent advertisement that directly commented on British imperialism. In 1884, a rebellion against the British occupation of Sudan, then administered by Egypt, led to the decision to withdraw from that country. The British government sent Major-General Charles George Gordon to help evacuate Egyptian soldiers and administrators. Gordon disagreed with the plan, so he organized the soldiers to fight the rebels instead, but was unsuccessful and became trapped. In early October, Britain sent a second general, Garnet Wolseley, with Canadian and British troops on an expedition up the Nile River to rescue Gordon and his men.[57] The cigarette ad published on October 18 in the *Illustrated London News* featured a drawing of Wolseley arriving, but Gordon, happily smoking Richmond Straight Cuts and wearing a fez, sends him away, saying, "You can go away again, Wolseley, I'm quite comfortable" (fig. 1.2).[58] The joke, complete with pyramids in the background, inserted bright leaf cigarettes precisely into the circulating news and imagery of empire that had given Egyptian cigarettes an edge.

Another Allen & Ginter advertisement showed the Old Smoker character drawn from the Joseph Ducreux self-portrait "Introducing to the Nations Allen & Ginter's choice Cigarettes & Tobaccos" (fig. 1.3). The Old Smoker presides as world leaders sample Allen & Ginter's cigarettes from a table piled with boxes of the company's most prominent brands. The figures, save one, resemble actual state leaders; London clubmen could enjoy identifying them all.[59]

The figure of the Sultan anchors the ad squarely in imperial tropes: unlike the other leaders, the figure does not resemble the current Sultan, but is a generic stereotype, invoking the Sultan without legitimating the actual leader as an equal among the rulers of the West. Even the Sultan eagerly receives the Allen & Ginter bright leaf cigarette. The ad thus displaces the Egyptian industry's recent

Fig. 1.3 Allen & Ginter advertisement, *The Illustrated London News*, December 1, 1884. From the figure standing at the left and moving clockwise the figures are: British Prime Minister William Gladstone; Russian Czar Alexander III; US President Chester A. Arthur; the Old Smoker; German Chancellor Otto von Bismarck; King Umberto I of Italy; Prussian Crown Prince Frederick; the Prince of Wales; and the Sultan.

success in defining taste in Britain with an image of Allen & Ginter cigarettes as the global leader and tastemaker.

Ginter repeated this tactic in the US in a marketing story that he released, not as an advertisement, but as an item of news. When Congressman A. M. Hewitt was in Turkey on official business, the story went, he accepted an invitation to dinner with that "great connoisseur of Oriental smokers," the Sultan. Hewitt "was surprised when a box of 'Richmond Straight Cut, No. 1' was handed him. 'Why,' said he, 'this is unusual. I had expected, of course, that His Majesty would smoke his far famed domestic brand. Is this done as a compliment to me?' 'I much prefer the Richmond Straight Cuts,' replied the Sultan. 'They are my favorites.'"[60] The story appropriated the authority of the Sultan over cigarettes while offering an image of Western industrial primacy. Ginter, then, attempted to combat one imperial fantasy with another.

In Britain, despite the inroads of the Egyptian industry, the 100 percent bright leaf cigarette remained "the British cigarette" for decades, but it never became "the American cigarette" for US consumers. The Egyptian trend certainly changed the game in Britain, though. The Allen & Ginter and W. D. and H. O. Wills companies maintained their bright leaf cigarette sales, but Duke, who began exporting cigarettes at this moment, could not get his brands established there. But the effect of the Egyptian industry on the bright leaf cigarette may have been even more profound in the US, where "the Virginia cigarette," as it was called, held no foreign cachet. Indeed, there would be no clear "American cigarette" until the 1920s, when RJ Reynolds's Camel cigarette achieved unprecedented sales. The American cigarette became Camel's distinctive blend of Turkish, bright leaf, and burley tobaccos, soon copied by other companies.

It is only possible to understand why RJ Reynolds named its brand "Camel" and included expensive Turkish tobacco in the blend by acknowledging the long-lasting presence of the Egyptian industry in the US. Nestor Gianaclis was likely the first Egyptian producer to export directly to the US, beginning as early as the 1870s. He had a well-established export business based in Cairo before the boom of the 1880s and was especially prepared to take advantage of it. In addition, Sotirios Anargyros emigrated from Greece to New York City in 1883, when the Ottoman state tobacco monopoly displaced his business. Anargyros imported the Egyptian-made brand, Egyptian Deities, and produced several successful brands, including Helmar, Murad, and Mogul, the last of which carried the slogan "Just like being in Cairo."[61] Within a decade, there were at least ten Greek-owned companies making "Egyptian" cigarettes in New York City, with additional factories in Boston, Philadelphia, Chicago, Milwaukee, and other cities. Brands like Fatima, Mogul, Ramesis II, Contax, and Skinazi "were in circulation [in the US], whose producers were Greeks, whose names were Egyptian, and whose tobacco was from Macedonia, Smyrna, Aginion and Samsun." Especially in New York City, but soon in many major cities, the Egyptian cigarette—imported, made by a branch of a foreign company, or by an immigrant company—soared in availability and popularity.[62]

The Egyptian cigarette gained in popularity quickly in the US in part because it circulated not as an isolated product but as part of a public scene that celebrated imperialism and the exoticism of the East. People encountered and participated in Egyptomania within international expositions, amusement parks, restaurants, and the smoking rooms of elite private homes. Participatory concessions at fairs, such as the Streets of Cairo at the 1893 World's Columbian Exposition, introduced Egyptian cigarettes as part of a much more comprehensive sensual experience. Developed and managed by an Egyptian entrepreneur, the Streets of Cairo included a mosque, coffee shops, camels, snake charmers, and dancing women.[63] Sights, sounds, and smells of Egypt enveloped the consumer while rugs and fabrics begged to be touched and Egyptian cigarettes and food engaged the taste buds. So popular was it that an entrepreneur established a copy at Coney Island that remained open for ten years. Why not smoke a Mogul brand cigarette that carried the advertising slogan "Just like being in Cairo" when stopping at a coffee shop in the Streets of Cairo concession on Coney Island?

Seemingly private spaces also become part of the public scene of the Egyptian cigarette, as well-off urbanites used imports to decorate smoking rooms and parlors in fashionable Oriental style. Import firms like Vantines, with a warehouse on Broadway in New York City, offered Egyptian rugs, furniture, clothing, and household items to US retailers and consumers. Smoking rooms for men expressly made cigarette smoking the purpose of the space. Oriental rugs, reclining couches, curtains, and other textiles encouraged luxurious repose. Men might retire with guests to the smoking room after a meal for male camaraderie and cigarettes. Women also decorated parlors or "cozy corners" in Oriental style, using lushly colored pillows, rugs, and draped wall curtains to evoke the female space of the harem. Here too, Oriental divans invited a reclining pose rather than the upright Victorian posture expected of respectable women.[64]

Smoking rooms and cozy corners could evoke both British sophistication and an imperialist vision of the opulence and sexual privilege of the Ottoman man and his harem. British imperial thought displaced sexual aggression onto Ottoman men in order to define British culture as rational and sexually ordered; such a system unintentionally invited sexual longing to flourish within fantasies of otherness. Both British imperialism and the Ottoman male elite conjured images of privileged masculinity that could be enacted in the US smoking room and ritualized through the smoking of an Egyptian cigarette.

Despite the advent of the cigarette machine in the 1880s and the 1890 formation of the ATC monopoly, Egyptian and immigrant cigarette producers continued to expand and define US tastes into the twentieth century.[65] Likely to evade the US's high tariff, Nestor Gianaclis opened a branch factory in Boston, which he soon moved to New York City. He often advertised his brand "Nestor" with the slogan "The Original Egyptian." Another well-established producer in Egypt, Melachrino, migrated to the US and began production in New York City in 1904.[66] One observer noted that by the turn of the twentieth century, in addition to "big Turkish cigarette factories, such as Surbrug, Schinasi, the Khedivial, Monopol, and Anargyros," there were also "numerous small factories founded on the East Side by Russians, Hungarians, Syrians, Greeks, and Armenians for making the cigarette of their own lands." In 1906, this observer counted "over one hundred small factories where Russian cigarettes are made. The owners and operatives are Russians, Roumanians, Slavonians, and Hungarians, of whom a majority are Jews." These immigrant producers blended tobaccos "from Turkey, Syria, Greece, Herzegovina, and Persia, not to speak of Virginia and North Carolina."[67] From a strong dominance of 90 percent of the market in 1895, the ATC dropped to 75 percent of the market by 1904.[68]

Egyptian and immigrant producers could compete against the powerful domestic monopoly in part because they used a separate distribution system. Native-born competitors cried foul when the ATC used its monopolistic power to assert new control over previously independent tobacco distributors by requiring them to sign exclusive agreements with the company. Competitors suddenly could not get product onto the shelves. Larger Egyptian producers, however, simply opened their own shops in major cities, as Ginter had done in London, and capitalized on the exotic and novel reputation of their goods. Immigrant producers reportedly sold one half of their output on New York City's

Fig. 1.4 The ATC bought Anargyros in 1899 and made this advertisement in the 1910s, using long-lasting orientalist themes. From the collection of Stanford Research Into the Impact of Tobacco Advertising (tobacco.stanford.edu).

Fig. 1.5 Egyptian producers built US factories to evade high tariffs. *New York Times*, February 14, 1906.

Lower East Side, one third in other immigrant neighborhoods in New York and other cities and, in an intriguing mix of old sales methods and new geographies, peddlers hawked the remaining one sixth "in the stores and business offices of the metropolis."[69]

Failing at beating the Egyptian industry through advertising, the ATC purchased the S. Anargyros Tobacco Company in 1899 and also attempted to gain control over tobacco sources in the East.[70] The fastest-growing cigarette brands in the US between 1900 and 1910 were Turkish–bright leaf blends with Egyptian-style branding made by both Egyptian companies and the ATC.[71] A 1904 elite club banquet held in New York City and attended by John D. Rockefeller included a cigar and a "Turkish cigarette" at each place setting.[72] One 1906 observer remarked that the Turkish tobacco cigarette "is largely replacing the Virginia cigarette, especially in the large cities and summer resorts. At nearly all formal dinners and supper-parties, Turkish cigarettes are served with the coffee."[73]

The Queer Cigarette in Britain and the United States

Another blow to US and British marketers that had a lasting impact on the character of the cigarette was the Anglo-American view of the cigarette as effeminate, a view that grew in intensity with the Egyptian cigarette craze and the Oscar Wilde homosexuality scandal that cast suspicion on the homosocial ethos of the gentlemen's clubs. The imperial discourse of civilization had long ascribed greater manliness to northern white races and effeminacy to the men from countries in southern Europe, the Middle East, and Asia. Cigarette smoking became tied to this worldview. *The Smoker's Guide, Philosopher and Friend*, a book published in the West End the same year that Ginter placed his first cigarettes in London, explained the connection. "[The] Man of the North, with his coarse and dull faculties, rejoices in strong Tobacco. The Man of the South, more sensual, more effeminate, prefers mild and aromatic Tobacco—'Turkish,' and that sort of thing, which is only Tobacco in name and pretensions." *The Smoker's Guide* was a light, entertaining book, likely meant for perusal in the gentlemen's club smoking rooms. In other words, it targeted the same market as Ginter did. The author waxed eloquent about pipes and men but averred, "In truth, there is nothing manly in the cigarette."[74] Ginter's success in entering this market, then, may have rested on offering a "northern" and Anglo, and therefore more manly, cigarette.

British gendering of colonial men only intensified in the 1880s, especially men in the Middle East and India, all understood in general terms as cigarette smokers. Bengali men who migrated to the immigrant neighborhoods of London became newly targeted as effeminate when they gained education and positions in the civil service, threatening notions of primitiveness that justified colonial rule in India. The Egyptian craze was even more threatening. While the Sultan surrounded by his harem was a highly stereotypical image of masculine sexual prowess, homosocial male cultures in the Ottoman Empire came under increasing scrutiny for being effeminate and sexually deviant.[75] Such imperial discourse about effeminacy served to reinforce, in contrast, a bourgeois British nation based on the British man as head of a normative family. "To us," wrote the London journal *Tobacco* in the late 1880s, "the smoking of cigarettes savours of the effeminate, and is not suited to the English nation."[76] The Egyptian cigarette could enable men to embody the prowess of the Sultan, but what if it had other effects?

In the 1890s, the rise of sexology joined with bourgeois imperial thinking and generated a sex panic surrounding Oscar Wilde and the elite male culture of the gentlemen's clubs, the very culture that had contributed to the branding of Allen & Ginter's bright leaf cigarettes. In 1895, Oscar Wilde was convicted of gross indecency for his relationships with young men. Had Ginter learned of this before his death in 1897, he certainly would have had reason for personal alarm. Though the clubs were not explicitly implicated, their homosocial culture did indeed enable same-sex relationships of many kinds, including sexual ones, though these developed in other locations as well. Until the 1890s, the clubs, aestheticism, and intense relationships between men that were common to club culture were beyond reproach and unquestionably manly in both Britain and the US.

Ginter's experience in Richmond was a case in point: no one ever questioned his aestheticism or his relationship to John Pope. Indeed, his obituaries celebrated these elements that would soon be signs of deviance. The *Richmond Dispatch* recalled that "he went to work as soon as he was able to earn money, and a natural instinct and inborn taste for the artistic led him to seek employment in stores where fancy articles and art fabrics were for sale."[77] Ginter's stately home included an extensive library, lush smoking room, and massive dining room. "From his early manhood he was fond of a nice table," enthused the *Richmond Times*. In addition, "The tapestry and bric-a-brac in his home show itself wherever one turns. Hundreds of beautiful and valuable souvenirs he picked up in foreign climes during his worldwide travels, and these show the love of the man for the beautiful and aesthetic." Ginter's embrace of an aesthetic sensibility went hand in hand with his often-voiced disdain for "stiff-necked" Puritans and their descendants.[78] As in the British clubs, Ginter's was an elite embrace of pleasure that valued aesthetics and did not fully conform to bourgeois values of work and thrift.

Ginter and Pope spent about twenty-five years together, parted only by Pope's death. When Allen retired in 1883, Pope became a partner in the company and Ginter and Pope gained recognition as life companions. The *Richmond Dispatch* celebrated the emotional bond between the two men, writing "Mr. Pope never married, but lived quietly with Major Ginter, for whom he possessed the most ardent affection."[79] In 1894, the *New York Times* society page mentioned Ginter and Pope's vacation, taken with other leading Richmond families, at the Hot Springs resort, noting, "Major Lewis Ginter and John Pope of Richmond have taken possession of one of the Virginia cottages."⁸⁰ The *Richmond Times* confirmed Ginter's orientation toward men, noting, "Major Ginter never married. His manner towards women was tender, considerate, at times gallant, though he never pointedly sought their company."⁸¹ Ginter's intimate dedication to Pope was clear to all and utterly respectable. If he learned about Wilde trial or heard Wilde's defense of the beauty of relationships between older and younger men, he would have been shocked that such a defense would have to be made.

Wilde, too, assumed that his relationships with men would be understood to be a noble part of British elite masculine culture, while the sexual license accorded to elite men would protect his private sexual relations as long as he followed rules of decorum.⁸² He correctly read the codes in the 1880s but by the 1890s, increased scrutiny about "perverse" sexualities in London's commercial districts and rising sexologist theories caused greater attention to fall on elite men's activities.⁸³ Wilde's scandal, and his genuine surprise about it, is indicative of the transition to a more binary understanding of sexual identity—gay and straight—that would characterize the twentieth century.

If this homosocial, elite male culture came to carry the markers of gayness or camp—the attention to aesthetics, especially—that is simply because it was the first public staging of masculinity that included same-sex desire to become known, studied by experts, and declared deviant. The cherished aesthetic flourishes that the *Richmond Times* celebrated as Ginter's particular gifts came only later to seem gender-deviant and queer, and indeed, to define queerness. In other words, the first public scene of the bright leaf cigarette also became the first public scene of homosexuality and the codes of masculinity associated with both became redefined and stigmatized as "effeminate."⁸⁴ As elite male culture came under suspicion, Allen & Ginter's branding of the bright leaf cigarette within that culture became regrettable.

The growing gendered and sexual stigma in Anglo-America about the cigarette may explain the slowness of many US men—even men in the cigarette business!—to adopt cigarettes. At an early ATC stockholder's meeting, chaired by James B. Duke, the *New York Times* reported, "Not one of [the stockholders], either in person or by proxy, uses his own product. The air became more or less clouded at the meeting, but it was from the smoke of cigars, which were smoked man-fashion [chain-smoked], as if the gentlemen were addicted to the weed in that form."⁸⁵ Duke continued smoking his cigars man-fashion throughout his life. "Take care, my Boy," wrote a friend when Duke was in his sixties, "twenty-four cigars per day is too much for anyone to smoke, and it will get you."[86] Though Duke sold cigarettes for thirty years, he never, ever smoked them.

The Western Cigarette in China

The 100 percent bright leaf cigarette set the standard in China in a way it could not in the US and Britain. Though the 100 percent bright leaf cigarette never became widely popular in the US, in China it became both "the American cigarette" and "the British cigarette," as well as more generally known as "the Western cigarette." By the early twentieth century, Chinese companies that made cigarettes mostly offered bright leaf cigarettes in order to compete with the Western cigarette, and they used brand names written in English lettering to give their products a foreign air. The cigarette that boomed in the 1920s, Ruby Queen, was a 100 percent bright leaf cigarette. Like everywhere else in the world, other kinds of cigarettes were available, but the bright leaf variety became the tastemaker, a huge boon for BAT. It remained dominant through the height of foreign influence and the Mao era. Only when China opened up again to foreign trade in the 1980s was the bright leaf cigarette's dominance significantly challenged.

Bright leaf cigarettes were just one type among many available in China in the late nineteenth century. Cigarettes likely entered Chinese coastal markets sporadically as soon as they did New York City, that is the 1850s, but as in the US a steady stream began later. By the late 1870s, cigarettes flowed to South China via British and American merchant houses in the northern Philippine island of Luzon, particularly cigarettes produced of Luzon tobacco by Mestizo and overseas Chinese companies, including the brand Bronze Drum.[87] In 1885, Allen & Ginter signed a contract for its bright leaf brand, Little Beauties, with Mustard and Co. commission agents in Shanghai. Chinese historian Fang Xiantang claimed this as "the formal introduction of the foreign cigarette to our country."[88]

Duke also may have exported to China in the late 1880s. In June 1882, W. Duke and Sons sent salesman Richard H. Wright on a nineteen-month tour to place the company's smoking and chewing tobacco—but not yet cigarettes—in key export centers. Wright traveled to Europe, South Africa, India, the East Indies, Australia, and New Zealand to establish connections with agents. The Dukes had begun cigarette production on a small scale in 1881; by 1885, production had increased enough that they could be added to the company's offerings.[89] Soon after, the Dukes' Pinhead brand appeared in many countries, possibly including China. In 1890, when Ginter and Duke joined forces in the ATC, the new company maintained the contract with Mustard and Co., and Pinhead cigarettes certainly flowed to China (fig. 1.6).

By the 1890s, the field became more crowded and more production occurred on Chinese soil. The Japanese Murai Brothers Tobacco Company offered Peacock cigarettes and the major Egyptian producer M. Melachrino and Co. had a depot in Shanghai.[90] Russian exporters and small Chinese producers also entered the fray.[91] Several American, Japanese, Russian, and Egyptian entrepreneurs or merchant houses founded small factories, including the American Trading Company (owned by Mustard and Co.), the Murai Brothers Tobacco Company, the Russian A. Lapato and Sons (in Harbin), and the Taipei Tobacco Factory, also known as the Egyptian Cigarette Manufacturing Company.[92] Chinese entrepreneurs also opened up factories, including the short-lived Maoda Rolled Tobacco Manufactory (in Hubei Province) and the Fanqingji workshop in Shanghai.[93] The British W. D. and H. O. Wills exported bright leaf cigarettes through a British commission house, Rex and Co. The ATC's commission agent, Mustard and Co., worked with the Chinese distributor, Wu Tingsheng, while Wills' agent contracted with Zheng Bozhao. When the ATC and Imperial (which included Wills) joined to form BAT, the major bright leaf exporters to China became consolidated in one company. BAT maintained the contract with Mustard and Co., rather than Rex and Co., but it arranged to maintain a relationship with Zheng Bozhao.

Though US and British companies began work in China much as did others, their bright leaf cigarette became dominant after 1900 for a number of reasons. After 1902, BAT put more resources on the ground in Shanghai, including foreign personnel and advertising, than any other company. By 1907, the Taipei Cigarette Company moved to India, reportedly because it could not compete with BAT in China.[94] Chinese companies also struggled, in part because BAT's foreign representatives took them to court for trademark infringement. Because the cigarette was a foreign novelty in China, Chinese companies copied foreign packaging practices, including using Western lettering and Western-style pictures. This was not so different from US producers copying Egyptian brand names and package imagery. If Chinese companies came too close to copying an established Western brand, however, CE Fiske would sue them for trade-

Fig. 1.6 Shanghai BAT cigarette vendor, 1905. Advertisements for Pinhead and Atlas cigarettes still used Western imagery. Duke University Archives.

mark infringement in the foreign court in the International Settlement, where Chinese producers invariably lost.[95] This use of the foreign court is an excellent example of the kinds of advantages available to foreign companies within corporate imperialism.

These moves certainly helped the bright leaf cigarette get established, but they did not make it the standard bearer. Cigarettes became defined as Western before BAT began its significant buildup to being a major producer in China. BAT's bright leaf cigarette became the standard bearer among Chinese people because it circulated in two significant public scenes: via the figure of the high-class courtesan and through the anti–American goods movement of 1905. Ironically, it was not just imperialism that defined the cigarette in China, but also anti-imperialism.

Merchants from Guangdong, such as Zheng Bozhao specifically, became the most important innovators in shaping the emerging character and charisma of the cigarette in China. Very few Chinese people adopted cigarettes in the first decades of their limited availability. Companies distributed cigarettes almost exclusively in the treaty ports, and a large part of their market was foreign business and military men, a fact that underlined the foreign reputation of the goods. In this way, the early introduction of cigarettes in Shanghai bore a resemblance to their introduction in New York City and London, where foreign residents were the first smokers. In contrast to cigarettes' prevalence, pipe smoking in China was nearly ubiquitous. Locally grown pipe tobacco came in a number of grades, the cheapest of which was vastly less expensive than foreign cigarettes. Most Chinese people who did begin smoking cigarettes lived in the treaty ports, had expendable income, and had contact with Western businessmen.

Fragmentary evidence suggests that members of the famous merchant guilds of Guangdong were among the first to smoke cigarettes. Presbyterian missionary Fred Porter Smith recalled seeing Guangdong people smoking cigarettes from the Philippines in the 1870s, and Tianjin resident Zhang Tao recalled seeing Guangdong merchants smoking cigarettes while in the city on business in 1884.[96] Zheng Bozhao's trading company fits precisely this description. The company reportedly owned a cigar factory in the Philippines (which may have also produced cigarettes).

Guangdong merchants handling cigarettes undoubtedly sold them to courtesans in Guangzhou, the capital city, and it made good business sense to do so. High-class courtesans became important cultural intermediaries in the promotion of the Western cigarette precisely because they interacted with the international business community in treaty ports. Pipe tobacco had been part of courtesan practices dating to the seventeenth century; around the turn of the twentieth century in Guangzhou and Shanghai, cigarettes came to replace the pipe in courtesan practices. Furthermore, Guangzhou's courtesans were early adopters of Western products. Many Shanghai courtesans came from Guangzhou and others often took cues on Western appropriations from Guangzhou's courtesans.[97]

Shanghai was a unique, imperial space by the turn of the twentieth century precisely because hundreds of foreign businesses from Western Europe, the US, and Japan made their China headquarters in Shanghai, and thousands of foreign employees lived in the French and International Settlements, where they enjoyed extensive privileges. Western architecture dominated in the business district, or Bund, as well as the foreign settlements. Stores on Nanjing Road offered foreign and Chinese commodities in the latest fashions. Foreigners had such a strong stamp on Shanghai that, as historian Antonia Finnane put it, "In the eyes of both Chinese and foreigners, Shanghai seemed a foreign city. To the Chinese it was 'the ten-mile Western quarter.'"[98] Indeed, Shanghai bore some resemblance to world fairs in its monuments to Western industry and its cosmopolitan mix of people.[99] Shanghai courtesans played a very public role in this context. Their houses were located in the foreign settlements and they built their public images among Chinese people through their engagement with Western material culture, including cigarettes.[100]

Courtesan houses and courtesans themselves became so important to the early scene of the cigarette in China because they played a critical function in Chinese business culture, particularly through the business banquet, an indispensible aspect of Chinese business. An ambitious Chinese businessman would host a banquet to honor superiors, make new business connections, and announce his own status through the generosity of the food, drink, and entertainment and the prestige of the guest list. Banquets carried a reciprocal expectation: if one had been a guest at a banquet, eventually he must reciprocate the favor. In this way, banquets were critical to the formation of Chinese business networks and the lubrication of business deals.

Elite courtesans threw such banquets as one of their main services and sources of income. A Chinese businessman contacted the female proprietor of a courtesan house and told her the numbers of guests and his price range; she took care of everything, including location, food, drink, and entertainment.

A large house might have its own banquet room; smaller houses would contract with restaurants. Sex acts were not part of the package; rather the courtesans contributed refined beauty and artistry—they often sang and danced for guests—along with glamour, companionship, and an aura of desire.[101] The long-stemmed pipe or water pipe was a longstanding component of the banquet; when courtesans incorporated cigarettes, they gave cigarettes an important public role in Chinese culture.

Foreigners also attended such banquets when doing business with Chinese men. Indeed, it is possible that business banquets with cigarette company representatives were one way that courtesans first encountered the cigarette, for representatives brought their wares to all social occasions. James Hutchinson, a BAT employee from North Carolina in the 1910s, explained that "the Chinese men never received their friends in their homes. They gave each other dinner parties, and none was complete without the sing-song girls. A girl sat behind each guest and acted as companion and entertainer."[102] Lee Parker, who attended a Chinese cigarette dealer's banquet in Wuzhou, noted that "sitting back of each man was a sing-song girl robed in bright colored silk. Her training from infancy had been to make men happy. She smiled, she laughed, she lighted cigarettes and cracked melon seeds, all the while she was chattering and giggling." The girls sang singly and in groups and danced for the guests. "It was all graceful, charming and very beautiful,"[103] recalled Parker.

Courtesans were important intermediaries for the incorporation of many Western products, including furniture, decorative objects, fashion, and cigarettes. Courtesan Hu Baoyu created an entire Western room in her quarters with all imported furniture where she entertained both Chinese and foreign clients. Courtesans could buy Western-style lamps, furniture, and knickknacks at a variety of stores on and near Nanjing Road. Some shops also offered Guangzhou-made furniture in Western styles. When an elite courtesan was building a more intimate relationship with a Chinese client, she would invite him to her quarters and there would share with him a pipe—or a cigarette.[104] The top courtesans were also fashion trendsetters. Known by name, their actions covered by the tabloid press, they were famous for being "innovative, even outrageous." A certain degree of iconoclasm was a long-standing privilege for these women, who stood outside the confines of Confucian codes for female movement and behavior, but their style took on new connotations in the rapidly changing world of Shanghai as "curious, fascinating and exotic."[105]

That cigarettes did worldly work is especially clear when we consider them in relationship to the couch. Courtesans had Western-style couches, but they looked remarkably like the couches that US customers saw as "Oriental" and placed in cozy corners and smoking rooms. The couch invited new, intimate postures in contrast to traditional furniture, which conveyed "distance, order, and hierarchy."[106] Some courtesans bought rattan couches that were made in Guangzhou expressly for export to Western markets. The identical couch could be an "Oriental" good in an import store on Broadway in New York City, and a "Western" product on Nanjing Road. Likewise, the cigarette was "Oriental" in the US and Britain, but "Western" in China.

Courtesans also served as cultural intermediaries by smoking cigarettes—and cigars—on the street, usually in their carriages but sometimes when walking. The newspaper *Shenbao* complained about this practice: "I have seen Western-clothed prostitutes in carriages with cigarettes in their hands, spewing smoke incessantly along the journey."[107] Appearing in public was new in itself for courtesans in the late nineteenth century, as they began attending public entertainments and taking early evening carriage rides through Shanghai, acts that increased their notoriety.[108] When newspapers criticized courtesans' use of cigarettes, the transgression was not smoking itself, for pipe smoking had long been commonplace among both women and men, but doing so in public, outside the Confucian courtyard house or the courtesan house that kept women out of sight. Courtesans also occasionally smoked in public while attired in Western business suits made to fit them by their expert tailors. In doing so, they staged a carnivalesque gender and cultural inversion, irreverently donning the costume of male Western business power, complete with cigarette.[109] Through all of these uses, courtesans gave Chinese affects and cachet to Western cigarettes, while underlining their Westernness.

Perhaps even more important to the emerging definition of the cigarette as Western than courtesan's uses, however, was the anti–American goods movement of 1905, which particularly targeted BAT cigarettes. As in Britain, imperial events created unanticipated problems. The anti–American goods movement was a global Chinese protest against US immigration policies. The US had passed Chinese exclusion laws in 1875 and 1882, the first and only federal laws to deny entry to a particular group explicitly on the grounds of race. Between 1898 and 1904, deportations and harassment of Chinese people entering the US increased in response to fears that Chinese people would enter through the newly acquired US protectorate of the Philippines. In 1904, the exclusion law and corresponding treaty with China were up for renewal, and the Chinese government hesitated to sign, encouraging activism already brewing in China against the exclusion policy.

Through the summer of 1905, a boycott of American goods spread through ten provinces and was especially strong in Guangdong, where most migrants to the US originated, and in the city of Shanghai. BAT felt the hit. Support for the boycott spread through overseas Chinese communities in Havana, San Francisco, Seattle, Vancouver, Manila, Penang, Bangkok, Saigon, Yokohama, Nagasaki, and Kobe. These communities put pressure on the US government and sent monetary support to the boycotters. In China, the boycott crossed class lines and is said to have launched the anti-imperialist National Products Movement, which promoted Chinese-made commodities. The boycott of BAT cigarettes lasted through the spring of 1906 in Shanghai and the end of that year in Guangzhou.[110]

Protesters taught consumers that BAT cigarette brands were American goods, identifying them as such on hundreds of posters. Agitation against the American cigarette spread through Guangzhou in part through love songs, with lyrics modified to signal a breakup with the American cigarette. One such song, "Elegy to Cigarette," positioned the male smoker and the cigarette as estranged lovers. "I thought our love affair would remain / Unchanged until earth and sky collapsed / . . . Then this movement against the treaty got underway / And spread everywhere / Because America mistreated our Overseas Chinese / . . . . Ah cigarette, / You have the word 'American' in your trademark for everyone to see / So I must give you up along with my bicycle." Such popular songs gained rapid circulation in poor urban neighborhoods as well as among the more wealthy patrons of high-class courtesans.[111] Protesters helped define the cigarette even as they urged people not to smoke them.

Zheng Bozhao advised BAT on its response to the boycott and suggested that BAT deny that its products were American goods and promote them instead as British. (In truth, of course, they were both.) Accordingly, Zheng Bozhao gave Ruby Queen cigarettes a Chinese name, Da Ying, which means Great Britain. The company issued a new cigarette package with both brand names—Ruby Queen in English lettering and Da Ying in Chinese characters—and advertised them as British cigarettes.[112] This was among BAT's earliest changes in branding to satisfy Chinese customers. Eventually, the boycott ended and the company continued to grow. However, BAT's short-term solution would prove to have long-term consequences, for Ruby Queen/Da Ying cigarettes boomed in the 1910s and even more in the 1920s, strengthening the foreign—and specifically Western—identity of cigarettes. The Western cigarette in China, then, explicitly engaged with China's struggles with Western imperial powers. BAT's tactic would come back to haunt the company in the May 30th Movement, when British goods in particular were under fire (see chapter 5).

Just as orientalism invoked for Westerners an East that was fantastical, capacious, and shifting, so was the Chinese courtesan's West much more than a simple reference to certain countries or geopolitical alignments. Especially in Shanghai, Chinese people encountered a mythical notion of the West, not only because their cigarettes and couches came to them largely waiting for their stories to be written, but because the Bund and international settlements themselves were rather spectacular manifestations of foreigners' fantasies. In addition, not all foreigners were from Western Europe or the US: Japan also had a significant presence in early twentieth-century Shanghai, especially in the ownership of massive textile factories, and would come to have an even greater one after military invasion and occupation in the 1920s. For Shanghai people, commodities from the West may well have helped create an East/West binary, but the Chinese East and West were not the same East nor the same West as it was for US and Western European consumers.

Conclusion

In the US, Britain, and China, the early cigarette became central to the very construction of notions of East and West. The cigarette's varied international career and bidirectional flow confounds notions of Western progress and primacy associated with modernity. Nevertheless, the history of the US industry has been told until now as an archetypical story of the triumph of Western technology and marketing, spreading the cigarette from West to East.

While machine-made cigarettes have been the focus of industry history, the hand-rolled cigarette reveals a different, richer story. The Egyptian industry's taste-making power challenged Ginter and Duke in the US and Britain throughout their careers. In addition, it was Ginter who built a very extensive foreign export business by the mid-1880s, when Duke's cigarette business was just picking up steam. Historians have dismissed rather than investigated this early public scene of the cigarette, in part because of the focus on the machine-made cigarette, but also because the elite, masculine culture crucial to Ginter's early cigarette market lost the prestige it had carried in the 1870s and '80s as it came to seem suspiciously effeminate and marked with the stigma of homosexuality. The early hand-rolled cigarette saw a related fall from grace: it became seen as an elite aberration before the machine-made cigarette captured a truly popular and significant consumer base. As a result, this early but critical history of the cigarette has been almost entirely repressed.

The industry did mechanize and consolidate, as Ginter's bright leaf tobacco cigarette became the foundation of the ATC and BAT and grew to have dramatic global consequences. Ginter tied the bright leaf economy, a foundation of Southern racial capitalism, to the cigarette, an international product that circulated and gained resonance through multiple imperial contexts. When the "Western" bright leaf cigarette circulated in China, it gained the ability to define the reputation of the cigarette in a way it had not in the US.

As significant as Ginter was to the early history of cigarettes, Duke did indeed eventually displace him and gain control of the ATC. Yet Duke's triumph was not due to disruptive innovation, but to his ability to align his personal fate to the spectacular empowerment of the business corporation that was underway. Chapter 2 tells a new story of this dramatic moment in the history of cigarettes and of capitalism itself.

2

Corporate Enchantment

Let us begin our story of James B. Duke's unpredictable rise to power in the cigarette industry at a crucial but rarely mentioned moment of his change in fortune. On December 19, 1889, Duke's family company joined a new corporation called the American Tobacco Company, which was headed by Lewis Ginter, chartered by the state of Virginia, and headquartered in Richmond. This new corporation linked the five most significant companies run by native-born cigarette manufacturers in the US: Allen & Ginter, W. Duke and Sons, F. S. Kinney, W. S. Kimball, and Goodwin tobacco companies. The corporation was capitalized at $1 million with the right to expand that capital at will. The corporation did not fully merge the five companies; rather, the former company owners owned the new corporation's stock.[1] The incorporators, however, were all tobacconists from Richmond. The new corporation gained immediate publicity because it suggested that the native-born cigarette producers would function as a monopoly.

In a dramatic series of events, however, the Virginia legislature erupted in controversy and rescinded the charter only one week later. Their plan scuttled, the five manufacturers quickly sought a corporate charter from another state. They turned to New Jersey, the apex of rising corporate empowerment in the US, and there gained a corporate charter in January 1890; this time, significantly, the corporation was headquartered in Newark, with Duke replacing Ginter at the helm. This corporation did merge the five companies, all of which ceased to exist except as factory branches of the ATC. The New Jersey charter resembled the rescinded Virginia charter in its division of stock between the former owners—Allen & Ginter and W. Duke and Sons received an equal share—and by the former owners' membership on the board of directors.[2]

Duke had enjoyed earlier successes, but this was the turning point that portended his coming power as a global capitalist. His head must have been spinning. By a fluke event, Duke was now positioned, at the relatively young age of thirty-three, to direct the brand new cigarette monopoly. Furthermore, the ATC had just acquired the most lenient corporate charter possible in US history. Recent reforms of New Jersey's incorporation laws created new entitlements for corporations and put few limitations on monopolistic practices.

Yet Duke immediately faced challenges to his power from within and without the company. New Jersey's laws created widespread controversy; no consensus existed on capitalism's rightful direction. A growing antimonopoly movement organized against the ATC, challenging it on multiple legal fronts. Even within the ATC, Duke faced significant opposition to his leadership from Ginter, who had other ideas about how to run a business. By the end of the century, however, Duke had prevailed and enjoyed unprecedented power. How did Duke so completely overcome opposition?

The answer does not lie in innovation with the cigarette machine, as so many have claimed, but in Duke's ability to align himself with a dramatic process of corporate empowerment. Duke had not gained clear domination over the cigarette trade when the ATC formed; nor was he the corporation's inevitable leader. The important story of Duke's power begins with the formation of the corporation in New Jersey, on the cutting edge of the transformation of the corporate form itself. For the ATC, the combination of New Jersey's incorporation laws and the Fourteenth Amendment was to prove to be almost magical. With the ATC's triumph, corporate personhood emerged in a newly animated—indeed, we might say enchanted—form that wrested free from the traditional regulatory function of states. In addition, corporate empowerment resulted from the growing importance of the stock market and its creation of enhanced forms of property. Duke did not pull himself up by his innovative bootstraps; rather, he rode the coattails of the rising corporation.

The ATC's dramatic empowerment also dovetailed with waxing US imperial might, making possible the development of a novel kind of corporate imperialism. By the 1870s and '80s, the US had made a series of incursions in the Pacific, staking claims to ports in Hawai'i, Samoa, and Korea. With the War of 1898, the US colonialized the Philippines and declared its presence as an imperial power in Asia. Military and diplomatic muscle created opportunity in the Pacific, but the only way to siphon resources from imperial outposts to US pockets was to build institutions there. The newly empowered multinational corporation would become one such institution, but its status as a "private" business enterprise obscured its role.[3] Significantly, the ATC and BAT, under Duke's leadership, were at the cutting edge of these developments as well.

Corporate personhood itself was not new: the corporation had long melded a large set of business agents and interests into a single legal entity, as both US and British systems took the individual as the unit of law. In the late nineteenth century, however, business corporations successfully extricated themselves from longstanding tethers of government control and gained protections as "private" entities that many human persons concurrently struggled in vain to claim.[4] Brands loosened from their previous status as symbolic elements of ownership and sale and became aspects of stock assessment, enumerated for the residue of past value and their potential for future profit.[5]

It may seem facetious to refer to the change in the business corporation as magical or as an enchantment, but I mean this quite seriously. Our long reliance on ideas of modernity and capitalism has made it difficult to see how deeply our economic system is simultaneously a belief system. Westerners have long considered the modern world disenchanted; that is, they believed that industrial and postindustrial societies operated according to a scientific rather than religious cosmology. Max Weber's *Protestantism and the Spirit of Capitalism* famously distilled this idea. Weber wished to explain why some countries appeared to be more advanced economically than others. Writing in the era of corporate empowerment, he argued that countries in North America and Europe had a culture particularly conducive to the development of capitalism, whereas countries in the East did not. In the West, he argued, the social world became disenchanted: specifically, superstition declined and scientific rationality took hold in most aspects of life. The desire for salvation did not disappear, according to Weber, but Protestants sublimated that desire into an experience of a calling to a job and, as a result, worked with particular fervor. In the East—China was Weber's favorite example—superstition and senseless proceduralism blocked the development of such ambition.[6] In this view, the corporation is the opposite of enchanted: it is based on rational investment, management, and bookkeeping.

Weber had a profound and lasting influence on the way the story of the corporation has been told. Key twentieth-century historians such as Alfred Chandler and Robert H. Wiebe praised the corporation's rational management and the enumeration possible with modern business accounting.[7] Since the rise of postcolonial studies, historians have roundly criticized Weber's orientalism but have left many of his linked presumptions about "the West" in place.[8] Few have considered the ways that Western countries remained enchanted and that new forms of enchantment arose flying under the banner of "science" or "economic common sense."[9] But there were magical transformations within the process of corporate empowerment, from the formation of the ATC in 1889, through multiple challenges to the new corporate form, and culminating in the expansion of the China branch of BAT, beginning in 1905. These transformations inaugurated a new era of the corporation and the cigarette.

Forgetting Ginter

When Lewis Ginter (fig. 2.1) decided to join forces with James B. Duke in seeking a corporate charter, he was sixty-five years old and reputedly the wealthiest man in the South. One might expect him to be ready for retirement, but he had his legacy to protect. The cigarette industry was changing rapidly as multiple cigarette machines appeared on the market. In the preceding few years, the five major manufacturers shifted increasingly to machine production and the W. Duke, Sons and Company had become his most significant US competitor. Having built his company on the hand-rolled cigarette, Ginter took steps to adapt to the new conditions and secure his considerable domestic and foreign investments.

Ginter placed great hope in his business partner and life companion, John Pope (fig. 2.2), whom he had groomed to succeed him in the business. Pope was exactly the same age as Duke—both were born in 1856—and he brought a younger man's energy and ambition to the business. Much like Duke knew his father's company, Pope knew every aspect of the Allen & Ginter Company's business. He began working in the factory as a teenager and became a full partner in 1883 after Allen retired.[10] He managed the factory and traveled with Gin-

Fig. 2.1 Lewis Ginter. Courtesy the Valentine.

ter across the company's vast sales territories. In addition, Ginter and Pope had established a family relationship that gained both informal and legal recognition. Ginter legally adopted Pope, thus designating him as an heir and ensuring that Ginter's will could not be challenged after his death.[11] There are other historical examples of wealthy, childless men adopting younger male adults in order to control the inheritance of their assets; some but not all of these were clearly life partners making their relationship legally binding through the mechanism of adoption.[12]

Ginter actively negotiated the Virginia charter of the American Tobacco Company and saw it as a positive step for the future of his business and his hometown of Richmond. When the Virginia legislature revoked the charter, he and Pope publicly expressed their dismay. Ginter complained that newspapers had raised a "great hullabaloo" about their charter and "printed many foolish things," resulting in a serious mistake. Pope told the *Richmond Dispatch*, "We had made arrangements that would enable us to continue our business as it is now conducted and retain our identity with Richmond."[13] The wound was a lasting one. Ginter's obituary seven years later reported, "Not long before his death, Major Ginter referred with regret to the loss which this unwise course of the Legislature had inflicted upon Richmond and the Commonwealth. . . . He said that the city and State would have received taxes to the amount of $200,000 annually if that act had not been recalled. The result was only to carry elsewhere

Fig. 2.2 John Pope. Courtesy Cook Collection, the Valentine.

the benefits which had been secured for Richmond."[14] Though Ginter went along with the New Jersey charter, he never fully reconciled himself to its terms or to Duke's leadership.

None of these elements of Ginter's and Pope's experiences fit within the established story of the ATC, which holds that Duke leveraged the merger by controlling the cigarette machine, cutting prices, advertising heavily, and driving his competitors to the bargaining table. As a result, the Virginia charter and Ginter's role have been forgotten. The myth of Duke as destructive innovator fully flowered in the post–World War II era, when the influential theories of Joseph Schumpeter refracted the ATC story.[15] Schumpeter argued that the innovator was often a newcomer in an established field, and Duke's youth could make him seem so if you discounted his training in his father's business. The innovator used a new technology to create a cheaper, inferior product; in this case, the machine-made rather than hand-rolled cigarette. This allowed the innovator to catch his sluggish competitors unaware as his low prices captured and extended the established market. This disruption of business protocols thereby allowed the innovator to restructure the industry around his newly victorious model, giving him power over his competitors, who had no choice but to follow in his footsteps or fail.

One doesn't have to look far to discover why scholars applied Schumpeter's theories to Duke's story. Schumpeter himself suggested that scholars should look for cases of entrepreneurial innovation that might be preemptively "sifted and arranged with definite hypotheses in mind."[16] In the early 1960s, two business school professors did so, applying Schumpeter's theory to several entrepreneurs, including Duke.[17] A few years later, a scholarly article elaborated the story of Duke as destructive innovator.[18] Alfred Chandler, the father of business history, drew on this article in his masterpiece, *The Visible Hand*.[19] The story solidified and became iconic.[20]

The fly in the ointment of the Schumpeterian interpretation is the inconvenient fact of the generally overlooked first ATC charter of incorporation in Virginia in 1889. If Duke had disrupted the hand-rolling industry with his innovative use of the cigarette machine, if his profits had allowed him to wage price wars and gain the upper hand in the industry, if his prescience next had led him to force his competitors to the bargaining table for a merger on his terms, then why was the first ATC charter obtained in Virginia? Why did it name Ginter as president and have only Virginians as officers? Clearly, Ginter was not quite as defeated in 1889 as the Schumpeterian interpretation might lead us to believe. If Ginter was still a serious competitor, then exactly what was the role of the cigarette machine? And, if the Virginia merger was not a result of Duke's coercion but contained an element of a choice, why did Ginter choose to join in business with Duke?

The cigarette machine *is* important to this story but the critical struggle was over who would receive the exclusive patents for the use of the Bonsack machine in overseas markets. There are three points that need to be illuminated in order for the Virginia charter to make sense. First, there were several cigarette machines coming on line, not just the Bonsack. Second, Duke's advantage with the Bonsack machine was far less significant than has been believed; and third, the Bonsack Machine Company's own foreign expansion practices threatened international growth for all US producers, giving them ample reason to join forces to enhance their leverage. Global ambitions drove this monumental US merger, not Duke's purported competitive dominance.

The Cigarette Machine Revisited

The Bonsack machine very quickly achieved global significance but its origins are a Virginia story. Around 1876 or '77, Ginter announced a contest awarding $75,000 for a fully functional cigarette machine.[21] This was a massive sum — equivalent to approximately 1,730,000 in 2017 dollars.[22] Seventeen-year-old James Bonsack of Roanoke County, Virginia, started tinkering in the barn. Bonsack knew he had competition: other inventors were patenting cigarette machines and bringing them to market. In 1879, Charles and William Emery, owners of Goodwin and Co., patented the Emery machine and installed it in their factory, the first of the major US producers to use a cigarette machine.[23] The Emery was fully owned and controlled by Goodwin and was not available to any other producer. Meanwhile, Bonsack rushed to develop his machine. Who knew how many other Virginia and North Carolina boys had some invention in the barn that they hoped would be life changing?

In 1880, the young Bonsack presented a machine to Ginter. Though Ginter saw merit in the machine and installed one in his factory, he did not award Bonsack the $75,000 prize, for the machine still had too many glitches to put it into full-time production. According to the trade periodical *Tobacco*, "as late as 1886, [the Bonsack machine was] more of a failure than a success."[24] Had he awarded

Bonsack the prize, Ginter would have obtained an exclusive agreement for the machine's use, forcing other US producers to invest in other machines.

Significantly, however, because Ginter did install a Bonsack machine in his factory, Bonsack could not offer an exclusive agreement to any other US producer. Rather, the Bonsack Machine Co. formed to take the machine on the road, negotiating contracts on a royalty basis domestically and exclusive contracts abroad. Duke installed the Bonsack in his Durham factory in 1884 on an experimental basis and began using it on a significant portion of his production in 1885. The F. S. Kinney Company patented the Allison cigarette machine; both Kinney and Kimball used this machine in their factories. Other machines hitting the early market included the Everitt and the Cowman.[25] By 1886, all five major producers had machines—Bonsack, Allison, or Emery—in some of their factories, though all also continued to emphasize hand-rolled cigarettes.[26] When the Bonsack ultimately proved most efficient, Kinney joined Duke and Allen & Ginter in contracting with Bonsack. Only then—1887—did both Duke and Allen & Ginter begin converting their New York factories to machine labor.[27]

In order to understand the transition to machine production, we need to understand the way the hand-rolled industry was transforming. The conventional assumption that hand-rolling is primitive and machine production modern misrepresents the development of the industry in the US, Europe, and Egypt. In fact, there were three labor systems in the rapidly changing world of cigarette production: two handrolling systems—artisan and deskilled—and machine production. While artisan labor did decline, especially in factories run by native-born men, all three forms of labor remained significant to the US and global industry well into the twentieth century.

The first cigarette producers in the US employed immigrant, artisan hand-rollers. Most were Russian Jews and they quickly unionized, typically with the cigar makers union. The Kimball, Kinney, and Goodwin companies all operated on this model, as did most European and Ottoman producers. In the 1870s, however, employers developed a deskilled model of handrolling that divided the cigarette rolling process into steps and used molds rather than skill to ensure correct shape, which dramatically lowered labor costs. In the US and France, this model employed "girls" instead of skilled male artisans. Indeed, "girls" had emerged as a new, deskilled labor category defined as females as young as eight and as old as late twenties who worked before marriage, though actual labor forces that bore this label sometimes included older or married women.[28] Some element of skill with cigarettes remained, but the production process became much more like an assembly line: standardized, segmented, and highly supervised. Ginter pioneered this model in the US when he began cigarette production in Richmond in 1875.[29]

Ginter was likely inspired to employ girls by watching the much-larger cigar industry, which had already begun replacing artisan cigar rollers with European immigrant girls in New York City and Chinese men in California. This deskilling process prompted the rise of the five-cent cigar and the corresponding skyrocket in cigar sales — decades before cigar machines emerged. Cigar employers segmented and standardized the labor process so that rolling a perfect shape became a guarantee rather than an accomplishment. With skill less important, cigar employers increased the size of operations to take advantage of economies of scale. So did Ginter: he soon employed one thousand girls in his Richmond factory. Artisanal cigar workers fought back against this change and, in California, adopted virulent anti-Chinese tactics that directly influenced the development of the Chinese Exclusion Laws as well as the strategy and structure of the American Federation of Labor. Artisan cigar rollers maintained control over the high-grade cigars but lost the growth industry of the cheap cigar.[30]

Ginter encountered few labor conflicts, probably because he hired girls from the outset of cigarette production; in contrast, his competitors faced opposition when they attempted a shift from male artisans to deskilled female labor. The Kinney cigarette factory, founded in 1869 in New York City, initially employed Russian Jewish men; by the late 1870s, these artisans shared the workplace with newly hired Irish girls, ages twelve to twenty. In 1883, the company's 300 young female rollers and 150 seasoned male rollers struck to protest a rate decrease. The male rollers were unionized; they admitted nine girls into their union at the first strike meeting. Meanwhile, the company announced that it "did not intend to employ any more men at cigarette-making. Cigarette-making was almost essentially a woman's work, and the firm advised its men to seek some other and more remunerative business." No record exists of the conclusion of the strike, but two years later, Hyman Grossman, formerly employed at Kinney, shot himself in a New York hotel room, reportedly because he had "lost his situation and was despondent."[31]

Duke (fig. 2.3) began making cigarettes in Durham by hand-rolling on a small scale in 1881, six years after Ginter had started operations with girls, but he was beset with labor shortages.[32] Durham's population in 1880 was 2,041,

Fig. 2.3 James B. Duke. Rubenstein Library, Duke University.

compared to Richmond's 63,600 and New York City's 1,912,000. The immigrant population that fueled industrialization in the North was absent in North Carolina. The South's low agricultural wages depressed industrial wages and discouraged immigration to agricultural centers like Durham.[33] Richmond, as an industrial and port city, had grown to sizable proportions before the Civil War and again afterward, as industry and trade rebounded. In 1884, Duke responded to his need for labor by opening a factory in New York City where he employed 250 immigrant girls and "a few men," for a deskilled hand-rolling operation, and by recruiting 125 Jewish artisan hand-rollers to move to Durham from New York City.[34]

The Jewish artisans immediately rebelled against the Duke factory's "tyranous [sic] shop rules" and launched a new local of the Cigar Maker's Progressive Union of America. After over a year of pitched battle, they returned to New York.[35] James Duke's father, Washington Duke, reflected, "We have never had any trouble in the help except when 125 Polish Jews were hired to come down to Durham to work in the factory. They gave us no end of trouble. We worked out of that, and we now employ our own people."[36] The Dukes did advertise in the Durham newspaper for five hundred "girls and boys" to make cigarettes, but they remained desperate for workers. For example, while Kinney and Ginter hired girls as young as twelve, the Dukes employed Laura Cox as a cigarette roller at age eight. Despite raiding the local playground for workers, the company also had to recruit New York artisans again, this time workers recently laid off from the Goodwin Tobacco Company factory. Cox later recalled that "the majority of the employees were Jews from the North."[37] Like their predecessors, however, the Goodwin group lasted less than a year before returning to New York.[38]

Duke's motivation to shift to machine production in Durham was integrally tied to these persistent labor problems. He installed the machine in his Durham factory; in his New York factory, he continued to rely on deskilled hand-rolling. When the cigarette machine became effective in 1886, his sales of machine-made cigarettes rose quickly. W. Duke and Sons soon became Ginter's most serious US competitor, though the company did not match Ginter's overseas trade.[39]

The deskilled hand-rolled cigarette made by cheap labor remained the standard for several more years, even as all producers had their eye on improvements to the competing cigarette machines. Contrary to accounts, the other cigarette entrepreneurs' behavior, including Ginter's, was not dramatically different from Duke's. Duke and Ginter both maintained some hand-rolling until the formation of the American Tobacco Company, and perhaps longer. The Egyptian industry, however, maintained its deskilled hand-rolling system well into the twentieth century and competed very ably with European and US tobacco companies. The viability of hand-rolling in Egypt was likely linked to cheap labor costs and the success of the Egyptian industry in setting the standard for the elite hand-rolled cigarette.

Duke's success has also been attributed to the fact that he negotiated to pay a lower royalty rate for the Bonsack, but this is simply false. Duke did sign a secret contract with the Bonsack Company that locked in a royalty rate of twenty-four cents per thousand, rather than the standard thirty or thirty-three cents, and promised that W. Duke and Sons would always enjoy a rate 25 percent lower than any competitor. These seemingly superior terms were in exchange for using a large number of machines in Durham (which Duke did) and for using them on higher-priced brands, rather than only on cheap cigarettes (which he did not do until the rest of the industry shifted). But this was not, in fact, the best contract. The Lone Jack Tobacco Company secretly paid even less for the Bonsack machine: only fifteen cents per thousand.[40] As tends to happen with secrets, the varying royalty rates leaked out, and much wrangling ensued. The Bonsack Company had signed contracts that were impossible to honor—for example, promising Duke a lower rate than anyone else while promising Ginter that his company would never pay a higher rate than any competitor.[41] Some order had to be brought to the matter, and manufacturers began talking of joint agreements with Bonsack as early as 1887.

The Bonsack contract operated very differently overseas and should also cast doubt on the Duke entrepreneurial myth. Nothing prevented the Bonsack Company from offering overseas manufacturers exclusive agreements for the machines in a given country, posing another set of serious risks for US producers. The market in cigarettes, as we have seen, was already an international one. The Bonsack Company recognized this and marketed its machines internationally from the outset. When James Bonsack applied for his first US patent for the machine, for example, he simultaneously applied for patents in Canada, England, Germany, France, Belgium, Austria and Hungary, Spain, Italy, and Russia.[42]

The W. D. and H. O. Wills Tobacco Company of London, a bright leaf cigarette producer with an active trade in London's gentlemen's clubs, signed the first exclusive agreement with the Bonsack Company in 1883, giving Wills the sole right to the use of the machine in England more than a year before Duke first installed a machine at Durham.[43] Notably, Ginter's well-established hand-rolled cigarette market continued despite competition from Wills. In 1884, Ginter had the largest share of the market in the city.[44] Duke's machine-made exports, by contrast, proved unable to compete with Wills. Duke reported that he "received more objections to machine-made cigarettes from England than any other Country ... [and] did not succeed in making a market there for our goods."[45] As the cigarette machine—and the machine-made cigarette—grew more important, Wills' exclusive contract promised to hamper Ginter's business development as well. Ginter had established a branch factory making hand-rolled cigarettes in London, but no US producer would be able to manufacture cigarettes in London using the Bonsack machine.

Even more of a threat to US producers was that the Bonsack Company hired Richard H. Wright as its sole agent empowered to sell the rights to its machines in several Asian countries, South Africa, and Egypt.[46] Wright might well sign an exclusive agreement with another manufacturer, closing Duke or Ginter out of production in that country. W. D. and H. O. Wills posed a particular threat in this regard, as it was also expanding internationally. Alarmed, Ginter and Duke joined forces informally to pressure the Bonsack Company to keep Australia out of Wright's hands; Australia was already a critical market for both Ginter and Duke.[47] Though their companies' combined contracts were lucrative enough for Bonsack that Duke and Ginter won the Australia battle, they had cause to worry about their ability to access the machine in much of the rest of the world.

Duke had particular reason for alarm: Wright was his nemesis. For five years, Wright been a partner in W. Duke and Sons and had served as the company's principal salesman, making an international sales tour in 1882. In 1885, however, an ugly falling out left Duke and Wright at loggerheads and in repeated litigation. Furious, Wright joined the Lone Jack Tobacco Company of Lynchburg, Virginia, which had ties to the Bonsack Cigarette Machine Company of the same city. At least one of the officers of Lone Jack was also an officer of the Bonsack Machine Co., and Bonsack experimented with machine improvements at the Lone Jack factory. These ties explain why the Bonsack Machine Co. surreptitiously gave Lone Jack a better royalty deal than Duke had.[48] When Duke discovered the perfidy, he became enraged. From Duke's perspective, there could hardly have been a worse choice of men to control global access to the Bonsack machine.

Duke railed about Wright's new position to D. B. Strouse, president of the Bonsack Company, saying, "The world is now our market for our product and we do not propose sitting idly down and allowing you or anyone else to cut off any of the channels of our trade by establishing factories where there have been none and tying up your machinery with them, and afterwards you could say to us, 'you must manufacture your goods in the United States; this country and that country is taken.'"[49] Duke had no objection to the sale of Bonsack machines in Cuba, where manufacturers made cigarettes from a very different kind of tobacco. "But where mild [that is, bright leaf] cigarettes are sold or wanted," he insisted, "the US leads the world in their manufacture."[50] Duke urged the Bonsack Company to develop a policy that it would stand by, writing: "I think it about time for you to outline to us your policy regarding the future disposition of the Bonsack Machines in foreign countries."[51]

By creating the ATC, the major, native-born US cigarette manufacturers were able to negotiate an excellent and common contract for the Bonsack machine, shared rights to the Emery and Allison machines, and increased their ability to purchase the rights to competing machines. As a key participant in all negotiations, the Bonsack Company agreed to "restrict the machines to the large factories," as long as their combined sales exceeded 50 percent of the domestic market.[52] Even more importantly, joining together provided the ATC more clout in overseas negotiations with Bonsack. And not a moment too soon: Wright departed on his first overseas sales trip for Bonsack in 1889, the year of the first ATC corporate charter in Virginia.[53]

Duke's own words reveal that he did not believe he had more power than Ginter in the formation of an alliance. Strouse, the Bonsack Machine Co. president, wrote Duke in March about "the organization of a consolidated cigarette company." Strouse's choice of wording alarmed Duke, who replied, "I am perfectly willing to meet the manufacturers at any time to discuss any matter for our common good, but if you mean by consolidated cigarette company that there is to be another factory started or a trust formed, and want us to take stock in them, I am opposed to anything of the kind, as we want the full control of our own business, which we could not have with a trust. If there is any other proposition to discuss I will meet if called."[54] Duke here opposed the kind of consolidation that the Virginia ATC would create just a few months later. Though he worried about losing power, he must have felt it was the best he could do.

Indeed, Ginter had more reason than Duke to be pleased by the Virginia charter, not least because its headquarters were in Richmond and Ginter was president. The December 1889 charter secured the manufacturers' shared goals and gave the ATC an exclusive contract with the Bonsack Machine Co. At the same time, the manufacturers remained in charge of managing their own companies, which remained intact. The ATC would act on behalf of the five companies with regards to the purchase of materials, and it would be able to "buy and manufacture and to establish depots and agencies in other countries."[55] The incorporators included Ginter, John Pope, Milton Cayce, and Thomas F. Jeffress, all of whom were associated with Allen & Ginter, and James M. Boyd, another

Richmond manufacturer.[56] The other entrepreneurs would serve on the board of directors, but Ginter stood to guide domestic and, especially, international expansion.

As noted, this charter did not prevail; the day after its passage, Virginia legislators, many of whom were also tobacco farmers, returned to the capitol building horrified by the impact an ATC monopoly might have on leaf prices. They claimed they *had* read what they voted on, but the bureaucratic nature of passing corporate charters may have led them to merely glance at the incorporation bills. Now, however, "there is a fear that this may be a gigantic combination of the leading rich tobacco manufacturers of this country to control the leaf tobacco market." Farmers and leaf dealers across Virginia and North Carolina joined the call for a repeal of the charter.[57] On December 21, just two days after the charter had been granted, Representative Henry D. Flood of Appomattox introduced a bill for the charter's repeal, and it was immediately clear that he could marshal the votes to pass it.[58] Men in the tobacco trade successfully blocked the corporate charter.

The tobacco company owners quickly turned to New Jersey and secured a corporate charter on rather different terms. Capitalized at $25 million, the ATC's first action as a corporation was to buy all five of the cigarette companies. "The firm of Allen and Ginter no longer exists," announced the *Richmond Dispatch* with alarm.[59] In this merger, the five companies became one entity rather than existing as linked but separate legal entities, a detail that would soon have significant legal and practical consequences.[60] As president, Duke moved his New York City domicile across the Hudson to Somerville, New Jersey, in order to fulfill the residency requirement. John Pope of Allen & Ginter served as first vice president; the owners of the original five companies served on the board of directors.[61]

Trouble in the House of Capital

The New Jersey charter demonstrates what a big difference a seemingly small contingency may create, for New Jersey was the cutting edge of corporate empowerment. By incorporating in that state, the ATC gained freedom from regulations that had typically governed corporate actions. New Jersey had radically loosened its general incorporation laws throughout the 1880s, and lawyers representing corporations had a direct hand in the drafting of its new corporate law in 1889.[62] Emboldened by railroad corporations' judicial wins, the New Jersey incorporation law's basic premise was that business corporations should be able to do virtually anything they wished.[63] Of course, a charter established the right to do something—in this case, pretty much anything—and marked only the beginning of a battle. After securing the charter, the ATC manifested its new rights in its business organization and practices while many looked for ways to fight this new consolidation of power.

It would be difficult to overstate how much opposition greeted the new corporation: there was anything but consensus on its form and powers. Even within the ATC board of directors there were members, first among them Ginter, who openly stated that they did not see the corporation as legitimate. Farmers, leaf dealers, distributors, manufacturers, retailers, state-level regulators, and investors also opposed the ATC. People of the Upper South were especially angry. The *Reidsville Review* wrote, "That this soulless monopoly not only wants the control of the entire tobacco trade of the world, but the earth itself, we do not doubt for a moment."[64] These were the words that were fit to print but other expressions flew in private. One Winston-Salem smoking tobacco manufacturer friend wrote to Duke: "I have heard you cursed by farmers, warehousemen, leaf dealers, and manufacturers. We have been cursed because we did not curse the trust."[65] The business class was riled.

Alongside the particular passion that North Carolina and Virginia brought to the fight, opposition to the ATC raged nationwide and quickly took the form of court challenges. The company faced cases in New Jersey, New York, North Carolina, Virginia, Georgia, and Illinois; the last ruled the ATC an "illegal" corporation and prohibited it from doing business in that state. Some state legislatures, including North Carolina, Virginia, and Georgia, passed new antimonopoly laws so that stronger cases could be brought against the ATC specifically.[66]

What exercised opponents of the ATC monopoly was *not* a moral objection to cigarette smoking. Yet Duke's first biographer claimed that "long-haired men and short-haired women" wrote "lurid accounts" that absurdly implied that "the boy who hid behind the barn to sneak a smoke was taking his life in his hands."[67] There were indeed some reformers who saw both tobacco and alcohol as destructive drugs, but when Washington State outlawed the sale of cigarettes in 1893, "nine-tenths of the members who voted for the bill did not care a nickel about the reform of the cigarette fiend, but they were anxious to knock out the Tobacco Trust."[68] Likewise, the Iowa Wholesale Grocers refused to handle ATC products because of the monopoly's control over distributors, not because they saw cigarettes as immoral.[69]

Nationwide, antimonopoly sentiment developed into a conversation about the proper direction of the economy. Antimonopoly thinking had deep roots in US culture and was honed in railroad controversies. Most people believed that competition was necessary for the health of the economy; many worried that monopolies would allow destructive and corrupt economic practices to persist. Railroad corporations appeared to be proof of this. They seemed like the "embodiment of special privilege and discrimination," for they were granted charters that gave them wide latitude but could not be regulated.[70] At the same time, railroads discriminated, giving better rates to large contractors, with the result that they destroyed competition in a range of industries, such as beef. Their use of the courts was pioneering in setting federal protections against states' abilities to tax and regulate. Tobacco planters in Virginia opposed the ATC because they worried about the manipulation of prices paid for tobacco, but their wider concern was about the ability to discriminate in general: to pay and charge more to some than to others in order to control business and further monopolize. Such debates gained fervor because many believed that the "economic system of a democracy had itself to be democratic."[71]

Those who opposed the ATC sometimes couched their battle in melodramatic terms as the little people against the big monopoly, but avowed antimonopolists were not necessarily for racial equality or workers' rights. Rather, these were economically enfranchised people who wished to influence the means of exchange—how corporations would be empowered and regulated and therefore how business practice would proceed—more than the means of production.[72] The farmers and leaf dealers of the Virginia legislature, for example, were upset about the ATC's ability to control leaf prices and disrupt the means of exchange of tobacco through the auction system. These were not yeoman tobacco farmers on small plots of land—though the ATC hurt them as well and they also organized—but plantation owners who had farmed and cured tobacco with slave labor before the Civil War, and currently contracted with black and white sharecroppers. They had no problem with the idea that those who produced the tobacco would see very little of the profit and they surely thought that sharecroppers' condition of labor was irrelevant. They were long accustomed to thinking about economic fairness among propertied white men while profiting from slavery or oppressive wage labor systems.

When the ATC monopoly expanded from cigarettes to the chewing and pipe tobacco industries, it confirmed people's worst fears. Until this time, the cigarette industry had not greatly changed the bright leaf economy, but the ATC overhauled it. The ATC bought up hundreds of chewing and smoking tobacco companies and displaced thousands of employees. Beginning with no chewing tobacco manufacturing in 1890, by 1902 the ATC controlled 71 percent of the market; by 1910, it controlled 85 percent of the market and had absorbed approximately 250 companies.[73] In this way, the ATC transformed the tobacco South from an economy based on partnerships to a corporate monopoly. Companies that survived rallied their forces to compete directly with Duke. Liggett and Myers, for example, went into cigarette production for the first time "as a means of retaliation."[74]

Even within the ATC, though, all was not well. Ginter publicly opposed the New Jersey corporate charter and referred to it as a trust, a loaded term.[75] Congress had passed the Sherman Anti-trust Act the same year that the ATC formed. The Standard Oil Company was the most famous example of a trust: its trustees held the voting stock of all of the original companies. Critics claimed that New Jersey's incorporation laws simply created another avenue for a trust. ATC director William Butler, formerly of Kinney, publicly defended the company saying, "The American Tobacco Company is not a trust, but is a regularly-authorized company."[76] But George Arents, Ginter's nephew and treasurer of the ATC, said, "Although [Ginter] was opposed to the formation of the trust, he was not opposed to the formation of a legitimate tobacco company, and has been one of its main supporters."[77] Even tobacco executives disagreed about what constituted a legitimate business form.

The battle over chewing tobacco drew even more internal opposition. Duke, generally given to hyperbole, later recalled in a rare understatement, "I was the one that promoted the idea that we should go into the plug business and some [of the other directors] didn't seem to favor it very much."[78] Ginter and other directors argued that taking over chewing and smoking tobacco diverted attention from cigarettes; indeed, the diversion of profits toward buying chewing tobacco companies depressed the domestic cigarette market.[79] Meanwhile, Egyptian and immigrant competitors gained on the ATC. Though the ATC awarded dividends of 10 to 12 percent in its first several years, in February 1896 it awarded none.[80] For Ginter and others, the shift to monopolizing chewing tobacco was a risky move that jeopardized the business that the cigarette companies had built.

Ginter and his allies on the board of directors made at least one significant attempt to shift the direction of the ATC, to no avail, and soon after all internal opposition to the ATC collapsed.[81] Part of the reason could have been that John Pope died unexpectedly of an infected abscess in his esophagus at age thirty-nine on April 8, 1896, just two weeks after the *Wall Street Journal* had printed numerous reports of dissension in the ATC. Pope had become ill the previous year—the papers mentioned acute "laryngitis"—but had appeared to recover. Ginter was devastated personally and professionally, for he had lost his life partner, his plan to pass his power and the fruits of his labor to his heir, and his hope of checking Duke. Pope had been "one of the biggest men" in the company, having served as first vice president through the company's first five years, and had both the experience and the youth required to take the company into a new time. In a poignant gesture, Ginter buried Pope in the grave he had reserved for himself.[82]

Ginter's own health took a sharp turn for the worse after losing Pope. He retreated from the ATC and resigned from the board of directors in the Spring of 1897; he died just six months later of diabetes-related complications.[83] Ginter's *Richmond Times* obituary recounted: "The illness of Major Ginter may be said to have begun just after the death of Mr. John Pope, his protégé, business companion and friend. . . . there is no doubt that Major Ginter was prostrated, mentally and physically, at his death."[84] Ginter's and Pope's deaths removed the significant internal obstacles to Duke's power.

The external opposition, however, coalesced into a formidable offensive effort led by entrepreneurs who had formerly been artisan cigar and cigarette rollers. They formed a new corporation, the National Cigarette and Tobacco Company (NCTC), solely to compete with the ATC and bring lawsuits challenging its monopolistic practices. Certainly, these businessmen had not forgotten Duke's poor treatment of artisans at his Durham factory, which had often sparked into a power struggle. For example, when union representatives had approached Duke in his New York City office to discuss pay issues, Duke traveled to Durham in order to personally cut artisanal pay by an additional 60 percent.[85] These were no longer artisans, however. NCTC's managing director, Bernhard Baron, had once worked for Kinney but had recently opened a cigar factory in

Baltimore; Baron would later form the successful London-based Carreras Tobacco Company, maker of internationally famed Black Cat cigarettes.[86]

The NCTC was not clinging to the artisanal past, then, but was staking a claim on how industrial capitalism developed, and it intended to draw the ATC into battle. Baron's first volley was to install in the NCTC factory his own invention, the Baron cigarette machine, which reportedly worked even better than the Bonsack. The ATC immediately initiated a patent infringement lawsuit to protect its exclusive access to the Bonsack machine's many patented components. The monopoly lost in court, which effectively ended its brief monopoly over the best-functioning machine.[87] The ATC's opponents celebrated this significant success.

The NCTC's greatest challenge to the ATC came in the form of two linked lawsuits, brought in New Jersey and New York, both of which tested the validity of the New Jersey's laws of incorporation. The court cases represented an alliance between the NCTC, distributors, and the attorneys general of New York and New Jersey. The ATC hired high-powered lawyers and prepared to defend itself on multiple fronts.

Corporate Enchantment

The coordinated cases brought by the attorneys general of New Jersey and New York, in particular, had the potential to jeopardize the ATC's corporate charter or state certificate and halt the advance of the ATC. The New Jersey case, brought in equity court, asked that the company be prevented from using its charter; if successful, the case would essentially shut down operations. The New York case could not affect the corporate charter, but that state had a very strong antimonopoly law, and the case there, brought in criminal court, could cancel the certificate allowing the company to do business there. Because New York was both the largest domestic market for cigarettes and the location of the majority of the ATC's factories, losing New York would cripple the ATC. That the ATC saw these cases as serious threats is reflected in the annual salary of $50,000 provided to the ATC's top lawyer, W. W. Fuller, a sum that converts roughly to $1.3 million today.[88] Curiously, previous histories of Duke have entirely ignored these cases.

Both cases rested on complaints brought by distributors, or "jobbers," that the ATC used its monopoly to destroy competition. Traditionally, jobbers were independent contractors. They purchased products from the tobacco companies and sold them to retailers at a profit. The ATC sought greater control over this process, so it required jobbers to handle no competitor's tobacco products, to sell the goods at a price set by the ATC, and to receive a sales commission from the company. As President Frank McCoy of the NCTC put it, "[The ATC] says to [jobbers], 'You can buy our goods, but you must sell them for exactly the sum you pay us for them. If at the end of six months you are a good boy, we will pay you a 35 per cent rebate.... If you are not a good boy, you get nothing.'"[89] In effect, self-employed jobbers became dependent on the ATC, with little capacity to influence the terms of labor.

The agreement also created obvious barriers for competitors who could not get their products distributed. One such competitor, McCoy, stated, "We have come into this business to stay, and no trust with illegitimate methods of doing business can drive us out."[90] In 1895, the NCTC signed a distribution contract with the Wholesale Grocers of New England, representing 133 jobbing houses, which represented the first breach in the ATC's monopoly over distribution.[91] The NCTC's Admiral cigarette brand would finally receive broad distribution. The ATC backed up its policy by refusing to pay 23 jobbers who carried both NCTC and ATC brands.[92] The ATC's manifest business practices became a matter of public record and the lawyers got to work.

Though the immediate issue was distribution, the New York and New Jersey cases more significantly arbitrated the ability of the states to regulate corporations in the wake of the revolutionary New Jersey incorporation acts. States historically had the legal right to regulate corporations through their delegated right to issue corporate charters. Charters traditionally laid out what a corporation could and could not do. If a corporation engaged in business practices that exceeded those granted by its charter, the state could charge it with being "ultra vires," or beyond the law. Reforms through the nineteenth century made the process more regularized and bureaucratic via state incorporation laws, which stipulated what rights charters typically granted. With the New Jersey laws, the notion of "ultra vires" became moot or nearly so, for the laws placed virtually no limits on corporate activity. Because the ATC was one of the first large companies to incorporate under New Jersey's laws, these cases tested the enhanced powers of the corporation and the states' ability to regulate them.

New Jersey Attorney General Stockton brought suit in equity court charging that the ATC engaged in actions injurious to trade and commerce, and requested that the ATC "be restrained from using the corporate organization of the said company . . . to stifle competition in the manufacturing and selling of cigarettes."[93] New Jersey state lawyer Benjamin F. Einstein argued that states chartered corporations for the public good and that states may restrain corporations that cause public injury. The ATC business practices, said Einstein, injured trade and commerce in the state.[94] By filing the suit in equity court, Stockton indicated that New Jersey retained the right to regulate a corporation that harmed trade in the state; if successful the suit would result in the effective revocation of the ATC's corporate charter. Beyond trying the ATC, then, the case tested the New Jersey incorporation law. A win for the state would have stabilized the corporation's status as public and created great pressure on the legislature to revise the controversial incorporation law.

In New York, Attorney General Hancock had more distant jurisdiction but a strong antitrust law. The New York legislature passed an antimonopoly bill in 1892 that made it illegal to conspire to stifle competition. New York had issued to the ATC a standard certificate authorizing the company to do business in that state. Hancock used this to claim authority over "foreign" corporations, meaning those chartered in another state. He asserted:

> I have no doubt that [New York state courts] have sufficient jurisdiction over a foreign corporation to restrain it from transacting an illegal business. . . . To hold otherwise would be to . . . place the State in the unfortunate position of having no power to protect its citizens against conduct of a foreign company which would not be permitted in the case of a domestic corporation.[95]

Hancock charged that the ATC was guilty of conspiracy; if the suit was successful, the case could be tried in criminal court.

ATC lawyers filed a demurrer[96] in the New Jersey case that used the notion of corporate personhood in order to boldly challenge the notion that corporations had any responsibility to the public good whatsoever. They openly asserted that the company had a legal monopoly granted by its charter, and insisted that it should be treated no differently by the law than an individual human in business would be. The lawyers maintained that the ATC could choose to consign or sell its cigarettes, or not, as it wished, even if it was injurious. They argued that the ATC "is a private manufacturing corporation, owing no duty to the public or any part of it. It has all the dominion over its property that an individual has over his, and the same right to do what it will with its own.... in all it does stand justified before the law as fully as if it were an individual carrying on the same business."[97] The corporation was a private individual, according to the ATC lawyers, not a public institution.

The lawyers also challenged the validity of the test case, arguing that the only way to regulate a corporation was through the ultra vires principle and that the case should not have been brought in equity court at all. They denied that any state law applied to them outside of their charter, except criminal laws under which an individual could be charged. The ATC lawyers insisted repeatedly that the ATC must be treated as a private and singular person, equivalent to a human person and with the same rights and protections. The ATC lawyers also held that there should be no such thing as regulations that apply to corporations specifically; they should experience the same property rights and other rights as an individual.

In a dramatic victory, New Jersey Vice Chancellor Reed accepted the ATC's demurrer—indeed the ATC lawyers could have written his decision, so closely did it parrot their arguments—and endorsed the claim that the complaint had been brought to the wrong court. The vice chancellor created a conundrum for anyone desiring to legally challenge a New Jersey corporation's powers. They could only file an ultra vires complaint that the corporation had exceeded its charter, but the 1889 law meant that charters no longer significantly limited corporate activities. The usage of ultra vires would obviously be futile but, according to the vice chancellor, no other avenue for challenging corporate power existed. Vice Chancellor Reed also declared that the corporation was an essentially private individual. Accepting the equation of the corporation with a person, he said, "A trading or manufacturing corporation, until its charter is annulled by such a proceeding at law, has the same authority as an individual trader or manufacturer to sell or consign its goods, to select its selling agents, and to impose conditions as to whom they shall sell to and the terms upon which they shall sell."[98] In other words, the fact that a corporation had a particular potential to concentrate wealth made no legal difference: it must be treated legally as though it were an individual making contracts.

In New York, ATC lawyers filed a very similar demurrer, but this time the judge rejected their arguments and ordered the ATC officers to stand trial in criminal court. In response to the specific charge of conspiracy, the ATC lawyers had argued cleverly that because the ATC was a single legal person, it could not be guilty of conspiracy. By definition, two or more were needed to conspire, they claimed. They also cited the Fourteenth Amendment's due process clause to argue that the ATC officers should be immune from criminal charges. Judge Fitzgerald was not impressed by either argument: "To rule that the officers and agents of a corporation are relieved from individual criminal liability for all they may do under the color of corporate acts would amount in many cases to a practical suspension of law."[99] Judge Fitzgerald clearly defended the right of states to regulate corporations.

The ATC officers faced a jury trial in 1897. If found guilty, the ATC would be restrained in New York State and the officers would face fines and imprisonment. The trial lasted five months, but the jury failed to achieve the unanimity required for conviction.[100] District Attorney Olcott immediately called for a retrial, saying, "When a jury stands ten for conviction and two for acquittal the people . . . are entitled to a decision on the law while the matter is still fresh in their minds."[101] During the summer court break, however, the ATC dropped its policy on distributors. The court cases changed an individual monopolistic policy, but failed to significantly constrain the ATC, cause revision of New Jersey's incorporation law, or check corporate empowerment.

Beyond enabling the ATC to do what it wished, the new empowerment of corporate persons had the stranger and ultimately even more significant effect of animating the corporation as an "autonomous, creative, self-directed economic being." The problem was not that the corporation was considered a legal person when it was not a person: the individual is the basis of British and American common law and corporations have long been seen as entities or persons, but as ones requiring special regulation.[102] In this era, however, the corporation shifted in the collective imagination from being a public to a private entity. Whereas the state had once regulated the corporate person through its charter as a matter of course—think of a puppet controlled by firmly attached strings—the ATC cut the strings and claimed protection *from* the state.

It is misleading to say, as some do, that corporate personhood was a "social fiction" that allowed an individual-based legal structure to handle collectivities; rather, it was a legal fact, for the law is a closed system of rules and procedures.[103] There was nothing fictional about the corporation's powers and status as a person under law, despite its obvious nonhuman nature. In this way, the corporation's new powers were not only symbolically or metaphorically person-like but literally and functionally so. When a metaphor and symbol becomes animated and materially efficacious, it has attained a kind of spiritual power. Most religions contain an element of this kind of transfiguration. But here we can see that such transfiguration exceeds religion and lies at the heart of US economic life. Weber's argument that science disenchanted the West made this spiritual power more difficult to see. To say that the corporation is enchanted, then, is to acknowledge that this transfiguration is part and parcel of its power.

The secret to how the ATC gained this empowerment is rooted in its use of the Fourteenth Amendment.[104] The Fourteenth Amendment awarded African American men citizenship rights, despite the opposition of the former Confederate states, and expanded the federal protections awarded to legal persons, including protections to private property and due process. The ATC was not the only company to defend itself using the Fourteenth Amendment in these years—the railroads had created the precedent—but it may have been the first since the more liberal New Jersey incorporation laws were passed.[105] When corporate lawyers successfully defended the corporation's new powers from states by using federal protections of private citizens, they animated the corporation, making it creative and self-directed—nothing less than "the quintessential economic man."[106] Indeed, because the corporation was an abstraction of myriad people and business practices into a single "entity," it was able to achieve the legal abstraction of pure economic motivation more nearly than could any human, but especially more than those who because of race, gender, or nationality appeared to lack the capacity for the rational, independent action required of the citizen in the marketplace.

The ATC's successful claim on the protections of the Fourteenth Amendment came in the context of highly contested struggles for legal personhood by African Americans, women, and Chinese people who sought redress from legal exclusions from citizenship protections. The fact that the US relied on the individual as the unit of law was deeply woven into cultural and political life; the individual's right to contract emerged as the single distinguishing factor between slavery and freedom in the post-Emancipation US. The right to form wage and marriage contracts, particularly, rested "on principles of self-ownership, consent and exchange," and defined the ideal of freedom.[107] Activists worked to extend the right to contract to all women and people of color, that is, to establish them as full legal persons. Overall, however, new local, state, and federal laws checked the gains women and people of color achieved in the wake of the Fourteenth Amendment.

For example, laws of coverture continued to position married women in a special mediated relationship to their own employment and entrepreneurial activities. In addition, even though they were nominal citizens they lacked the right to vote. Their association with the private sphere of the home accorded them none of the entitlements that corporations claimed as private businesses. Black men found their citizenship claims greatly strengthened because of the Fourteenth Amendment, but many states immediately acted to delimit those rights.[108] The system of Jim Crow segregation became foundational to capitalism by excluding blacks from particular job categories and schools; a combination of corrupt laws and violence curbed the franchise. While the attempt of states to delimit the rights of corporations largely failed, their attempts to delimit the rights of African American men mostly succeeded and were formally endorsed by the Supreme Court in the 1898 *Plessy v. Ferguson* case that legalized segregation. Not until the Civil Rights Act of 1964 would African Americans—and women—gain recognition as full legal persons in the economic realm of jobs.[109]

Likewise, in the era that corporations strengthened their claim to be rights-bearing legal persons, Chinese people in the United States found their claim to personhood weakened by the federal Chinese Exclusion Acts. The Supreme Court ruled in *Yick Wo v. Hopkins* (1886) that resident noncitizens, including Chinese men, were entitled to the protections of the Fourteenth Amendment.[110] In 1882, however, the Chinese Exclusion Acts introduced the first federal limitations on entry explicitly on the grounds of race. People arguing for exclusion described Chinese men not as legal persons making contracts but as undifferentiated "hordes" who took any job and lived in squalor. The racialized figure of the "coolie" was the diametric opposite of the "quintessential economic man," the status increasingly accorded the corporation.[111] As the corporations' claim to "private" legal personhood increased, then, African Americans, Chinese people, and women found their claims to personhood repeatedly undermined.

The Mystification of Property

The ATC also gained in financial power in these years through its formative participation in the redefinition of property in the stock market. The ATC became one of the first industrial manufacturing companies to issue stocks in the New York Stock Exchange. The way that its assets were enumerated, particularly its many brands, represented another transformation at the line between materiality and ephemerality, this time in the definition of property itself.[112]

Before the cigarette industry incorporated, property meant the "ownership, disposal and exchange value of tangible things." With the merger movement, *intangibles* came to be included, and protected, as property. When one company took over or merged with another, the former owners of the purchased company typically received securities or equity in the future earning of the new company in exchange for all buildings, equipment, and supplies—tangible property—as well as patents, trademarks, and brands. Considering intangibles as property raised stock values as "the consolidated prestige, goodwill, and reinforced market power, of previously separate enterprises, raised the putative earning power of the new corporation."[113]

The redefinition of property was a momentous shift for the ATC monopoly because it came to own hundreds, perhaps even thousands, of brands. The going wisdom in the post–Civil War tobacco industry was that if a company wished to expand into a new market, it would offer a new brand, distinctive by the blend of tobaccos, the formula of spices and sweeteners, or a new grade and price. A small town like Reidsville, North Carolina, had eight to ten tobacco companies, each producing four to twelve brands of chewing and pipe tobacco. One Winston manufacturer offered forty brands.[114] The ATC took over approximately 250 US companies, thereby making brands a significant portion of its property.

A brand's value on the stock market was based on an estimation of the brand's established customer base, considered in terms of its reputation or the goodwill that it had accrued in its career to date and the sales, therefore, it was likely to be able to sustain in the future. Brands had originated as trademarked names and glyphs that allowed customers to distinguish between similar goods and repeat a purchase. The concept of goodwill began with the notion that consumers valued a brand by its trustworthiness (would the tobacco be fresh, weighted as advertised, and of consistent quality?). The concept soon also referenced a brand's popularity quite apart from elements of consistency and reliability. A very successful brand, therefore, not only had a pleasing name and glyph but also had accrued a cultural value—goodwill—based on a combination of company reputation and popularity. A brand's prior purpose and value—already fundamentally symbolic—transformed in the context of the stock market, for stock market calculations of the value of a company's property put a number on the estimated value of a company's brands based in goodwill.

The brand, then, lifted from its commercial, symbolic role in the process of sales and came to have quite another function in the stock market as an indicator of future selling potential. Part ghost of past aggregate sales, part prophecy of future sales, a brand on the stock market was a strange sort of property. Endowed with a numerical value, it appeared stable and objective when in fact it was a highly volatile property composed of symbols and desires. The mutability between economic materiality and spirit was not entirely new—we could say it is as old as money itself—but it had a new and defining presence in the stock market and was foundational to corporate empowerment, especially in the case of the ATC.

The ATC valued its brands highly. When the company incorporated in 1890, it issued stock to the former owners in exchange for their properties. An investigation found that "about $5,000,000 worth of stock was issued for the live assets of the [former companies] and $19,990,000 for good will."[115] This is an extraordinary number: goodwill constituted almost 80 percent of the purported value of the corporation. Criticisms flew that the stock was watered and that intangible values were arbitrarily assigned. In 1908, after eighteen years of acquiring companies and hundreds of brands, goodwill accounted for 55 percent of all of the ATC's assets. The Bureau of Corporations determined that the value of brands had been overestimated by $85,000,000 but these calculations too were vulnerable to criticism.[116] At the time of its dissolution by the Supreme Court in 1911, the ATC's assets totaled $227 million, of which fully $45 million was attributed to the company's many trademarks.[117] The brand was magical property indeed.

By 1898, the ATC, empowered as a new kind of legal person with a novel sort of property, embarked on a more aggressive and imperial mode of global expansion. By then, the company had overcome both internal and external impediments and seemed unstoppable. The company war chest was flush with funds due in part to some shady stock market maneuvers on the part of the directors, actions that also lined their own pockets nicely.[118] Duke was the only original cigarette entrepreneur to remain on the Board of Directors; financiers drawn by the smell of profits held the rest of the positions.

Corporate Imperialism

The ATC developed new means of corporate expansion in dynamic relationship with US and British imperialism in Asia. The US had long desired to be a dominant power in the Pacific Basin. In 1844, Caleb Cushing traveled to China to negotiate a treaty for the US on the same advantageous terms that Britain had won in the wake of the 1842 Opium War, including extraterritoriality in the treaty ports. In 1854, the US led an expedition to "open" Japan to foreign trade, using the treaty with China as a guide. The US established several treaty ports there as well and in the late nineteenth century also established extraterritorial rights in Siam, Samoa, Korea, and Tonga.[119] In the late nineteenth century, the US repeatedly asserted itself in the Pacific, gaining control over critical ports in Hawai'i and Samoa. In the War of 1898, the US won the former Spanish colonies of the Philippines, along with Cuba and Puerto Rico, and set up an ongoing military occupation of the Philippines.[120] In doing so, the US clearly signaled to other imperial nations that it would be a factor throughout the Pacific. US companies immediately began to pursue expanded overseas opportunities.

Ginter and Duke and other cigarette manufacturers had a lively international trade before the formation of the ATC, but the corporation used new tactics for increasing its global reach, as did other companies. Ginter and Duke had primarily sold through commission agents in various countries around the world. Where they had a strong market, they would build a depot for coordinating distribution. In a few instances, Ginter started branch factories. After the formation of the ATC, the tobacco entrepreneurs continued to use commission agents, but they began a program of forcing a takeover of existing and functional foreign cigarette manufacturers. The ATC acquired companies in Germany, Australia, New Zealand, Canada, and Japan, which enabled it to evade high tariffs and to access the local domestic market with little expenditure of personnel, as well to obtain production and distribution centers for further export.[121]

In the wake of the War of 1898, the ATC's first major move in the Pacific was acquiring a controlling interest in the Murai Brothers Tobacco Company of Kyoto, Japan, in 1899. The ATC converted the Murai operation to bright leaf cigarettes and assigned Edward J. Parrish of Durham, North Carolina, to serve as vice president; Murai Kichibei remained as president. Parrish also oversaw the establishment of a Murai branch factory in Tokyo. Even before this move, the Japanese government was not particularly pleased with the ATC because it was trying to extricate itself from the influence of foreign imperialists. In 1895, Japan won the Sino-Japanese War, establishing itself an imperial force in East Asia. In 1898, Japan, finally freed of unequal treaties, implemented a high tariff that was reportedly directed specifically at the ATC. With the purchase of the Murai Brothers Tobacco Company, the ATC evaded the tariff and sent cigarettes to ports in China, India, and Southeast Asia.[122] Its purchase also became the largest investment by any foreign company in Japan and quickly aroused opposition. Political leaders in Japan described Duke as "the capitalist [who] is intending to monopolize the whole world"; in 1904, Japan nationalized the company, requiring Duke to find another East Asian production center.[123]

In the meantime, the ATC had made a bold move with the formation of BAT, creating a novel kind of international cooperation. The ATC began its incursion into Britain much as it had elsewhere, by purchasing a local company, Ogden's Limited of Liverpool, and using it to compete in the domestic market. In this case, eleven British companies merged into their own monopoly, the Imperial Tobacco Company, in order to fight Duke's price cutting. In 1902, the two monopolies created a historic agreement: they agreed to stay out of each other's domestic markets and jointly formed British American Tobacco, one of the world's first multinational corporations, with the explicit and sole purpose of pursuing foreign expansion. The ATC had many other joint ventures with foreign companies, but this was the first in which the goal was not to compete in the domestic market but to coordinate overseas expansion. The ATC had tapped the power and reach of the British Empire and Duke had become the planet's most powerful cigarette entrepreneur. That Duke saw this expansion in imperialist terms was demonstrated in his promise that BAT would "conquer the rest of the world."[124]

The structure of BAT as a novel cooperation between cigarette corporations of two nations had its echo in the multinational imperialism that had developed in China and other places in the Pacific. The roots of multinational imperialism lay in the treaties that established extraterritorial legal structures and privileges. Each time a country won a new privilege, other Anglo-European nations and, after 1895, Japan would claim the same rights. This highly racialized formation rested on the belief that the "Euro-American 'Family of Nations'" had developed civilized legal structures but that Asian countries required the imposition of foreign law.[125] Multinational imperial cooperation gained military expression in the jointly executed repression of the Boxer Rebellion of 1900. British, Russian, German, Italian, Japanese, and US military forces participated in putting down the rebels; the US sent troops from Manila, demonstrating the imperial benefits of its nearby colony.[126] Multinational imperialism gained expression in the built environment in the treaty ports' foreign settlements. Though this imperial formation was multinational, and therefore developed distinctive forms of governance and control, it is not best described as private, solely economic, or "informal."[127]

BAT developed an operation in China that integrated British imperial policies with the bright leaf network. China was different from all of Duke's prior expansions in that it lacked a productive cigarette factory to take over. Never before had the ATC or BAT had to develop management staff and factory workers from scratch outside of Europe and the United States. In addition, China's vast agricultural lands opened the possibility of sourcing some bright leaf tobacco, or perhaps something much like it, on site. BAT's East Asian managers in Tokyo and Shanghai weighed the options and recommended the risk, and Duke approved the shift in strategy.

Duke sent James A. Thomas of Reidsville, North Carolina, to head operations. Thomas had recently run BAT's business in India and in building personnel for BAT-China, he implemented the British Empire's contract terms: single men between the ages of approximately twenty and twenty-four signed on for four years, after which they received a vacation home. At the same time, Thomas sourced these men mostly from his own home region: the bright leaf belt of North Carolina and Virginia. BAT-China would become one institutional manifestation of a larger imperial story.

Conclusion

By 1905, Duke, the ATC, and BAT were enormously powerful—that much is true. However, the ATC's rise to power was not a result of Duke's brilliant entrepreneurial innovation with the cigarette machine. Rather, the ATC rode a wave of corporate empowerment that included the legal and social transformation of the corporation within the US as well as the new role that the corporation could play within rising imperialism in Asia. With every triumph of the corporation, Duke found his own fortunes spectacularly enhanced.

Corporations had long been legal persons but they acquired a status as private in these years, so that they were more successfully accepted as natural economic persons than many human persons on the corporation's payroll. Entrepreneurs like Duke accrued some of this power, as legal categories influenced social values within daily life and the lines between corporate and executive personhood blurred. One danger of the overapplication of Schumpeter's theory of creative destruction is that it focuses our attention overmuch on the entrepreneur, conflating him with the corporation itself and occluding the nature of corporate power. In addition, when corporate power is figured as the result of individual exceptional brilliance, it is inflated even beyond its considerable might, and it becomes more difficult to see the way that corporate power is still contingent, embattled, and partial, even when it is increasing.

This story calls upon us to rethink how ideas of public and private were organizing resources and obscuring certain interpersonal and institutional relationships. Consider that Ginter's entrepreneurial perspective was inextricable from his nonnormative intimate life. His relationships with men, and particularly with Pope, were not simply a private matter but influenced his marketing innovations and the structure of his imagination for company leadership succession. Creating strong business history explanations requires understanding the whole of life, including intimacy, desire, and sexuality, not just a falsely isolated notion of economics. At the same time, the business corporation shifted from being considered a public to a private institution, a shift that did not simply reflect its relationship to state and federal power but was the means by which that relationship was reconstructed. The results included a decline in regulation and the obscuring of the business corporation's role in imperialism. Schumpeter's theory of entrepreneurial creative destruction strips content from the story of change and replaces it with an archetype that can personify the corporation, deflecting reasoned ethical debate about the role of the corporation in US life.

In the case of the ATC, the power of Schumpeter's theory to distort is especially notable because the ATC's story is no backwater in the annals of history. As one of the nation's most powerful and consequential corporations, it has drawn the attention of myriad superb historians. Given how well traveled the pathway of ATC history is, one would assume that all relevant facts and sources would long ago have been excavated, itemized, and debated. Yet of Ginter's role in the ATC, the Virginia ATC charter, the role of the NCTC, and the New Jersey and New York court cases, none has ever before made its way into the story. The Schumpeterian thesis was so convincing, and the conflation between entrepreneur and corporate power so satisfying for admirers and critics alike, that no one looked further into the story for over fifty years.

With the formation of BAT, Duke did indeed assume an astonishing amount of power. He still, however, could not dictate what occurred on the ground in China, a place he never went to. That venture was made daily by hundreds of US and British and thousands of Chinese employees who built the company's operations from seed to smoke. With the decision to shift cigarette production to China, BAT staked its future on even greater imperial reach. Yet the multinational also remained closely tied to Virginia and North Carolina because it had built its fortune primarily on the agricultural resource of bright leaf tobacco particular to that region. What bright leaf tobacco would require, and how BAT-China's imperial business culture grew in relation to it, is the subject of the next chapter.

3

The Bright Leaf Tobacco Network

When the British American Tobacco Company decided to initiate tobacco agriculture and cigarette manufacturing in China, it set in motion a dynamic flow of people, plants, things, and ideas between the US Upper South and China. The bright leaf network, already developed in the US, now reached across the Pacific and experienced managers, bright leaf tobacco, bright leaf seeds, and cigarette machines arrived at the port in Shanghai. In time, the traffic increased in volume and variety, as a crescendo of entry-level white-collar employees, spiffy business suits, letters from home, record albums, the occasional wife, recipes for Southern dishes, and even Smithfield hams made the journey. Along with these people, plants, and things came culturally rooted knowledge about bright leaf cultivation and curing, cigarette manufacturing, and labor management, as well as expertise in the proper way to make Brunswick stew, manage servants, and maintain cleanliness and order. Always changing in composition, the bright leaf network was an active and generative phenomenon that served as the primary mechanism for the corporation's expansion.

The bright leaf network developed first in the 1890s, when the American Tobacco Company absorbed hundreds of North Carolina's and Virginia's bright leaf chewing and smoking tobacco companies. What had been a white ownership class rather suddenly found its enterprises, almost without exception, owned by the ATC. The owners called it a "war" (the "plug war"), and there was a brief period when common hatred of the ATC and Duke brought them together. Companies that fell to Duke encouraged those that held out against the voracious foe to stay the course. Once former owners realized that they had lost utterly, however, they found themselves consorting with the enemy. As the bright leaf economy restructured around the ATC monopoly, the ownership class went to work for the ATC, becoming a white-collar corporate network of white men who shared regional ties and an understanding of the industry.

The bright leaf network shaped the expansion of the industry, first in North Carolina and then elsewhere. Expansion depended on building factories and establishing new markets, but first and foremost it required more bright leaf tobacco. As we have seen, after the Civil War, the bright leaf agricultural belt spread southward from its origins in three piedmont counties at the border of North Carolina and Virginia to the region that included Winston-Salem, Reidsville, and Durham, North Carolina. During the 1890s and the early twentieth century, bright leaf spread to the eastern North Carolina tidewater region and would continue spreading southward to South Carolina and Georgia, to China, and eventually much further.[1] The bright leaf network organized this agricultural and corporate expansion for decades, directing new white-collar opportunities in the ATC and BAT to white farm boys and sons of the former owner class. The class basis of this network was porous—many an economically unfortunate white boy gained the patronage of a powerful man on the network—but the race basis was absolute.

While it is accurate to say that Southerners brought bright leaf to China, it perhaps captures more of the truth of the situation to say that bright leaf tobacco brought Southerners to China.[2] Because bright leaf had certain biological, environmental, and technological requirements, knowledge of bright leaf had evolved with the tobacco type only in the Upper South. That knowledge became especially valuable as demand for bright leaf grew. African Americans did not gain any positions with BAT-China, despite their expertise with bright leaf tobacco; their absence is one of the clues that the bright leaf network operated to produce a racialized corporate structure and culture.

The bright leaf network was a product of Jim Crow, and part of its mission in both the US and China was to manage corporate expansion while generating racial distinctions and hierarchy. Of course, white Southerners could not simply export Jim Crow segregation whole cloth to such a radically different place as China, but they necessarily drew upon what they knew in order to build new systems in concert with myriad Chinese people, from businessmen to farmers to servants. The bright leaf network, then, transmitted knowledge about race management forged in the Jim Crow South as well as knowledge that might not immediately appear to be racialized, such as ideas about debt, sanitation, or cornbread.³

The bright leaf network in China functioned as a particularly interesting meld: it was a product and agent of Jim Crow segregation; it was an active arm of the ATC and BAT corporations; and it manifested imperialistic privileges in China. Furthermore, there was nothing particularly abstract about its power. Though generated by segregation, corporate empowerment, and imperial entitlements, the bright leaf network did its work through daily contacts between BAT foreigners and Chinese people in the course of developing a bright leaf tobacco program, running a foreign business, and relaxing at company events.

This chapter explores how the bright leaf network expanded to China through the work of organizing the bright leaf agriculture program there, directing further opportunities in BAT-China to white Southern men, and developing BAT's foreign corporate culture around white Southern identity. North Carolinians Henry and Hattie Gregory anchor the chapter, much as they anchored BAT-China. Henry Gregory ran the agricultural department, working with Chinese and Southern employees to establish a bright leaf production system that was quite different from sharecropping but similarly based on producing debt. Hattie Gregory maintained the couple's home in Shanghai's French Settlement, where young employees from the South enjoyed Southern food and hospitality at weekly events, reinforcing regional bonds in a new context of global capitalism. The Gregory home was, like the middle-class home in the South, also a site of daily contact with servants that defined racial hierarchies just as surely as did contacts with farmers in the tobacco fields.⁴ In other words, Henry and Hattie performed consonant work producing and managing racial hierarchies in their respective arenas. Through this labor, the corporation took shape as a creative, productive, and imperialist organization.

The Bright Leaf Network in the United States

In 1905, R. Henry Gregory and James A. Thomas departed North Carolina by railroad for San Francisco, where they then boarded a steamship to China. There, the two Southerners would lead the BAT enterprise in its conversion from a sales depot to a corporate branch that produced cigarettes from seed to smoke. Thomas headed the operation and would first tackle the task of transforming a shell of a factory into a functioning operation. Gregory's charge was to develop the company's bright leaf tobacco agriculture program, even as no one knew if it could be done. China grew plenty of tobacco, but not the bright leaf variety. If Gregory succeeded in cultivating bright leaf in China, the dream of creating a full cigarette production system might also succeed. If he failed, the enterprise would forever struggle under the weight of high transport costs of raw material; consequently, cigarette sales to ordinary cash-poor Chinese consumers would almost certainly elude the company. As the two men boarded the steamship in San Francisco, they couldn't know that their work in China would ultimately define their careers. As it turned out, Thomas headed BAT-China for eleven years, during its period of most dramatic transformation, after which he served on BAT's London board of directors (fig. 3.1); Gregory ran BAT-China's agricultural department for fully thirty years, the longest tenure of any foreign executive.

The stories of how Thomas and Gregory came to be on that ship to China exemplify the consolidation of the bright leaf network. Thomas experienced firsthand the ATC's takeover of the chewing and pipe tobacco industry in North Carolina because, for a short time, he had been a partner in the AH Motley Tobacco Company of Reidsville. That Thomas had reached that status was surprising in itself, for his family was "as poor as church mice," according to his cousin.[5] Born in 1862, Thomas spent his first decade of life on the Rockingham County farm that his father and grandfather converted to bright leaf tobacco after the Civil War. Like so many others, the family manufactured chewing and smoking tobacco on site and sold it themselves through the Southern states. When Thomas was just ten years old, however, their business failed, and the family moved to nearby Reidsville. Thomas quit school for a job "in a [tobacco] warehouse at twenty-five cents a day, which . . . gave me an opportunity to . . . learn about leaf tobacco."[6] Thomas's future was not looking particularly bright, but his luck turned when, as a teenager, A. H. Motley, owner of the AH Motley Tobacco Company, offered him a job and subsequently took him under his wing, grooming Thomas along with his own son for a leadership role in the business.

As one of Reidsville's most successful chewing and pipe tobacco companies, AH Motley Tobacco Company had established distant markets for its growing products and aimed for further expansion. Motley sent Thomas and his son to a four-week course at Eastman Business College in Poughkeepsie, New York,

Fig. 3.1 James A. Thomas as pictured in a 1923 commemorative volume. Courtesy Harvard Yenching Library.

where the teenagers learned basic bookkeeping, how to write a professional letter, and other professional skills. Thomas became Motley's principal salesman, traveling first through the Southern states, then the West Coast, and finally to Australia, New Zealand, and Tasmania. With his savings, Thomas also became a partner in the business. In 1896, however, the ATC took over AH Motley Tobacco Company and Thomas's brief moment as an owner and his excellent job both evaporated.[7]

Thomas tried hard to avoid working for the ATC. We know this because he next took a job with the Liggett and Myers Tobacco Company of St. Louis, originally a burley chewing and pipe tobacco manufacturer and the largest remaining company that loudly opposed the ATC's business practices. In fact, L&M went into cigarette production solely to give the ATC significant competition. Scores of companies the size of Motley and smaller still held out against the ATC, but mostly because Duke had not gotten around to targeting them yet. If there was any hope of competing with the ATC, it lay in L&M, for the company had established a strong domestic and international market for its goods.[8]

L&M broadened Thomas's experience by sending him on a sales trip to "the Hawaiian Islands, Japan, China, the Philippines, Borneo, Straits Settlement, Java, Sumatra, Siam, India, Burma and Ceylon." In 1898, however, financiers took over L&M and delivered the company to the ATC, and Thomas lost his job once again. At last, Thomas sought a position with the ATC; for the kind of work he did, it was virtually the only game left. The ATC sent him to the Philippines and to Singapore; after the formation of BAT, he served as the head of BAT-India. He left his position in Calcutta when he contracted malaria. After his recuperation in North Carolina, the company assigned him to build up BAT-China.[9] Thomas personally experienced the dislocations caused by the ATC, but eventually became part of the ATC and BAT's bright leaf network.

R. Henry Gregory was fourteen years younger than Thomas and entered the bright leaf network through his role in the expansion of bright leaf to eastern North Carolina. Gregory was born on a bright leaf tobacco farm in Granville County, in the first zone of bright leaf's expansion. When Gregory was thirteen years old, his mother died. The farm could not function properly without her agricultural and domestic labor, so Gregory's father sent the children to live separately with nearby relatives. Gregory quit school and moved to Warrenton, North Carolina, where he worked in a general store. There, Gregory met George Allen, an associate of James B. Duke who took the boy under his wing and helped him land a better job in the ATC's extensive structure, specifically in a Durham leaf redrying factory, where workers prepared cured tobacco for manufacturing.[10]

From there, Gregory gained incrementally better jobs in the newest bright leaf district, North Carolina's tidewater region. In the 1890s, farmers in the tidewater began growing bright leaf tobacco, and new leaf factories and warehouses sprung up to accommodate the growing trade. Gregory accepted a supervisory job at a leaf factory in Rocky Mount, and then at an export factory in Kinston, both along the border of the new bright leaf belt. Gregory had officially entered the small white-collar tobacco class. He married and settled down. In 1904, tragedy befell his young life a second time when his wife died of tuberculosis. The following year, at age twenty-nine, he called upon his friend George Allen, now at the ATC's New York City headquarters, and said he was ready for a change. Allen offered him the job in China establishing a bright leaf tobacco agriculture program.[11] When Gregory and Thomas set out for China, Thomas had already traveled the world, but Gregory was leaving home for the first time.

Establishing Bright Leaf in China

Thomas and Gregory arrived in Shanghai on June 2, 1905, but Henry Gregory's plan to begin work immediately was foiled.[12] "There was right much excitement when I first reached here on account of the riot," Gregory wrote home, for the anti–American goods movement was in full swing, with BAT cigarettes a main target.[13] If Gregory had harbored assumptions that he could simply impose an agricultural process on passive Chinese people, those surely wavered as he witnessed the dramatic protest. Shanghai merchants inaugurated the movement and donated their supplies of BAT cigarettes to large bonfires in the streets.[14] As chapter 1 explained, the movement protested the US's 1882 Chinese Exclusion Act and its corresponding Chinese treaty, then up for renewal. The anti–American goods movement delayed Gregory's first trip into the countryside to investigate tobaccos for almost a full year.[15]

Gregory had two strategies for developing a bright leaf source in China, each born of his prior experience with bright leaf cultivation. The first was to find a light tobacco already grown in China that might "be improved," as Thomas put it, to become something like bright leaf. That is, BAT employees would work with established Chinese tobacco farmers to modify their cultivation practices and teach them the bright leaf curing method. The other strategy was to use bright leaf seed from North Carolina and seek soils similar to those in the Upper South.[16] One of the challenges was that one could not simply distribute seeds to Chinese farmers and return later for bright leaf tobacco because tobacco seeds are very sensitive to soil and climate conditions. The same seed that produces bright leaf in the North Carolina piedmont, for example, might produce a heavy cigar tobacco in Wisconsin's soil. Even on the same farm, different soils had different effects on tobacco leaves.[17] Seed, soil, cultivation practices, curing tech-

Fig. 3.2 R. Henry Gregory inspecting Chinese tobacco, 1906. Courtesy Rubenstein Library, Duke University.

niques, and climate were all variables that Gregory and his staff had to juggle to obtain a suitable leaf.

Gregory finally departed for Sichuan Province in June 1906 in hopes of finding Chinese farmers who already grew a light tobacco for the large Chinese pipe tobacco industry. Gregory described the trip to his friend Kate Arrington back home: "I am going from here to Chung King [Chongqing] about 2,000 miles in the interior near the Tibet line ... It will take between two and three months to make the trip as I will have to go on a houseboat from Ichang [Yichang]."[18] As Gregory traveled, he eyed Chinese farmers' tobacco fields for similarities to North Carolina conditions or tobaccos. He mapped the types of tobacco he saw, the quantities farmers produced, and their prices (figs. 3.2 and 3.3). In Jiangxi, for example, he recorded, "This district raises two grades of tobacco, the light colored the same as the Huang Kong [Huanggang] district ... The other grade is a heavy coarse tobacco ... No good for our use."[19] Gregory also speculated on how bright leaf artisanal cultivation and curing methods might affect the tobacco he found. In Shashi, he noted, "tobacco is grown all along this canal; a large rich colored ripe tobacco the color of the R grade. . . . This tobacco would be fair if it was properly cultivated and cured."[20] Gregory grew excited when he arrived at Dongzhou and discovered that "some of the [tobacco] fields look as

Fig. 3.3 North Carolina workers dry tobacco, Miscellaneous Subjects Image Collection (P0003). Courtesy North Carolina Collection Photographic Archives, The Wilson Library, University of North Carolina at Chapel Hill.

well as I have ever seen them at home. The tobacco planted about two feet apart, topped and suckered and would make very fine tobacco if allowed to get ripe and cured properly." Gregory bought two boatloads of this tobacco so that he could cure it using bright leaf methods.[21]

The results were promising and Gregory proceeded with plans to develop certain local tobaccos into something like bright leaf. "After some investigation," Thomas recalled in his biography, "I found that the quality of native tobacco could be improved to such an extent that it could be used in the grade of cigarettes we sold in China. I next brought out from North Carolina some men who were thoroughly familiar with growing cigarette tobacco. They taught the Chinese farmers how to cultivate it."[22] After a few years, Gregory and his staff of North Carolina and Chinese tobacco farmers achieved an acceptable quality of bright leaf–like tobacco from Chinese light tobacco.[23]

Gregory simultaneously pursued the second strategy, one virtually identical to how tobacco experts expanded the bright leaf zone in North Carolina: he matched seeds imported from North Carolina to particular soil and climate conditions by running field trials and intentionally selecting the most successful seeds.²⁴ This method sent Gregory looking for soils, not tobaccos, like those in North Carolina. Thomas explained, "The land in that part of China [where we introduced tobacco] is very much like that in some parts of North Carolina"²⁵ (figs. 3.4 and 3.5). This method required developing the appropriate seeds and recruiting farmers who worked soils deemed compatible with BAT's seeds, whether the farmers previously had grown tobacco or not. These farmers, too, would need instruction in bright leaf cultivation and curing methods. Ultimately, Gregory's staff found that their experiments with imported seed yielded a larger and heavier tobacco leaf than those they had achieved with their best Chinese-seed tobacco. Thomas recalled that "within seven years we had produced a good hereditary tobacco seed in China."²⁶ BAT phased out its efforts in Sichuan Province and focused instead on districts with appropriate soils in Shandong, Anhui, and Henan Provinces. Gregory had succeeded.

In 1912, then, BAT entered a new phase of growth based on Gregory's achievement of a viable bright leaf tobacco program. Gregory continued seed development, began building the extensive infrastructure necessary for bright

Fig. 3.4 R. Henry Gregory standing in a bright leaf tobacco field. Courtesy Rubenstein Library, Duke University.

Fig. 3.5 Farmers in a bright leaf tobacco field near Wilson, North Carolina. Commercial Museum (Philadelphia, Pa.) Collection of North Carolina Photographs (P0072). Courtesy North Carolina Collection Photographic Archives, The Wilson Library, University of North Carolina at Chapel Hill.

leaf production, and recruited Chinese farmers. Thomas pursued a significant expansion in factory production and sales and stepped up the hiring of entry-level recruits from the US. The corporation grew in size after 1912, as well as developing more capacity and complexity.

Also in 1912, Thomas and Gregory brought Wu Tingsheng and Zheng Bozhao, BAT's two highest-ranking Chinese merchants, as their guests on a trip to the United States. Both Wu and Zheng had worked with BAT longer than Thomas and Gregory, having begun contracting with the ATC and Wills, respectively, before BAT's founding. Thomas and Gregory treated Wu and Zheng to a tour of BAT's New York City headquarters and the North Carolina and Virginia bright leaf region in preparation for a new level of sales collaboration that would give Wu and Zheng considerably more decision-making power (discussed in depth in chapter 5).[27]

The return voyage back to China also marked a significant development, for Gregory had married Hattie Arrington while in North Carolina, and his new wife accompanied the four men to Shanghai. No action could have testified more strongly to Gregory's confidence in his future in BAT-China than this. Hattie would live in Shanghai for twenty-three years, bear two children, lose one to diphtheria, and make the Gregory home a significant location for BAT's foreign business culture. The bright leaf network was enabling the reproduction of something like a North Carolina home in China.

Developing a Chinese System of Bright Leaf Production

Upon his return to China, Gregory and his staff turned to creating a bright leaf infrastructure and recruiting Chinese farmers into a workable system of bright leaf production. BAT needed experimental stations, leaf factories, and leaf buying stations, not only in Hankou, Hubei Province, where Gregory had headquartered his agricultural work, but in locations that would be accessible to farmers in Shandong, Anhui, and Henan. BAT also needed to interact with Chinese farmers' material and social lives and priorities and come up with incentives for producing bright leaf.

The Southern staff of BAT's agricultural department applied their knowledge about bright leaf forged in the Jim Crow South to their new setting whenever possible. Their shared background certainly greased the wheels within the foreign parts of BAT. When writing to BAT's London headquarters, for example, Gregory dealt with director A. G. Jeffress, a Richmond, Virginia, native who once worked for Allen & Ginter. Gregory explained that his staff had taught Chinese farmers to build tobacco barns for curing. These barns, he wrote, "are built just like our flue curing barns at home, only they are made of mud bricks instead of logs." He also assured Jeffress, "I believe the Chinese will learn to flue cure even quicker than the average farmer will at home," countering the widespread belief among Westerners that Chinese people were primitives that lacked technological capacity.[28]

Despite Southerners' common framework, Gregory and his foreign staff had to negotiate with Chinese businessmen, farmers, local officials, and gentry to make the venture succeed. High-ranking Chinese employees were indispensible partners who mediated between foreign employees' visions and local conditions. Chen Zifang was in charge of this process in Shandong; Wang Yanzi in Anhui; and an employee identified only as Compradore Ou in Henan.[29] Though BAT provided its foreign employees with basic tutoring in Mandarin, no foreign employee spoke any Chinese dialect well enough to negotiate with local farmers, gentry, or officials. Nor did they possess the necessary cultural competencies that would allow them to navigate such business interactions. Indeed, foreign employees were utterly dependent on Chinese translators and servants for obtaining food, water, and shelter, arranging transportation, and all business communication. In addition, foreign employees relied on businessmen like Chen, Wang, and Ou and other less high-ranking Chinese employees to provide them with a picture of local social structures so that foreign and Chinese employees together could envision a bright leaf infrastructure and social structure in Chinese villages.

The bright leaf system BAT's foreign and Chinese employees built was something new, neither sharecropping nor quite like the way Chinese farmers produced other crops. BAT's foreign and Chinese agents added a component of debt to existing agricultural systems that, though different in details, created investments and dependencies analogous to tobacco sharecropping in the US South. Most Chinese farmers that BAT targeted were poor, as in the rest of China, but the US's system of debt tenancy would not have worked. Some Chinese farmers were tenants, but many of those who owned their own land were not better off economically than renters because their acreage (in China, *mu*) was so low. There was a small class of farmers who owned enough land that they regularly made ends meet and a small class of farmer-landlords who rented a portion of their land to tenants.[30] This system offered very little social mobility, and the vast majority of farmers remained highly vulnerable to risks of drought or other causes of low yield. BAT did not change this basic system but, through the enticement of cash and the loan of items necessary for growing bright leaf, created a new subsystem of dependence and auxiliary profit within the established structure.

Chinese farmers' need for cash drew them to bright leaf, but the high cost of growing it created dependence on loans. Farmers' needs for cash were increasing across much of rural China because of rising taxes and increased availability of goods such as yarns for weaving and oil for cooking and heating. Crops other than bright leaf brought lower prices and were often paid for partly in barter or in installments.[31] Chen, Wang, and Ou offered cash in one payment for the harvest and promised (at first) to take the entire harvest regardless of quality. Though bright leaf brought a high cash price, the need to purchase seeds, fertilizer, flue pipes, and coal for curing made it five times more expensive to produce than gaoliang (a grain) and twenty-six times more than soybeans, and farmers sometimes struggled with the costs of those cheaper crops.[32] In 1915, Gregory estimated that Fangzi farmers spent three-quarters of a cent per pound of tobacco on coal. Farmers also scrambled to build tobacco barns for curing. By 1915, Fangzi farmers had built 568 barns, "and as soon as those who had barns finished flue curing, their barns were run to their full capacity by other farmers." Farmers who lacked access to barns air-cured their American seed tobacco, but sold it at a lower price. All of the Chinese farmers hoped to flue cure their tobacco the following year.[33]

Chen, Wang, and Ou set up a system whereby Chinese employees and local propertied men loaned fertilizer and coal at substantial interest rates, reducing farmers' profits and making bright leaf a new source of revenue for a privileged class of men.[34] This move generated local investment in the success of the bright leaf economy. At first, the company gave recruited farmers seed and fertilizer, loaned them the thermometers and flue pipes for curing, and provided instruction, but soon a system of debt developed. In Shandong, Tian Junchuan, successor to Chen, bought bean cake fertilizer and coal in the off season, when prices were low, loaned them to farmers at the going rate at peak season and charged interest. Local landowners and wealthy individuals also made loans of fertilizer or coal to farmers, either directly or through their oil or coal companies. In Anhui, Wang even began a fertilizer production company that allowed him to profit from the loan system. Wang did not loan directly to farmers, however, but to local propertied men and officials.[35] In both of these variations, BAT employees directly profited from the system that they devised while also involving local influential people in profiting from bright leaf; both siphoned profits away from small farmers, who had to pay fixed costs regardless of the size of the harvest.

In the US South, the loan of such items was part of the sharecropping contract and was one way that farmers entered a cycle of debt, where their profits were so compromised by costs of land rental and supplies that they made next to nothing, or even were forced to sign a new contract for the next year in order to delay payment on their debts. Southern landowners initially resisted sharecropping because it required them to forfeit direct supervision of farm labor on their land. In 1873, however, new crop lien laws in North Carolina and Virginia guaranteed that the first share of the crop went to the landowner, not to the farmer, effectively stabilizing the downward distribution of risk and upward direction of benefit.[36] Like sharecropping in the US South, then, a bright leaf system developed in China where the local landowning class consistently profited while risk shifted down the economic ladder to be borne almost entirely by poor farmers.[37]

The new infrastructure required to support bright leaf agriculture in China was extensive. BAT's bright leaf program required building leaf factories and leaf purchasing stations. Leaf factories, sometimes called redrying factories, further prepared tobacco leaves for storage, shipment, and manufacture. Gregory found that redrying factories on site were necessary because "a good deal of [the tobacco] was too soft or too dry for packing in hogsheads" for shipping to the Hankou factory. At first, Gregory experimented by redrying the tobacco in the company's six curing barns on the BAT compound and re-sorting it in the warehouse. Gregory estimated that redrying the tobacco on site saved the company $2.50 per hogshead and "put up our tobacco in better condition." He requested $2,000 to build a proper redrying plant at Fangzi, acknowledging that the venture was "entirely an experiment."[38] These new properties were illegal, because foreign companies were entitled to build factories only in treaty ports, all of which were in urban centers. High-ranking Chinese employees were the official buyers of the properties, though the true ownership was an open secret.[39]

The redrying factories became large operations: by the mid-1930s, the leaf factory at Mentaizi in Anhui Province employed 500 workers and 60 Chinese and foreign staff. Tobacco purchasing stations were typically located adjacent to leaf factories and both operated seasonally. BAT built seven purchasing stations in Shandong; by one estimate, the company employed 1,600 Chinese workers in the large factories and buying facilities in Nianlibao, Shandong.[40] Henan and Anhui had additional redrying and purchasing stations. Though BAT's official position was that it would not pay off warlords, the company built local support in Ershilibao, Shandong, by paying a local militia leader monthly to support his private army and operate his coal mine.[41]

Though the introduction of bright leaf tobacco to rural China drew on a logic of debt found in sharecropping, BAT employees could not directly replicate the tobacco culture of the Jim Crow South. Land rental, for example, was foundational to sharecropping but not to BAT's system in China, where BAT worked with small landowners and tenants on much the same terms. Likewise, Chinese farmers owed cash payments on debts but not a lien on their crops. Debt was one of BAT's agricultural products, for it allowed a range of others to participate in and benefit from the introduction of bright leaf to the region.

However, debt was simultaneously a labor control strategy, as in the US South, designed to keep farmers' need high and pay low.

The benefits of the bright leaf program to BAT were palpable. Sherman Cochran estimated that after 1919, BAT paid an average of $0.08 per pound for Chinese-grown bright leaf and $0.44 per pound for imported leaf, which does not include the cost of transport or losses due to spoilage for imported leaf.[42] As one BAT director wrote to Thomas from London, "This [Chinese-grown bright leaf] is very cheap and will be our salvation in China."[43] By 1915, Chinese farmers already produced 10 percent of the tobacco used in BAT's huge cigarette factories; Gregory hoped to expand bright leaf output three to five times in Fangzi alone over the following year.[44] By 1918, the company used more Chinese-grown bright leaf than imported tobacco in its factories in China. By the mid-1930s, approximately two million Chinese farmers grew bright leaf tobacco in the three main tobacco provinces, Shandong, Anhui, and Henan.[45]

The Southern Identity of BAT-China's Foreign Staff

The growth of the agricultural program solidified the Southern predominance in BAT-China's foreign staff. Thomas later recalled:

> Most of the Far Eastern representatives of the company in the early days were recruited from North Carolina and Virginia. There was no rule about this, but as most of the company's men at home came from these two states, they knew where to find assistants who from infancy had cultivated, cured and manufactured tobacco, so that it was a second nature to them.[46]

Men who knew bright leaf tobacco guided cultivation and curing and were as important as knowledgeable tobacco buyers. BAT's main competitor in China, the Nanyang Brothers Tobacco Company, lacked people who could discern bright leaf from a cheaper light-colored leaf that could carry a bitter flavor. "We were swindled because we did not know the leaves," remarked Jian Zhaonan, one of the leaders of Nanyang.[47] Buyers also had to be able to grade bright leaf in order to assign a price to it. Fangzi alone produced twelve to fifteen grades of tobacco by 1915, all of which brought a different price at BAT's purchasing station.[48]

Zhang Jiatuo, who worked for BAT's agricultural department in Shandong, described the preponderance of Southerners similarly, though he emphasized experience with labor control as the significant factor. "The foreigners who were in charge of buying tobacco came from the southern states of the United States.... Many of their fathers were capitalists, landowners, and rich peasants who were engaged in the tobacco business in the US and who had experience in managing tobacco companies as well as laborers and slaves." Zhang's term "rich peasant" referred in China to small landowners who rent some land and farm themselves as well; "landowner" referred to people who only rented to others. Zhang's use of the word "slave" is jarring, as slavery had been abolished for six decades, but does indicate an understanding that the US bright leaf economy was racially divided.[49]

BAT exercised tight control over the class and race composition of recruits by advertising openings and the application process exclusively through word of mouth and by using established white-collar employees on the network to vet applicants. In 1915, Gregory requested more men from Jeffress, the BAT director from Virginia:

> We will need eight or ten more men who understand flue curing. As we will not require these men more than two months in the year ... Messrs Thomas and Cobbs ... are of the opinion that ... the Sales Department will be able to use them and let us have them during the curing season.... We believe Mr. J. W. Goodson and some of your other leaf buyers will be able to recommend bright [leaf] farmers that they know personally who will make good salesmen, and at the same time could be used during the curing season.[50]

By asking for men who have a personal connection to tobacco buyers, Gregory established a word-of-mouth network that would confine opportunities to certain class and race circles.

In 1916, Lee Parker got his job offer in just this way, by showed up at the Wilson, North Carolina, tobacco market and approaching Mr. Currin, a tobacco buyer for the BAT-owned Export Leaf Tobacco Company. Parker had heard from friends that Currin could hire for BAT, and indeed, Currin offered Parker a job on the spot. Likewise, James Hutchinson acquired his job with BAT-China by speaking with Mr. Toms, the head of the Durham ATC factory. A dean of Trinity College (now Duke University) had told Hutchinson, a recent graduate, that this was a route to a position. Trinity College had close connections to the bright leaf industry because its founding endowment came from Washington Duke, father of James B. Duke. In fact, Hutchinson referred to Trinity as "the 'tobacco' college." Like Parker, Hutchinson got the job offer on the spot; he left for China ten days later.[51]

Though BAT needed some people with bright leaf cultivation and curing skills for its agriculture department, for most entry-level jobs no particular knowledge was required. Parker and Hutchinson, for example, never worked for the agriculture department; rather, both men worked year-round in sales. In fact, applicants were not quizzed about their prior experience. Parker's interview with Currin lasted "anywhere from thirty seconds to two minutes," and demanded no evidence of skill with bright leaf. The only evidence Parker had to provide before departure was a character reference from "my preacher [or] my banker or some prominent citizen . . . to let him know . . . whether I would lie or steal or if I was alright."[52] At his intake interview in Shanghai, a BAT manager asked Parker if he knew how to grow tobacco. "Of course, I knew how to [grow] tobacco, but my daddy told me I didn't, so. . . . I said, 'No, sir.'" No one raised an eyebrow; Parker simply gained an assignment in the sales department. When pressed on the importance of bright leaf knowledge to BAT, Parker said, "I don't think [knowing tobacco] was worth a nickel. I really don't. . . . Knowing tobacco didn't mean a thing in the world."[53] Farming seemed far from Hutchinson's mind: his 418-page memoir about his years with BAT-China never mentioned the agriculture department. Many Southern recruits did not work with the bright leaf program; others worked in jobs in the agriculture department that could be learned quickly, with no prior experience.

The preponderance of Southerners in BAT created a foundation for the company's business culture. Men from the South often traveled to China together, as had Thomas and Gregory, and found easy connections with other Southerners once they arrived. Most were young: BAT liked to hire single men between the ages of twenty and twenty-five. Additionally, most were rural or small-town men who had just left home for the first time. Irwin Smith recalled that he and his traveling companions were "four of the greenest nuts you ever saw in your life." Their train ride to Seattle, where the men would board a ship, was already "more fun than any trip we ever had." Smith explained the active sense of connection on the bright leaf network: "The people out there, if you didn't know their family, you knew where they came from."[54] The network gained personal salience with the two-way flow of letters, gifts, and greetings. Thomas recalled: "When [a BAT employee] returned [to China] from his vacation, the boys from his part of the country gathered around to hear the news. The boys always took something from China to the home folks and brought back from home something to the others. This fostered a sort of community interest among the men in China."[55] BAT's business culture built upon an ever-renewing latticework of Southern connections.

The bright leaf network helped new recruits to cope with two very distinct kinds of cultural difference that they encountered in BAT-China: the ways of Chinese people of various classes and the ways of British people, mostly of a higher class background than US employees. New arrivals from the South had to interact daily with Chinese businessmen, interpreters, and servants. Perhaps even more disconcerting, however, were their British coworkers, most of whom had elite educations and accents. Shanghai was a huge, diverse city thronging with people. The Southerners' ability to adapt to their Chinese and British coworkers and the new context of the treaty port was critical to the overall success of the company.

Chinese people initially seemed mysterious, alien, and even frightening to many new US recruits. Raised on a mixture of Marco Polo, Fu Manchu stories, and fears of a "yellow peril," some new arrivals also had learned that "not so many years ago an army of natives calling themselves the Boxers had tried to slaughter all the foreigners."[56] During his first weeks in Shanghai, James Hutchinson hovered within a few blocks of the Astor House Hotel in the foreign-dominated Bund. He explained, "The slit-eyes of the Chinese looked so sinister, and tales of their mysterious ways were still so fresh in mind, that I was squeamish about venturing too far into their midst."[57] New employees encountered scores of Chinese employees at BAT headquarters, however, and immediately had to learn to communicate with their English-speaking translators. The company put new recruits up in the Astor House Hotel, where they could eat foreign food and socialize with other foreigners.

Other foreigners posed their own challenges to US recruits, however, because Shanghai's treaty port culture was elite, cosmopolitan, and dominated by unfamiliar British colonial practices. Likewise, BAT was a large multinational corporation with elaborate hierarchies and formalities, some of which bore marks of the British Empire. Contracts with BAT, for example, were drawn according to policies for British agents in the Empire, namely that new hires must be unmarried, no more than twenty-five years old, and willing to sign a four-year commitment after which they would have a four-month home leave.[58] British and US employees did not necessarily find immediate common ground. Most immediately, US employees were aware that their British counterparts, in general, came from a higher class background and had more formal education than they did.

Virtually all Southerners struggled to adapt to British ways in the treaty port or the company or both. Gregory wrote home shortly after his arrival: "Yes, Shanghai is certainly the Paris of the East and not a bad place to live. The greatest objection I have to it is everything is so English and stiff. You know how much I like formality. I certainly enjoy shocking some of these 'blooming' English ladies. You know they try to be very formal and I never could be and would not if I could."[59] Likewise, when Lee Parker and two Virginians had recently arrived, they misunderstood a BAT executive's invitation to a dinner "at the boss's home" that would be "white tie." "We thought he was trying to be funny. We all laughed heartily." But when the executive came to pick them up, he would not take them without formal attire. "We ate a lonesome meal in the hotel dining room," said Parker. Parker subsequently ordered formal clothes from C. R. Boone, Clothier of Raleigh. "This [requirement] seemed very strange to me," said Parker, "in what I had thought of as a backward country."[60] Parker's change of clothes signaled his own uncomfortable induction into the imperial multinational corporation.

Some young recruits had an inclination to work out conflicts with the British by themselves, as Hutchinson witnessed on one occasion. A North Carolinian, identified only as Bartlett, stopped in the Astor House bar on the birthday of King George. Hutchinson recounted that Bartlett "heard behind him some one exclaim, 'God save the King!' Turning to find three Englishmen in dinner clothes standing with their heels together and glasses raised, he hoisted his own and shouted at the top of his voice, '— queen, jack and ace!' Then the fight started." The melee had to be broken up by the police, and Bartlett spent the night in jail.[61]

The success of BAT-China, then, depended on creating an orderly business culture that managed these sizable differences and within which employees could thrive. The home of Henry and Hattie Gregory became especially important in creating BAT's business culture along Southern lines. The Gregory home hosted weekly events that welcomed even the greenest new recruit from the

South and created a familiar Southern culture to which any British guests would have to adapt. These events did not include, however, even the highest-ranking Chinese employees. In addition, the Gregory home provided this corporate function through the labor of Chinese servants, making the Gregory home itself not so much a refuge from China as a place where Chinese difference was carefully constructed and controlled.

The Southern Home in China

In 1912, newly married Hattie Gregory departed Raleigh, North Carolina, for Shanghai, knowing that, as an executive's wife, she would set up and manage the couple's home. She also knew that the home would function both as haven for her family, including children that she hoped to have, and a public showcase for the company. Hattie was young and thousands of miles from female relatives who ordinarily would have mentored her in a range of domestic management tasks, such as furnishing and decorating the home, determining the family's daily menu, sewing, child care, and putting on social events. Her success or failure would reflect on her new husband and had both private and public repercussions. Each of her new duties, furthermore, involved supervising Chinese servants. Hattie's endeavor was to make a recognizably Southern home, but she did so under radically different conditions than in the South, many of which were beyond her control.

Hattie entered a very different servant system in the treaty port than in the Jim Crow South. Hattie's family of origin had employed one domestic servant, an African American woman she referred to as Ella. Foreign homes in the treaty port typically employed four to seven servants. As C. Stuart Carr said, "Nobody worked very hard. Everybody had plenty of servants . . . number one boy and number two boy and a cook and a gardener. You name it, they had it. The exchange rate was so favorable. You could have anything that you wanted. It was a glamorous life."

What was ease for Carr produced work for Gregory. She remarked, "Sometimes I feel almost overwhelmed with the number of servants that I have, a boy, cook, coolie, two amahs [female maids, usually for childcare], in addition to the gardener and the washman, but it is necessary to have them all."[62] The treaty port servant system was a meld of British colonial practices—the word "amah," for example, comes from British colonial India—and Chinese conventions of servitude based in the wealthy Confucian home.[63] Foreigners' complete reliance on Chinese servants' language, cultural, and bartering skills to procure provisions and perform myriad other essential tasks meant that Hattie could neither dispense with the servants nor profoundly revise the way the system worked.

Just as her husband's first experience of China was the "riot" of the anti-American goods movement, so did Hattie immediately encounter resistance to her endeavor. Hattie's daughter Jane recounted:

> She came into a household staffed by servants who had been with her husband for a long time and had established ways of doing things. Mother started making changes. Before long the head servant went in to see father. He said that "everything fine," I "like everything," but "I can no stand missy." He resigned. A number of others also left.

Gregory hired a new staff and, according to her daughter, "had to establish herself first as mistress of the house."[64] Like white homes in the South, Hattie's home became a site of power struggle, labor management, and the ritual enactment of hierarchy.

Indeed, Hattie and Henry Gregory were engaged in analogous and interrelated tasks of cultivating bright leaf tobacco culture in China, Henry in the fields and Hattie in the home. In the South, bright leaf tobacco culture did not stop at the edge of the field or the tobacco warehouse door, but included the farmhouse and the white-collar home. Likewise, labor hierarchies in the fields found their analog in the home, making both sites important for the production and maintenance of the system of Jim Crow segregation. African Americans' lack of land ownership and depressed agricultural wages compelled black women to take jobs as domestic servants in white households. So little were black women paid that even many white cigarette and textile factory workers could afford to hire one. The white home thus became a key site for the daily production of race and management of the distinction between white and black. Similarly, in Shanghai, the imperial treaty port home also became an important location for drawing and managing a line between foreign and Chinese.[65]

The greater number of servants meant that Hattie enjoyed more assistance than she had come to expect in the South, including forms of service that African American women collectively refused to do. In North Carolina, white women worked alongside their servants to bring off any social gathering. In Shanghai,

Gregory delegated preparation for a meeting of her bridge club: "I simply told the boy to set the tea table for twelve people . . . [and] told the cook how many people were coming, and what I wanted him to prepare for them. . . . Isn't that different from the work we would have to do at home?"[66] In addition, when Hattie first arrived in Shanghai, the personal service possible in China struck her as funny. Gregory wrote home, "You would all laugh to see me sitting up here in the hotel reading or writing with two servants—the boy and amah, awaiting my orders. . . . Between them, I only eat and sleep without assistance, it sometimes seems, and the boy waits on the table while we are eating."[67] Under slavery, such service had existed in wealthy households, but it was unheard of in the New South.

Limits on personal service existed in the South because African Americans had, in the decades since Emancipation, reshaped the contours of domestic service. Black family economic strategies included, by necessity, domestic service jobs, but largely rejected them for men. Black women took such jobs, but refused to do tasks that smacked of slavery and would quit if employers asked too much. Black domestic workers shopped for food, cooked, and cleaned but balked at serving white adults in personal or intimate ways. They provided primary care for children but refused to "live in"; no matter how long the day, they returned to their own households and families at night. Black women's collective desire to dignify their employment became an effective, informal labor action. Whites struggled against these limits, contrasting "rude" and uncooperative servants with the fantasy ideal of jolly and willing Mammy, but because black women shared a collective willingness to quit rather than comply, whites had no choice but to accept the new norm.[68]

Gregory was thrilled to discover that there were few limits on how much child care she could demand from a servant. "I have a very good amah," she wrote, "and am satisfied to have her do almost everything for the baby."[69] The amah slept on a camp bed in the baby's bedroom so she could care for him when he awoke at night. She fed him four bottles per day while Hattie nursed him four times per day. As the baby grew, the amah's role only increased. Indeed, Gregory hired a second amah as a "wash amah" because "an amah that stays with him day and night cannot do his washing or anything else that I would like for her to do." While black women insisted on returning to their own homes and families at night, Hattie explained, "There is no foolishness about either of [the amahs], they do not mind work, and do anything that I want them to do."[70] Hattie enthused: "We are to be envied on the servant question. We pay much more for our servants than most people at home think, but we get the service as a rule."[71] For Southerners, Chinese servants would do precisely what employers could not get black women to do.

Though Chinese servants would perform tasks that African Americans refused to do, Gregory found her power significantly limited in other ways, especially in her dependence on the head servant. The head servant's role was closer to a British system than domestic service in the US. He supervised the staff, managed household items, and accrued personal profit through commission, barter, and pawn. When the head servant bought food and supplies for the household, for example, he charged his employer a commission of his choosing. Gregory recounted:

> [Servants] take good care of their master's belongings, but consider a certain amount of squeeze perfectly legitimate. This is decided by the master's income, manner of living, etc. We do not lock our pantries, for instance. Rather the boy who is No. 1, keeps the keys. My silver is in his care, and he locks it up every night. They love show and a dinner party or a big tea party is their delight. In the first place it means purchasing more supplies, on all of which they get a commission.[72]

Number-one "boys" also traded household items between houses for special events or even sold them temporarily to pawn dealers for cash. Clyde Gore recalled, "You knew that it was happening, and it was perfectly all right. It was the routine. You went ahead and paid whatever [the number-one boy] told you you owed him."[73] Head servants had considerably more economic authority over household resources than did African American servants in the US South.

In addition, while African American domestics in this period largely refused to live in, asserting their right to maintain their own families and households, a Chinese head servant asserted his right to move his entire family into servants' quarters in the house, and his employer was often expected to provide his wife with employment as well. Hattie Gregory explained, "Now we have the boy's wife here as a No. 2 amah, to wash, iron, mend and help out generally.... I did not want a second amah, we have so many Chinese around ... [but] the boy's wife is going to be here ... anyway so it does not really add another Chinese to the establishment, only recognizes her."[74] Thus, Southern employers in

China could not fully control the living situations of their servants in China any more than they could in the US, and often found themselves compelled to hire more servants than they wished. Supervision of servants in Shanghai, then, was a daily, intimate experiment in racialized labor management, as it was in North Carolina, where neither employer nor employee could dictate terms.

In the Gregory treaty port home, as in white North Carolina homes, segregation did not mean an all-white space; rather, it meant ordering intimate spaces in particular ways to create racial hierarchies.[75] Gregory's daughter, Jane, later stated that her earliest memories were of her amah, not her mother. The male servants also loomed large:

> I was a girl child in a house with five to seven male servants: two chauffeurs, cook, head servant (the "boy"), gardener, and the coolie who brought in the coal, took out the ashes and did other such jobs. The only other female servant (besides amah) was the woman who came in to do the wash. So we had a constant stream of men going and coming. The amah stayed with me twenty-four hours a day.[76]

Like white children in the South, Jane and her brother, John, bonded first with the domestic servant.[77] John visited family in North Carolina for a year when he was three years old, during which time he missed his amah very much. Gregory wrote her sister, "I heard John tell amah this morning that he had a bank full of money in America and would give her some if she would go to grandmother's house with him some day."[78] John was learning very personal lessons about his racial power as well as the limits of relationships with servants in the racial order of the home.

Hattie created a kind of labor relationship with Chinese servants similar to what she had with African Americans at home, and in turn she came to believe that Chinese servants were in fact similar to African Americans, especially through their daily indirect resistance to her authority. She wrote home:

> The Chinese are better servants than the Negroes, but very much like them in many ways. To get the simplest thing done properly it is necessary to stand right by and watch them — certainly until they get your ideas into their head. As with the Negroes, too, it is very necessary to show them that you are the mistress.[79]

Likewise, Gregory's daughter Jane recalled many years later that:

> [Mother] told me that there were similarities between dealing with Chinese servants and Negro domestic help back home. She thought that Americans from the South sometimes had an easier time in such relationships than Americans from other parts of the United States. Southerners were accustomed to dealing with servants, people who were in and out of their houses.[80]

Other foreigners likely had an idea of the bourgeois home as a private space defined by the absence of the market, but Hattie was accustomed to the home being both a place of intimate care and paid labor, and she consciously developed her role of race manager.

Like Henry, Hattie cultivated order in China using some of the principles of Jim Crow segregation, in her case, sanitation rather than debt. For Hattie, African American and Chinese people were both primitives who did not understand modern sanitation. Again she racialized Chinese servants by comparing them to African Americans:

> The Chinese have no conception of cleanliness as we understand it. A towel, for instance, that has no smudge on it, is clean enough to stay in use, regardless of the number of persons it has been used by or how long it has been in use. But all the servants take baths. There is an odor about them as peculiarly their own as that of the negroes, but nothing like half so strong. Therefore not so disagreeable.[81]

Though Jim Crow segregation is popularly associated with separation—separate drinking fountains, prohibition against eating at lunch counters reserved for whites—in fact, segregation operated through the idea of modern sanitation.[82] The point was not full separation but race management through particular forms of sanitation.

Segregation ordered space by separating clean white spaces from the supposedly inherently primitive and dirty bodies of African Americans. An African American woman working in a white home did not integrate the space. The very intimate labor of cooking, cleaning, laundry, and childcare sanitized the home under close white supervision. African Americans were present in white schools, stores, and churches as well, but as servants. A black woman could clean a public white bathroom, for example, but not use it herself. Southerners believed African Americans' inherent difference was so bodily that they smelled different from whites.[83] Whites especially policed locations where nonlaboring black bodies' fluids might come in contact with white bodies: swimming pools, drinking fountains, restaurants, bathrooms, waiting rooms, hospitals. Through sanitation, segregation "became, to whites, a badge of sophisticated, modern, managed race relations."[84] Hattie applied this logic in her daily management of the home in China.

Hattie's comparison between Chinese and African American servants led her to see Chinese people as halfway between black and white, a construction found in the US as well. "The Chinese look very much like mulattoes," she wrote to her cousin. "In fact, I can't get any other idea into my head when I am with them—even such as Mrs. Hawks Pott for instance."[85] Francis Lister Hawks Pott was one of the most important Americans in Shanghai. An Episcopal missionary, he was the head of Shanghai's St. John's University for over three decades. He married a Chinese woman, Huang Su'e, and the couple had two children.[86] Mrs. Hawks Pott had a higher class standing than Hattie, but Hattie's racial ordering created unresolved contradictions for dealing with Chinese class diversity.

BAT's Southern Company Culture

Hattie's imperial management of racial difference in the home played a corporate function because the Gregory home was a site of weekly BAT social events. In meals for her family and for corporate social events alike, Hattie provided Southern fare. The Gregory home became a pillar of BAT's business culture and a place for young Southern employees to gain an identity within the global company. Hattie's selection of menus and supervision of the chef infused BAT with a white Southern character and translated Jim Crow hierarchies into the corporate business culture.

From the earliest days, Thomas set the tone for BAT socializing by inviting other executives to dinner at his home one or more times each week and by hosting a weekly Sunday morning breakfast for a large group of foreign staff. He also threw parties regularly for a wider community of friends and associates in the foreign community in Shanghai, including the entire foreign BAT staff.[87] The Gregorys also hosted regular events and took over Thomas's tradition of Sunday morning breakfast after he resigned from BAT. Irwin Smith, recalled, "Well, you know Mr. RH Gregory, he was out there. He always thought a lot of the young fellows. If you were in Shanghai . . . you were always invited out to his house for Sunday morning breakfast about eleven o'clock."[88]

Among Southern hosts, there was a competition to serve familiar Southern food.[89] Thomas remarked on this, saying, "Out in China, when I thought of North Carolina, I would have been willing to pay any price within my limited means for corn bread, salt herring, and black-eyed peas, the food I was accustomed to eat at home. Later on I brought these out with me." Indeed, Thomas shipped many different Southern foods to Shanghai, including Smithfield hams for one large Christmas dinner. The Gregorys also shipped Southern foods to Shanghai, including black-eyed peas. Of course, they did not make these foods themselves. Thomas, the Gregorys, and many others trained their Chinese cooks to make Southern fare, including outside of Shanghai. Hattie Gregory traveled to Hankou and was pleased to find the men "comfortably fixed" and enjoying "delicious food," all distinctively Southern dishes. She explained, "The boys always claim credit for the dishes they have taught the cook to make. Mr. Wright from Virginia taught him to cut up and fry a chicken properly. Mr. Covington taught him to make Brunswick stew and Mr. Whitaker cheese and macaroni. I enjoy going up there."[90] Food from home gained heightened significance in China as a way to self-consciously stage Southern identity, giving the regional identity a new resonance in the global corporation.

Hattie Gregory quickly realized that teaching Chinese chefs to cook Southern fare would be one of her principal tasks in running an executive's household. Gregory wrote home to her cousin Kate asking for recipes:

> By the way, please write out for me some simple receipts for everyday N.C. and southern dishes — waffles, battercakes, both kinds of egg bread, corn meal and flour muffins, molasses pudding, smothered chicken, etc., anything you think of that we have at home. . . . Put in too such things as corn pudding, corn cakes, cheese and macaroni and anything you can think of that is made of canned vegetables and cooked mild.

Gregory was unsure of how to teach her cook, since she herself had never made these dishes: in North Carolina, Ella had made her family's meals. Accordingly, she asked Kate to be detailed in her instructions: "I have a general idea how they are made, and think I could make them myself after a few trials, but it is so much better to have reliable and explicit directions to give these Chinese cooks for a new dish."[91] Gregory's supervision of her Chinese chef carried echoes of similar relationships with domestic workers in North Carolina. One could say that Hattie taught her chef to cook in the style of an African American woman.

Most Chinese chefs in Shanghai who worked for high-ranking foreigners already knew how to cook British food and all could cook Chinese food; having a chef who made excellent Southern food carried cachet. Hattie regularly reported on the menus of dinners at others' homes, and she reveled in an entertaining success. She reported home that she succeeded in teaching the cook to make beaten biscuits, "which are something that I have not eaten at any other house." She also held "a stew"—a Brunswick stew dinner, which was a tradition in North Carolina. She wrote home, "Our Brunswick stew dinner was quite a success. . . . Mr. Covington came out and showed the cook how to make it. I went out and looked on, so now I am sure that I can make one, or have it made, equally as good."[92] Southern food thus played a unique role in creating a BAT executive domesticity and carried significant public functions, including solidifying the Southern dominance of BAT in a context where British colonialism seemed to set cultural terms and Chinese employees far outnumbered all foreigners.

Food and diet, then, served to underpin racial difference in Gregory's Shanghai household and in BAT-China's business culture in a way that it certainly did not in the US South. The Southern food that Hattie and the others desired was itself a hybrid of black and white culinary traditions. Generations of black women, during and after slavery, created Southern cuisine as they cooked for both their own and white families. The Southern diet, then, was shared across race precisely as a legacy of racialized and oppressed domestic labor. In China, however, the Gregorys explicitly resisted incorporating Chinese food into their diet, despite the skills of their chef. Their daughter Jane recalled that the family never ate Chinese food in the home, but always ate North Carolina food. Nor did the family patronize Chinese establishments. "I can't remember eating in a Chinese restaurant, ever," she said.[93] The refusal of Chinese in favor of Southern food was one way of maintaining a line between the races and keeping Chineseness in its place.

In addition, weekly social events at the Gregory house powerfully signaled lines of ownership and belonging in BAT-China by who was and was not habitually invited. While high-ranking Chinese employees only rarely gained admittance to the Gregory home, the greenest "griffin" (or newbie) from the South enjoyed a standing weekly invitation. Indeed, there is no evidence that Chinese employees were ever invited to Sunday morning breakfast. In this way, the home acted as a sieve, screening out Chineseness from the family's—and company's—central social activity, even as Chinese servants were present at all hours. The significance of the home as a location for company events thus becomes clear. Because the twentieth-century home signaled a private, homogeneous space, BAT company events in the Gregory home positioned Southern employees as family and Chinese employees as outsiders.

BAT's Chinese Business Culture

There was another business culture in BAT-China, of course, and that was the separate and occasionally intersecting Chinese business culture. Chinese BAT businessmen did sometimes invite foreigners to their social events when appropriate; indeed, on those occasions they tended to make foreigners the guests of honor. However, most Chinese events did not include foreigners, partly because the sheer number of Chinese employees dwarfed those of foreign representatives. The vast majority of the company's daily business transactions occurred between Chinese people, with foreigners nowhere in sight. In addition, BAT's Chinese business culture created and maintained a different set of relationships than did BAT's Southern business culture. BAT's Chinese business culture included many kinds of events. Every business transaction in China, for example, began with tea and tobacco. When Chinese businessmen wished to throw a social event to build business ties among Chinese men, that is, an event equivalent to the gatherings at the homes of BAT executives in Shanghai, they hosted a banquet.

The banquet was the central social form for making and maintaining business contacts in China. A Chinese businessman might host a banquet for a large number of reasons, including forging a new relationship with merchants, welcoming a new employee to the fold, or honoring a visiting company elite. Banquets were typically male-only events, held in restaurants, courtesan houses, or other public halls, and included many courses and plentiful drink, punctuated by toasts. As chapter 1 noted, high-class courtesans offered banquet planning as one of their services. Throwing a banquet allowed the host to demonstrate his status and generosity while building relationships of reciprocity with others, a process crucial to how Chinese business culture operated. Chinese business networks thrived through webs of favors given and owed and cultural practices of conferring and receiving honor. "Guanxi," a term with no direct English equivalent, referred to the credit and honor that one accrued through hospitality and favors and that later would require reciprocation. This process was compatible with hierarchy — indeed, it was the stuff of hierarchy within the company. A very large banquet with high-ranking company representatives in attendance, for example, would announce and raise one's status in the corporate structure. Much of the business of BAT occurred through these networks, hazy when visible at all to foreigners.[94]

Foreign employees did receive invitations to some banquets hosted by Chinese employees of BAT, and these events were the most common type of company-related social event where foreign and Chinese employees mixed. Early in his tenure in China, Frank Canaday recalled, "[one] evening provided me ... with my first experience of ... Chinese male dinner parties which I was to encounter in all my business contacts throughout the land."[95] Entry-level BAT employees attended banquets primarily when they went out on sales trips with a team of Chinese employees to promote a particular brand. Very often, a Chinese BAT dealer stationed in the vicinity would throw a banquet for the visiting team as well as the established and potential retailers in the area. Parker noted, "Though these affairs appeared to be purely social a great deal of business was transacted."[96] Hutchinson recalled:

> On the smallest excuse our Chinese dealers gave dinners. They were formal affairs ... The host started the dinner by filling everyone's cup with hot wine and toasting the guest of honor. The guest of honor drank the health of the host, and the dinner was on. When wine was poured a guest held his cup outstretched in both hands and murmured polite refusals, but always accepted, and a cup was never allowed to stay empty.[97]

Because foreigners carried status in the company, hosts often made them guests of honor, bringing status unto themselves as well.

At more elite levels, banquets were absolutely essential for Chinese businessmen to establish their place in the company and its hierarchy in relationship to other Chinese men. For example, Zheng Bozhao threw a very large banquet in December 1912 in Hankou to mark the birth of his first grandson, which occurred in the year that he assumed a new position within BAT, namely when he took charge of the distribution of the brand, Ruby Queen/Da Ying. While the banquet nominally celebrated a family event, he made Henry and Hattie Gregory the guests of honor and invited a number of other high-ranking BAT foreigners. This was an extremely large and elaborate banquet as befit a man of Zheng's stature, with two hundred Chinese guests. Chinese men arrived at 10:00 AM, foreigners joined them at 2:00 PM and the dinner lasted until 5:00 PM; after dinner there was a theater performance. Though the couple would have an interpreter, Hattie could expect much of the fast-paced conversation to be lost on her because of her lack of Chinese language skills. In addition, she would likely be the only woman guest in attendance, other than courtesan-entertainers. Hattie developed a headache and skipped this event, "but Henry and the others had to go," she wrote.[98] Zheng Bozhao used the banquet in a typical way to announce and reinforce his exceptional position in the company as well as the status that a new grandson brought him.

Segregated business cultures reflected and created corporate hierarchies, but they did not put all of the power into foreigners' hands. Chinese employees controlled the business of banquets, and foreign employees attended when invited, accommodating the way that banquets functioned to create BAT networks. Banquets were the key place where foreign and Chinese employees socialized, but they were not a location of foreigner control. Indeed, they signal just how much of the company's business was in the hands of Chinese employees. Perhaps it seemed natural to Southerners that Chinese businessmen maintained a separate business culture, as separate institutions for African Americans was the law of the land at home.

There was one exception that proved the rule in the Gregory household: Zheng Bozhao regularly paid calls upon Hattie. Zheng knew Hattie because they had made the long journey from North Carolina to China together in 1912. Paying afternoon social calls was a common practice in the foreign community as well as among middle-class Chinese in the Qing era, but in Chinese culture men typically called only upon men or mixed-gender groups. "The old compradore comes to see me quite often," Hattie told her cousin. Zheng always brought a gift, customary for Chinese social calls. He gave her a gold ring, sheepskin for a coat, and, upon the birth of her first child, a baby cap.[99] Zheng successfully melded Chinese and foreign social customs to make himself a guest in the

Gregory home. Nevertheless, when Henry Gregory reciprocated to Zheng, he neither made him guest of honor at a Southern dinner at the Gregory home nor threw a large Chinese-style banquet, but rather took him to a restaurant with Hattie and two other North Carolinian employees.[100]

Conclusion

As BAT employees and their families created corporate structures in China, the bright leaf network, founded in segregation, manifested corporate and imperial power. In the US South, the bright leaf network grew out of and served as a functioning part of Jim Crow segregation because it operated to direct white-collar opportunities in an expanding economy along hierarchies of race rather than expertise. In China, the bright leaf network continued the work of managing economic expansion along racial lines. Jim Crow segregation could not be imposed in China, but Henry Gregory drew upon it in cultivating a bright leaf agricultural program by creatively using debt to build farmers' dependency as well as widespread investment in the new system. Jim Crow also influenced the ways that Hattie used notions of sanitation to manage servants and order race in her home in the French Settlement. African Americans, then, were an absent presence in China. Southerners could not simply place Chinese people in the categories that African Americans occupied, but they used the racial hierarchies they knew so well as a resource in cultivating new social forms in China, forms that renewed the corporation's investment in race as a foundation for hierarchy.

In the process, BAT Southerners repurposed white Southern identity as a corporate imperial identity. The bright leaf network was imperial not just because BAT reaped the benefits of unequal treaties but because the network functioned as a mechanism for ordering the long-distance connections through which corporate power operated. It also generated the specific distinctions between foreign and Chinese that lent shape to the vertical hierarchies of the company and of empire.[101] Of course, large organizations like BAT have functional hierarchies but the organization of BAT's foreign business culture, especially, ensured that the vertical gradations within BAT-China were not determined simply by expertise or experience but by race.

It would be possible to tell the story of the expansion of bright leaf tobacco from the US to China through the mythos of modernity. Bright leaf developed in the US, required certain technologies, and spread to the East, to China. Indeed, Southerners in BAT seemed to see it that way, and spoke of "improving" Chinese tobacco farming practices and Chinese tobacco itself. The problem is both the imposition of the idea of progress and, even more importantly, the idea that modernity came first to the West, which makes it difficult to see connections and continuities. Bright leaf expansion was not completed when Gregory came to China but continued in the US South as well. The significance of the bright leaf network lies partly in its active and productive function in both places to transform local economies according to corporate imperatives, but it could not simply scale up an already existing system intact. The spread of bright leaf was part of an unfinished corporate transformation that occurred in the US and China at the same time.

The bright leaf network also coordinated the organization of new cigarette factories in the US South and China at nearly the same time. Of course, there were already cigarette factories in the US, but the ATC moved the Southern factories to the North in the 1890s. Only in the 1910s, the same years that Chinese factories grew to mammoth sizes, did the cigarette industry shift southward. Though the process occurred in both places, neither factory management system was simply an imposition of the techniques of the other. Nevertheless, in both cases, factory management shaped workers' movement through space according to race and gender and combined legal and extralegal modes of violence to keep order, albeit not always successfully. We turn to this story in chapter 4.

4

Making a Transnational Cigarette Factory Labor Force

In the 1920s, Ruby Delancy and Qian Meifeng both gained employment in the packing departments of cigarette factories that were linked by the bright leaf tobacco network. In 1925, thirteen-year-old Delancy began work at the Reidsville, North Carolina, branch of the ATC. Her first job was filling "fifties" (tins that held fifty cigarettes) by hand. In 1929, Qian, age nine, began work at BAT's Yulin Road Factory, in Shanghai. Her first job was distributing empty ten-piece packs to the people who worked the packing machines. In both factories, these entry-level jobs held status. Reidsville's packing department hired only whites like Delancy; black women worked primarily in the stemmery, stripping the stems from the cured tobacco leaves. Shanghai's packing department hired predominantly women and girls from Zhejiang Province; Qian hailed from this high-status province to the south of Shanghai, while women from more distant rural areas worked with tobacco leaves in the stemmery.[1] Both girls, then, had reason to feel relatively fortunate to have cleaner, easier work for higher pay.

Delancy and Qian participated in a similar factory choreography and danced a similar dance of labor. Both found their labor indirectly paced by the cigarette machine, which set the rate of cigarette output for the whole factory.[2] They each knew tobacco intimately—its pungent smell, the color and texture of the shredded leaves, the sleek slimness of the finished cigarette—and could perform the repetitive motions of their labor in their sleep. They entered the factory through particular doors and not others, wore uniforms that signaled their status, and did or did not receive free cigarettes, all of which conferred value or stigma upon their bodies. Both girls contributed their income to a larger family economy. Neither was the very first generation of young women in their region to earn a wage, but both were still seen as doing something quite new for girls when they went to work in the factory.

The commonalities between Delancy and Qian stemmed from the bright leaf network's role in recruiting, shaping, and disciplining a cigarette labor force in both the US South and China. For example, Ivy G. Riddick of Raleigh, North Carolina, worked in the management of both leaf preparation and cigarette manufacturing in China at the same time that his counterparts did the same in North Carolina. Riddick went to China in 1914; in 1919, he became the manager of the stemmery at BAT-China's factory in Hankou, and later served as factory manager there. In 1931, he became a special labor consultant at BAT's massive factory complex at Pudong, a job he held until 1939.[3] The bright leaf network transformed both societies, albeit not identically, by prompting labor migration, reorganizing local economies to accommodate industrial production, and shaping workers' bodies and subjectivities through disciplinary structures.[4]

Delancy and Qian's bosses used similar strategies in managing their workplaces, but factory policies and structures developed within vastly different contexts. Local political hierarchies inflected factory order in divergent ways. In addition, workers like Delancy and Qian themselves had profoundly different life experiences, and brought particular expectations about supervision and hierarchy to the workplace. They also answered first not to high-level managers but to immediate supervisors who loomed large in their daily experiences, though quite differently in each place. All of these factors governed how workers responded to management and how social interactions at the factory unfolded. Indeed, were Delancy and Qian somehow magically transported into each other's workplaces they would almost certainly be lost. Delancy and Qian both formed a part of the emerging global cigarette workforce, but they experienced and shaped the corporation within local contexts.

This chapter recounts the expansion of cigarette manufacturing as a story of the bright leaf network managers' nearly simultaneous constitution of a corporate labor force in two distinct sites. Though the US industry predated the one in China by two decades, US cigarette factories moved south, reconfigured the labor force, and began significant expansion after 1911, just a few years into the buildup of factories in China. The ATC's US cigarette factories had been in the urban North, where there was an abundance of immigrant female workers, the company's preferred labor force. The precipitating event for the southward shift was the Supreme Court's dissolution of the ATC monopoly in 1911, creating four successor companies: the new American Tobacco Company; Liggett and Myers; RJ Reynolds; and Lorillard. The new era of competition prompted these companies to open cigarette factories in Winston-Salem (1912), Reidsville (1916), and Durham (1917 and 1924).

The reconfiguration of the industry in the South united in one location the leaf preparation department, which employed mostly African Americans, and cigarette manufacturing, which employed mostly whites. In giving shape and different economic and cultural value to these positions, the cigarette companies participated creatively in the broader system of Jim Crow segregation, dividing jobs much more sharply by race and gender than had the New York City factories. Manufacturers had previously used categories of gender, age, and ethnicity to shape the hiring process, but created a relatively homogeneous workforce of urban immigrant girls. In North Carolina, managers created job categories that animated race and gender in hiring but also in the organization of daily workplace identities and hierarchies.[5] Leaf preparation remained in African American hands, but employers reassigned the operation of cigarette machines from white girls to adult white men. The factory — and the machine — newly took up a place in the political economy of the South in ways consistent with and productive of Jim Crow hierarchies.

BAT built up the industry in China at much the same time. The company began with new factories in Pudong and Hankou in 1906 and 1907. In 1909, the company added a facility in Manchuria at Mukden, and in 1914, at Harbin. By 1916, just over a decade after Thomas arrived, between one-half and two-thirds of the twelve billion cigarettes that BAT sold in China were supplied by its China factories. In 1917, BAT built a second large factory in Shanghai, followed by factories in Tianjin (1921), Qingdao (1925), and Hong Kong (1929). By the end of World War I, twenty-five thousand Chinese people worked in BAT's cigarette factories. As in the US South, Chinese workers filled job classifications sharply designated by gender, as well as by native place (the rural origins of one's family and ancestors).[6]

The bright leaf network generated and managed these massive labor force mobilizations in both places. As a corporate imperial structure, the network coordinated the development and management of a cigarette labor force by animating certain kinds of differences and muting others, by directing workers through factory spaces in some ways and prohibiting others, and by drawing on particular kinds of workplace discipline.[7] Discernable in both systems is the imprint of race management knowledge forged in the Jim Crow South, though policies based in this knowledge did not always make intuitive sense in the Chinese factories and therefore were sometimes less effective. In addition, both systems drew creatively on legal and extralegal forms of coercion through factory floor supervisors, who had enormous power in both places, as well as on external policing. The simultaneous development of these linked labor systems thus represents an effort on the part of the multinational corporation to create a global labor force and reveals its role as a site of imperial governance. At the same time, neither industry was merely an echo of a process that had already been completed somewhere else; rather, both unfolded through daily life in unpredictable ways.

Why Cigarette Factories Moved to China and the US South

Though the US South and China became huge centers for the production of cigarettes, neither had been a first choice for such development. China came onto BAT's radar as a production site only once Japan had nationalized BAT's Murai Brothers Tobacco Company in 1904; the venture seemed untried and risky, even to those who promoted it. In the case of the US, the ATC had actually shifted cigarette production *away* from the South, mostly to New York City. Moving factories to China and to North Carolina had the clear advantage of tapping low-wage workers but also required companies to prompt rural-to-urban migration streams, integrate with local economies and hierarchies, and work with a preponderance of new industrial workers, most of whom came from agricultural backgrounds. These factories would eventually move to the cutting edge of the industry, but at their inception there were already bigger, more established and successful factories in the northern US and Japan as well as throughout Europe and the Middle East.

There were a number of reasons that China would not necessarily seem like a good place to make cigarettes. Until this point, the ATC and BAT built factories overseas not primarily in order to gain cheap workers, but to evade high tariffs on their exports. The tariff structure was an expression of imperial power: the US instituted high tariffs to protect US businesses, which is why Egyptian companies opened branch factories in the US. The ATC exported cigarettes to Japan until that country raised its tariff; only then did the ATC buy Murai. China's tariffs remained very low, mandated by coercive treaties with foreign powers.[8] In addition, the Chinese market for US goods had yet to live up to fantasies about China's four hundred million potential customers. The *North American Review* even warned US businessmen in 1902 that "the vast possibilities of Asiatic trade will not be realized in the immediate present or in the near future. Its growth will be slow and in some measure disappointing.... until [China] is gridironed with railroads."[9] It was bold, some would say foolish, to imagine large, thriving cigarette factories in China in 1905.

Despite the risks, BAT and other companies saw potential in manufacturing in China. In 1895, Japan won the Sino-Japanese War and extracted a treaty that newly allowed foreign powers to build factories in Chinese treaty ports. British companies built textile factories in Shanghai in 1895 and 1896, and Germany and the US followed close behind, though US textile companies did not succeed in China.[10] Japanese companies began building textile factories by 1904. In response to these new factories, labor migrations were already under way, especially to Shanghai. BAT was motivated to build an East Asian production center in China in 1904, when Japan nationalized Murai. Edward J. Parrish, the vice president of Murai from Durham, North Carolina, had previously argued against building a factory in China, but in May 1904, he wrote to Duke and the London board of directors from Tokyo that "it appears to us China is a great place for a great factory or factories.... Now is the time to move rapidly and on a big scale in China." Parrish suggested that the skilled Japanese supervisor from Murai's Tokyo factory could train Chinese supervisors in BAT's practices, and indeed, Shanghai BAT manager William R. Harris subsequently hired him to do so.[11] Duke approved the shift to China, saying, "If cigarettes can be made properly and more cheaply in China they should be made there."[12] Duke's comment revealed that China's inexpensive labor was a draw.

BAT executives evaluated the potential of Chinese labor in the context of an international debate about the controversial "coolie trade."[13] The multifaceted debate produced diametrically opposed information about the implications of Chinese labor for democracy and civilization, but the two loudest positions both cast Chinese workers as having different racial capacities for labor than those of whites from the US and Europe. A contingent of US labor leaders argued that Chinese men lacked the more refined needs and moral sensibilities of whites required to support domesticity and a civilized manhood.¹⁴ In 1902, Samuel Gompers, a cigar maker who became president of the American Federation of Labor, pitted American labor—free, virile, and meat eating—against "Asiatic Coolieism." "Which will survive?," Gompers asked. Chinese men, he argued, were rice-eating, opium-smoking, emasculated, unfree, diseased, and lived in "colonies," rather than families. They therefore would work longer for less pay than white workers.¹⁵ He echoed E. A. Ross, a sociologist at the University of Wisconsin, who argued that Chinese people had a "race vitality" that enabled them to "underlive" white people.¹⁶

Businessmen refuted the idea that Chinese laborers were diseased or dangerous, but their defense of Chinese workers' capabilities often endorsed similarly racially charged ideas. In 1898, British textile company representatives looking into building a factory in Shanghai compared "cheap, plentiful, submissive" Chinese labor to "dear, dictating" British labor and declared that Chinese laborers required "nothing more than a bowl of rice, or wheaten cakes, to enable them to work without intermission for eleven hours."¹⁷ James A. Thomas endorsed the idea that Chinese workers were submissive and had superhuman endurance.¹⁸ He crooned to his Reidsville banker, R. L. Watt, "Labor is a great commodity these days and as you know these people work eighteen hours a day without the assistance of labor unions."¹⁹ Such ideas were long lasting. In 1922, Julean Arnold of the US Department of Commerce effused, "The Chinese laborer is remarkably good-natured, patient, industrious, able to subsist on comparatively little, [and] possesses splendid endurance."²⁰ BAT drew on and contributed to these rather magical ideas about Chinese workers' capacity for labor when recruiting its Chinese labor force and implementing factory policies.

Duke was convinced by arguments that cigarette production should move to China, but he never moved cigarette production back to his home state of North Carolina. When the monopoly formed in 1890, factories in Virginia and North Carolina each contributed 23 percent of the company's output and New York factories produced 47 percent. By 1905, North Carolina made only 3 percent and Virginia 16 percent of the company's cigarettes, while New York's share had risen to 60 percent.²¹ In fact, the domestic cigarette industry languished under Duke's watch, mostly because the ATC was busy taking over the larger chewing and pipe tobacco industry, and the company lost domestic market share to competition from Egyptian and immigrant producers. Furthermore, cigarettes did not make significant inroads against pipe and cigar smoking in the US, unlike in much of the rest of the world. In 1912 and 1913, just after monopoly dissolution, cigarette consumption relative to other tobacco forms was dramatically lower in the US than in the UK, Spain, Russia, Austria, Germany, Japan, Italy, and France.[22]

When the Supreme Court divided the ATC into four successor companies, dividing the factory properties among the four, it created motivation to build new cigarette factories. Because factories specialized in certain kinds of products, no one company ended up with a full menu of product offerings. RJ Reynolds received the large chewing and pipe tobacco factory complex in Winston-Salem, but was the only successor company to receive only Southern property and to receive no factory that produced cigarette brands.[23] In order to have a full complement of tobacco product offerings, RJ Reynolds introduced cigarette production at its Winston-Salem plant in 1912, becoming the first big producer to locate a new cigarette plant in the South.

The industry only slowly followed Reynolds's lead, probably because the other companies already had functioning cigarette departments, but in 1916, Charles Penn of Reidsville, North Carolina, now vice president in charge of manufacturing for the new ATC, successfully lobbied the company to open a cigarette department in the Reidsville plant. Penn's motivation came from his ongoing ties to Reidsville. While Duke moved to New York City in the 1880s, Penn had stayed in Reidsville even after the ATC absorbed his family's business and, like Reynolds, had served as the manager of the factory that his family had once owned. When he became a vice president, he had to travel to New York City frequently on business, which he reportedly hated, but he continued to live in his Reidsville home and even kept his old office at the factory. Locating the new Lucky Strike factory in Reidsville would direct much-needed jobs to Penn's hometown. Penn's loyalty to Reidsville likely was spiced with competition with Reynolds to establish himself as a civic leader in the North Carolina piedmont.[24]

Creating a Transnational Cigarette Labor Force

The nearly simultaneous buildup of the cigarette industry in China and the US required managers from the bright leaf network to interact with and shape very different labor pools into a functional cigarette factory system. In both cases, they utilized gender, age, and race or native place to create job classifications and systems of status. In both cases, many of the hires were new workers making the transition from farming to industrial work, which also meant a rewriting of familial gender roles. Interestingly, managers in the US and China hired women in nearly the same jobs and at virtually the same ratios. In both places, women came to make up about two-thirds of the workforce. Though women did not perform all of the manufacturing jobs, the bright leaf cigarette industry continued to participate in the global deployment of girls as a deskilled labor force. As chapter 2 showed, the term "girls" masked the age and marital diversity of the labor force and thereby helped justify low wages and poor prospects for advancement.[25]

Locating cigarette factories in the small-town South posed labor recruitment challenges that slowly led to changes in job classifications. For the first several years, the RJ Reynolds factory in Winston-Salem and the ATC plant in Reidsville followed the accepted practice of hiring girls for cigarette manufacturing, likely because the work had come to seem naturally suited for women. In 1903, Duke stated that the ATC hired women to make cigarettes because it entailed "all light, easy work."[26] There were very few immigrants in the South, so Reynolds and Penn prompted a migration of white country girls for these new positions. In 1918, Reynolds bought a hotel to house young women who migrated to Winston-Salem for cigarette positions.[27] Likewise, Reidsville offered a dormitory for female employees.[28] In 1918, the ATC completed its new $500,000 cigarette factory and built one hundred houses for employees in Reidsville.[29]

Despite the continued reliance on white girls, Southern tobacco manufacturers worried about how the composition of the labor force would affect the local social order. When the ATC announced the opening of Reidsville's new cigarette factory in October 1916, the newspaper explicitly recruited "white young women" for the two hundred new jobs, and apologetically acknowledged that there were no jobs for men at present.[30] Hiring tapped the groundwater of gendered assumptions to the company's benefit but immediate solutions could create new problems. An industry that hired girls when there were many white men who needed work provoked unease because it upset the idea of the white male breadwinner. Though the work was billed as for young women, specifically, many women who gained their jobs as children in this era worked in the factories until retirement decades later.

If hiring white girls proved a sensitive issue, considerably more anxiety surrounded African American employment. In fact, town leaders, many of them tobacconists, had long fretted about how the tobacco industry affected the racial makeup of the town because African Americans performed most work in tobacco, and new jobs prompted black migration to towns like Winston-Salem, Reidsville, and Durham. In its early days, Reynolds attracted African Americans to Winston-Salem by enlisting employees to recruit relatives, promising to hire whole families, and sometimes paying transportation costs. The city's African American population boomed and in 1880 reached 47 percent, compared to North Carolina's overall black population of about 12 percent.[31] Neighboring towns hesitated to take a similar approach. In 1889, chewing tobacco manufacturer Decatur Barnes of Reidsville wrote, "We need especially manufactories or mills which will give employment to white people the year around. A business which furnishes labor only to negroes and works about half the year won't build up a town."[32] Reidsville and Durham tobacconists actively promoted the opening of textile mills. Textile mills provided jobs primarily to white men and women who worked the weaving and spinning machines, reassuring white town boosters that the ratio of black to white would not tip to Winston-Salem levels.

Around World War I, the four big manufacturers changed their cigarette factory labor recruitment and management system. They all used race and gender to segment the labor force, but now the companies redefined the job of operating cigarette machines as privileged white adult male labor. The system of job classification that they developed at this time would stay in place with only minor variations until the 1964 Civil Rights Act outlawed job discrimination based on race and sex. White women continued to perform key roles in cigarette manufacturing and packing but no longer operated the cigarette machines.

Ruby Delancy's experience as a cigarette factory worker was typical for white Southerners. Delancy grew up on a farm near Ruffin, North Carolina, but her family moved to Reidsville when she was thirteen years old because the Lucky Strike factory was "hiring young girls." Though the factory had explicitly put a call out for girls, Delancy saw herself as taking on a male role. Her father was unable to work and her older brother had recently died at age twenty-one. "I had to be the little boy in the family," she recalled. Delancy's employment was a violation of new child labor laws but "I put on my long dress that's down to my ankles, and I told them I was sixteen." Several of Delancy's friends at the factory lost their jobs a couple years later because an inspector came through and the company quickly fired everyone underage, but Ruby's real age was not recorded in the factory office. She was glad to have her better-than-average job in the packing department because she "was able to help Mom raise the kids." Delancy worked at the factory for forty-two years.[33]

Delancy's relationship to her work developed in response to company policies, contemporary ideas and styles, and her own personal flair. Though Reidsville had a population of only about eight thousand people, it felt like a city to her. She declared with laughter at age ninety-two, "Oh, I was one of [the flappers]! I wore short skirts; I was a flapper and a flirt." The gratis cigarettes she received from the factory enabled her to smoke, which was also new for women in the South. "I was raised on Lucky Strikes," she said. "I'd smoke those all the time. Cigarettes were free then." In addition to a weekly allotment of free cigarettes, Delancy could help herself from the box of cigarettes in the women's bathroom. The racial division of labor was an ever-present fact for Delancy. She never went to the leaf preparation side of the factory, where most African Americans worked, but reported that "we had sweepers and ladies who cleaned our bathrooms." At seventeen years old, Delancy and her boyfriend slipped off to Danville, Virginia, to get married. They came back to Reidsville and continued working at the factory, but Ruby no longer lived with or gave her earnings to her mother, with whom she had a conflicted relationship.[34]

It would be difficult to imagine two places more different than Reidsville and Shanghai. Qian Meifeng worked at a job similar to Delancy's, but joined a massive migration of hundreds of thousands to new jobs in domestic- and foreign-owned industries in cosmopolitan Shanghai. While BAT would become Shanghai's single largest employer, textile companies together employed even more people, and countless other industries also located there. By 1933, Shanghai was home to nearly half of China's factories.[35] Merchants flowed into Shanghai from the south, while laborers came from all directions, but especially from the north. Qian moved to Shanghai from Zhejiang, a high-status province just to the south of Shanghai.

Drawing on its high-ranking Chinese employees, BAT-China made native place the foundation of labor recruitment for its factories. BAT hired floor supervisors for each department, who then hired from their own native place. There was nothing radically innovative about this basic strategy. In China, native place was a significant identity category. In addition, different Chinese regions carried very different levels of status, with poorer and more rural regions at the lower end of the spectrum. It would be difficult to overstate the significance of native place for how Chinese people created business networks and recruited labor. Native place connections had long served as the foundation for guilds and other social affiliations, bonds that were reinforced by a common dialect and cultural background. Because Shanghai was a city of sojourners, one's native place often became the basis for new kinds of associations in the growing economy, though the traditional functions of native place ties also strained under the sheer numbers of new migrants.[36] Zhejiang Province was among the highest-status provinces, and Qian's privileged position in the packing room came from her good fortune of coming from that area.

Qian Meifeng and women like her lived in dramatically different ways than Delancy, but the factory figured into her story analogously. Qian, like Delancy, did not move far to work in the Shanghai factory but gained a privileged position once she arrived. Qian and her counterparts resided near their factories, typically with family members and among people from their native place. Women and girls from other provinces worked in the leaf preparation department in another building in the factory. Men dominated in mechanics positions and in the cigarette department. Qian and her counterparts spent, on average, 80 percent of their wages on food, making extra purchases extremely rare. In addition, she had to buy presents for her immediate supervisor. Workers often lived in the factory shantytown, with very little space per family (as little as one hundred square feet) and a single water faucet for ten families. While Qian was the first young girl in her family to work for an industrial wage, she had counterparts whose mothers had begun working for BAT two decades before as children. Like Delancy, she likely hoped for fun and romance even as she toiled under repressive controls.[37]

Delancy and Qian built their lives, then, within the abstract categories and strategies that white managers from the Upper South attempted to marshal in the US and China. Managers utilized social categories of gender, age, and race or native place to order the division of labor and create differential status and wages on the job. Managers did not simply replicate already existing categories, however, but infused these markers of difference with new purpose. Managers in both places created disciplinary regimes that directed workers' movements through the factories, marked different jobs with uniforms, and augmented the status of certain workers with free cigarettes.

The Creation of Workplace Status and Stigma

Managers created status and stigma at the factories first by creating a functional and spatial division between the leaf and cigarette departments. In the leaf department, workers prepared tobacco leaves for manufacturing by removing the stems, checking for impurities, treating the leaves to achieve an appropriate moisture content, and sometimes adding flavorings. They also chopped, shredded, and sorted the leaves. On the cigarette side, workers ran the cigarette machines, caught cigarettes as they came out of the machines, and weighed, counted, packed, and labeled cigarettes, packs, cartons, and cases. All of the jobs in both the leaf and cigarette departments could be learned quickly and none took exceptional strength, with the exception of moving tobacco around the factory. Though they often came to seem natural, status and pay differentials did not track to skill but were almost entirely arbitrarily assigned in the interest of labor control.

The arbitrary nature of workplace values becomes especially clear when considering the job of stemming tobacco. In factories managed by white Southerners in both the US and China, leaf department jobs carried the lowest status and lowest pay. In the US, such work had been associated with African Americans since slavery. In the US, African American women stemmed tobacco leaves; in China, BAT assigned women from a low-status northern province to stem tobacco leaves (figs. 4.1 and 4.2). However, BAT's most significant competitor, the Nanyang Brothers Tobacco Company, *only* hired workers from the company owners' native place, making native place the determining factor in hiring but not relevant to further workplace segmentation. Nanyang hired women as stemmers, as did BAT, but Nanyang paid stemmers *more* than women in the cigarette department because they determined the work to be more difficult and otherwise less desirable.[38] This difference signals that BAT's constructed work values may not have seemed reasonable to Chinese workers who did not associate tobacco leaf preparation with slave labor.

In China, foreign managers depended especially heavily on Chinese supervisors in creating a segmented workforce. In BAT's Chinese factories, foreigners served as overall factory head and as main workshop managers, but they hired Chinese "foremen" or supervisors[39] to manage workers on the floor. Following US custom, these supervisors had the power to hire and fire their own workforce. BAT hired supervisors from different native places for different work-

Fig. 4.1 Chinese women stem tobacco at a BAT-China factory. Courtesy Rubenstein Library, Duke University.

Fig. 4.2 African American women stem tobacco in Kinston, North Carolina. The North Carolina County Photographic Collection #P0001. Courtesy North Carolina Collection Photographic Archives, The Wilson Library, University of North Carolina at Chapel Hill.

shops, scaling the prestige of the native place to the status that they wished to give to the work.[40] As was typical in China, supervisors hired from their own native place, segmenting the factory workforce in the process. In addition, Chinese supervisors also almost exclusively hired people of their own gender. If BAT wished for the packing department to be staffed by women of a high-ranking province like Zeijiang, then management hired a female supervisor from Zeijiang. In this way, BAT created a functional hybrid, joining its desire to segment the workplace with normative Chinese business practice.

Managers from the bright leaf network created loose parallels between the two labor systems. In the US, white women packed; in BAT-China, women from higher-status urban areas packed. In both countries, men worked in the leaf department, though not usually as stemmers. In the US, black men performed leaf department work; in BAT-China, men from lower-status regions performed it. The main difference came in the cigarette-making department. In the US, white women performed all of the tasks through World War I, after which both white men and women worked in this high-status department, with white men working the machines and white women catching the cigarettes. In China, only men worked in the cigarette-making room, with adult men running the machines and boys serving as catchers. Considered apprentices, these boys were in line for the privileged positions of running the cigarette machines. Unlike the boys in China, women in the cigarette-making room in the US had no chance of advancement (figs. 4.3 and 4.4).[41]

Managers gave job categories status and stigma by marking workers' bodies with uniforms and shaping their movement through the factory. Management's tactics were impersonal and bureaucratic, but affected each worker on a personal level: by amplifying the abstracted differences of gender, age, and race or native place, managers rendered other differences between workers less visible and less relevant. In other words, managers used highly personal and bodily tactics to render workers as anonymous and interchangeable within their classification as possible.

Workers in BAT's Pudong factory (across the Wangpu River from Shanghai) found themselves sorted by gender and race or native place upon arriving at the factory. Workers entered the factory compound by one of two gates, the men's entrance or the women's entrance.[42] Gender, this practice declared, was not just important for family life but was of utmost significance in modern factory life as well. Chinese workers next traveled within the compound to their

Fig. 4.3 Cigarette-making room, BAT factory in Pudong. Frank H. Canaday Papers, Harvard Yenching Library.

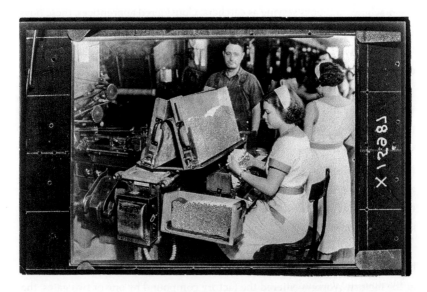

Fig. 4.4 Cigarette-making room, ATC factory in Reidsville, North Carolina. The Miscellaneous Subjects Image Collection (P0003), North Carolina Collection Photographic Archives, Wilson Special Collections Library, UNC-Chapel Hill.

department, typically segregated by native place. Through these movements, workers' investments in these categories deepened, especially because their job and pay were also scaled to differentials of gender and native place.

At the factory gate, each worker presented an identity card with their employee number. Assigning workers a number depersonalized them and also allowed for record keeping in English in the factory office, accommodating foreign managers' varying levels of skills in Chinese.[43] However, workers used their anonymity in the eyes of foreign company executives to manage their family economies. Mothers, especially, would loan their identity card to older daughters so that they could stay home with new babies. Chinese supervisors could be offered a gift in exchange for turning a blind eye. It would be difficult to call this resistance, as it meant that the life of a young girl was subsumed to capitalist goals, but it was an example of harnessing a disciplinary technique for the goals of the family.[44]

In North Carolina, workers' movements through the factory were no less choreographed than in Pudong, but the regulatory mechanisms were less visible. There were no gates shielding the factory property and workers did not gain entrance via a card or number. Nevertheless, black and white workers divided by race as they approached the factory and divided again by gender when they went to their department, where they universally worked with others of their race and gender. By assigning white men and women to different job categories, bright leaf managers did more than simply follow local custom. In the Southern textile factories in the same towns, managers hired both white men and women for the critical job of weaver. Depending on the factory, 25 percent to 50 percent of textile jobs could be performed by either men or women.[45] In tobacco factories, workers and managers like understood the compound to have a "white side" and a "black side." Ruby Jones recounted what happened when she went in the wrong door: "Come a thunderstorm up in the summertime and Mr. Lackey, my supervisor, and I got off [the bus] . . . and went in the door on the white side and he made me go back out there in all that rain and come in that black door around there."[46] Tobacco managers' distinctive sorting by gender and race amplified and gave new meaning to local hierarchies.

Within the leaf and cigarette departments, workers encountered dramatically different conditions. US Department of Labor inspectors noted that the "black" leaf department was extremely hot, dirty, and dusty, poorly ventilated, and had insufficient toilets and sinks. The department lacked chairs for female stemmers, who sat on barrels or boxes or stood while working. In contrast, cigarette departments were newer and cleaner and often served as models of modern factory facilities. White women on the cigarette side always had chairs available.

Black and white workers also ate in separate cafeterias designed to reinforce the higher status of whiteness. The white cafeteria was above ground and had a piano for workers' recreation; Lib Chaney recalled that she had a friend "who could play the dickens out of that piano." Though the lunch period was short, many workers would gather around and, "mouths full and what not, we would sing." In contrast, William Davis, who sang in his church choir for more than forty years, remembered that the black cafeteria was in the basement and remarked drily, "We didn't have no piano."[47]

In both the US and China, managers accentuated job classifications with uniforms for women only. Managers in the United States instituted uniforms in the mid-1920s; by the mid-1930s, managers in China implemented a remarkably similar uniform policy. At the ATC, white women wore uniforms in brand colors (green and red) in the 1920s, and white uniforms trimmed with green by 1930, while black women wore blue uniforms. In China, women in the packing department wore blue jackets with white aprons, and women in the lower-paid stemmery wore blue jackets with blue aprons.[48] In both countries, uniforms made job classifications visible on the body: they simultaneously abstracted workers and attached them even more tightly to gender and race or native place categories. In large factory compounds, the uniform was the first thing that any coworker or supervisor saw. If someone were out of place, the uniform would announce it.

The color differences in uniforms carried a powerful distinction in the US. White women wore uniforms associated with brand colors and the finished commodity of the cigarette. Ruby Delancy recalled, "We had to wear green uniforms with the American [ATC] emblem." Likewise, Evelyn Farthing recounted, "When I first went to work [in 1924], we had to wear solid green uniforms. The packs were green and red. And we had a red emblem on a green uniform with 'Lucky Strike' on the emblem."[49] The army green of the Lucky Strike brand proved unpopular with customers—and clearly with workers as well—so the company changed the pack to white trimmed with red. White women's uniforms changed to white trimmed with green. These uniforms drew much more positive commentary from workers, who universally described them as "pretty." White connoted purity, cleanliness, and white-collar or professional jobs. The redundancy of layering associations of white uniforms with white women's jobs built on deeply established meanings, while dressing white women in brand colors transferred some of the value of the commodity to them.

Blue uniforms, in contrast, widely signaled blue-collar work in the US, but also specifically smacked of domestic service in the South. White journalist Jonathan Daniels assumed that tobacco and domestic workers wore identical uniforms: "The blue and white uniforms for servants, sold everywhere by the chain stores, seem almost the uniform of Negro women in Durham, where they work in the rough preparation of the tobacco while the white girls turn out the endless tubes of Chesterfields and Lucky Strikes."[50] African American tobacco workers certainly registered the status difference between the uniforms. Anne Mack Barbee attached her resentment about job conditions to white women's uniforms, saying, "On that side where we were working, black women did all the hard and nasty work, that's what I say. On the cigarette side, where they wore those white uniforms ... [they] made sure no blacks worked over there." Marion Troxler recalled, "[Our uniforms] were blue trimmed in tan and .... the white ladies wore the white uniform that was trimmed in green. They wore beautiful white uniforms and ours was blue."[51] At the factory, the uniform policy required women to daily mark their bodies with their differential status.

Within African American community contexts, however, the blue uniforms of tobacco workers could hold more positive meanings. Tobacco factory jobs, though segregated and poorly paid, represented a step up for black women, who worked overwhelmingly in agriculture and domestic service. They also offered more autonomy than service in a white woman's house. Though Daniels saw no difference between uniforms for factory and domestic service, African Americans could distinguish them. Patsy Cheatham enjoyed wearing her uniform in Durham's vibrant black business district: "Well, you know that was the big thing, L&M and American, and if you get a job there and you have a blue uniform on, you thought you were really, that was envy. Clerks in the stores would stop and meet you half way, can I wait on you. You were somebody."[52] The company could not fully control the meanings of uniforms.

The color differences between uniforms in China would not carry such strong resonances as they did in the US. Blue was widely associated with labor in China's twentieth-century treaty ports, a fact likely related to the strong presence of British and US industry. Women in BAT's leaf department wore blue jackets with blue aprons; while the more privileged cigarette packers wore blue jackets with white aprons. In China, however, white did not signal purity and race privilege; in fact, Chinese cultures associate white with funerals and death. Managers' uniform policies, though similar in the two countries, gained an emotional charge from US race relations that did not translate to China.

There is fragmentary evidence that the uniform policies may have originated in both countries to control sexual expression and prevent women from dressing up at the factories. Some female workers in both Reidsville and Shanghai participated in consumer culture and creatively engaged with popular styles. Even in Reidsville, a small Southern town, workers shopped locally and through mail-order catalogues to achieve a stylish look.[53] In Shanghai, despite the twelve-hour shifts and the extremely low wages, cigarette workers from Jiangsu Province were known for their stylish clothes and boots.[54] If muting individual sexual expression at the workplace was one motivation for the uniform policies, it had a paradoxical effect of making gender especially visible.

Companies in the US and China both gave free cigarettes to higher-status workers, underscoring race privilege in the US and gender hierarchy in China. In the US, the ATC dispensed two hundred cigarettes per week to white workers regardless of gender, but not to black workers. BAT-China gave male workers fifty to one hundred cigarettes a week regardless of their native place, while women received only fifty cigarettes per year.[55] In both China and the US, this practice likely grew out of the desire to manage workers' ability to pocket cigarettes on the job. Cigarettes were small, easy to lift from the line and conceal on the body. As early as 1880, managers of cigarette factories in New York City struggled with workers who pocketed cigarettes; when they attempted to institute a search for cigarettes as workers left the factories, workers struck.[56] In China, managers regularly dismissed workers caught taking cigarettes. Managers allotted free cigarettes to the workers who had the most access to manufactured cigarettes in the cigarette-making room, which meant whites in the US, and men in China. Free cigarettes decreased the number of people who would lift cigarettes from the line, making it possible for supervisors to define the practice as aberrant and punishable when discovered.

Free cigarettes served as a management strategy beyond simply staunching the flow of pilfered cigarettes, however, because managers gave them out not by department but by identity category. In the US, black men and women worked as sweepers and bathroom attendants in the cigarette department, where they would have opportunity to pocket cigarettes that had fallen to the floor, that were not yet packed, or that were available in the bathroom, but they received no cigarettes. In China, women in the packing department had easy access to cigarettes, but their fifty cigarettes per year were certainly not enough to deter them from helping themselves. The differential distribution of free cigarettes, then, acted as another disciplinary device that constructed status at the workplace and built investments in race and gender identities.

Free cigarettes had significant value in both places whether or not workers intended to smoke them. Both urban China and small cities in the US South still operated partly as barter economies. In Reidsville, people sometimes paid the dentist or for their newspaper subscription in firewood rather than cash.[57] Nonprofessionals traded skills and resources even more informally. In Shanghai, the inflation rate was so volatile that the company paid workers in a combination of cash and rice so that people could eat even if the cash lost value.[58] Workers often sold or bartered their cigarette allotment, benefitting materially from their workplace privilege.

Ruby Delancy revealed a complex identity as a worker-consumer through her use of the commodity when she said, "I was raised on Lucky Strikes. I'd smoke those all the time. Cigarettes were free then." Since Delancy began working and smoking as a child of thirteen, she aptly and without rancor articulated management's paternalistic relationship to her, though the idea of a father raising a child on cigarettes is dark and strange. Even in Reidsville, parents tried to delay children's smoking, fearing it would stunt their growth. After a lifetime of smoking, Delancy was shocked to discover that she would no longer receive free cigarettes from the company after retirement. She promptly quit smoking in protest.[59] For Delancy, smoking free cigarettes enacted a privileged relationship to the company that she had viewed as personal and permanent.

For workers who did not receive free cigarettes, pocketing unattended cigarettes could seem like just compensation for inadequate wages. Riddick, head manager of the Pudong plant, noted in 1932 that their losses from cigarette pilfering were considerable and attributed it to the fact that the factory failed to pay a living wage to a large number of its workers.[60] Supervisors regularly fired workers caught with even a single cigarette in their possession.[61] Chinese women in BAT's packing department proved particularly difficult to control. In one skirmish, a foreign overseer caught a woman in Pudong taking cigarettes out of the packing department, and management fired her on the spot. When word reached the packing department, the women immediately struck. Management somehow convinced the workers to cease their strike, probably by threatening their jobs. However, at the end of the workday the women of the packing department made their point. They concealed and carried out hundreds of cigarettes and destroyed them on the street in front of the factory in protest.[62] By destroying the commodity in front of the factory, rather than smoking or selling it, they asserted their ultimate power to define its meaning. In this situation, they made the company's cigarettes into a symbol of women workers' power and defiance.

Factory Discipline and the Struggle over Workplace Governance in China

As with job classification, managers of the bright leaf network created parallel but not identical factory disciplinary systems in the US and China. In China, managers from the upper South depended heavily on Chinese supervisors, who hired for and ran their departments like small fiefdoms. Workers' close and demanding daily relationship with supervisors likely outweighed their contacts with foreigners in shaping their experience of labor, though oral interviews with workers also reveal a keen awareness of foreign authority in the factories. Supervisors shared dialect and cultural commonalities with workers in their departments and often belonged and answered to Chinese organized crime as well as foreign bosses. They exercised significant power that was often out of the reach of foreign control, with both positive and negative effects for workers.

In order to obtain a job, workers typically offered gifts or cash or hosted a banquet for the supervisor. Zhang Yongsheng began working at BAT's Pudong factory at age nine after his father paid seventeen yuan (a very high fee) to the supervisor. Another worker paid cash and hosted a dinner at Da Hong restaurant. Workers remained in good standing with additional periodic gifts, especially at holidays. "Those who gave generous gifts," recalled one worker, "would be sent to perform work at a higher piece rate, or be given more and better raw material." Supervisors "found fault in everything" when workers did not offer gifts. One supervisor gained the nickname "Invitation Party Queen" because she required many events to be thrown for her.[63] The costs could be significant. In 1920, approximately three hundred workers from the Pudong factory "invaded" BAT's headquarters in Shanghai to demand either a raise or the dismissal of a supervisor who "had too keen a sense of his privileges in the matter of 'squeeze.'"⁶⁴ Supervisors' power, then, operated separately from foreign managers and was arbitrary and coercive.

Many of BAT's Chinese supervisors also brought gang affiliations into the corporate structure. The Green and Red Gangs were secret societies that operated extensively as organized crime in Shanghai. They controlled the opium trade and participated in politics (including supporting the 1911 revolution). By the 1910s, "virtually every line of unskilled labor came to be ruled by native-place barons with gang connections—beggar chiefs, brothel madams . . . wharf contractors, and factory foremen." Migrants from the countryside "discovered that employment opportunities in the city were dependent on criminal ties."⁶⁵ Qian Meifeng worked in a cigarette department in which the supervisor would only hire male rollers who were part of his gang. "He was their old guy and adopted father," she reported, using terminology used by the secret societies and gangs. She also reported that he was a "bully and a local emperor" who made money on the side by running a gambling business that he expected workers to pay into.⁶⁶ Supervisors utilized gang connections to maintain status and control in their shops, but also could be pressured by gang leaders who questioned their loyalty: did they support their Chinese countrymen or the foreign devils?⁶⁷

At BAT-China's factories, supervisors essentially ran a system of social order that existed alongside factory rules and worked largely outside of foreigners' control. Supervisors maintained their own negotiations with workers about daily policies, creating great variation in actual practices within the factory. Foreign managers depended on Chinese supervisors for accurate information about workers' attitudes and grievances. Even the few foreigners with excellent language skills could understand only one or two of the many dialects spoken in the factory. This dependency proved to be a liability, especially in times of workplace conflict. Factory manager I. G. Riddick, from North Carolina, noted that "schroffs" occupied the "'No Man's Land' of Pootung [Pudong], between the foreigner and the labour. . . . Practically all of our shroffs are worthless as interpreters, and otherwise, in matters of discipline."⁶⁸

Factory rules were stringent and the daily power of supervisors to enforce factory rules and assign punishments was enormous. The most common points of conflict were around workers' bodily needs: lunch and using the bathroom. In Pudong, the huge factory compound had no lunchrooms. Workers purchased their lunches outside the factory gates from vendors who offered fried wonton, congee, noodles, sticky rice balls, soybean milk, vegetable rice cake, and big bing (probably a wheat flatbread), and had to consume their lunch before reentering the compound. However, the thirty-minute break minus travel and purchase time meant "in reality there were only fifteen minutes," according to one worker. In addition, many workers were paid by the piece, creating an incentive to send a friend to smuggle lunch to the workroom.[69] BAT's guards confiscated any food that they discovered. Workers' memories contradict the myth that Chinese workers could toil for eleven hours on just a bowl of rice. Li Xinbao recalled, "Workers whose snack was discarded had to work in hunger from morning, when they ate a little, to night, when the work ended. They were so starved that their heads were dizzy and their eyesight blurred." Li also recalled another worker being forced to stand on a desk in the factory in front of everyone and eat a big bing that she had successfully smuggled as far as the workshop. Ma Wenyuan recalled a supervisor throwing away his steamed cornbread and a jar of water.[70]

Bathrooms were another point of conflict, with employers assuming that workers were escaping to the bathroom merely to take a break. Each workshop had its own bathroom and regulations varied. One workshop locked the bathroom for an hour, every other hour, to limit use. Another used a system of tags to limit the number of workers who could use the bathroom at a given time (three tags for more than three hundred workers).[71] Foreign management at Hankou reportedly entered the bathrooms and beat and cursed workers to force them out.[72] Huang Zhihao recalled that "foreign management used hoses to force workers out of lavatory," and that there were separate bathrooms for workers, Chinese supervisors, and foreigners.[73]

Stringent factory rules led to particular hardships for mothers, structuring life outside as well as inside the factory. Workers would lose their jobs if they missed too many days in a row, which led to pregnant women coming to work until the day they gave birth and returning immediately afterwards. Zhang Yongsheng recalled, "Some gave birth in the lavatory. Some on the road. Some on the little sampan. A female worker had a son who was born on the sampan and his nickname was 'little sampan.'"[74] Mu Guilan reported that family members would bring babies to the factory gate at lunch break to meet mothers. "I chewed dry bing and ran to breast feed," she recalled.[75] Women were two-thirds of BAT-China's workforce, so BAT's maternity policies affected a huge number. When she entered the Pudong factory in 1938, Zhao Qizhang became the third generation of women in her family to work for BAT. In response to major strikes, BAT did implement some maternity policies in January 1928, including a six-week leave without pay for childbirth and a flat payment of thirty dollars.[76]

Though workers had far more contact with supervisors than with foreigners, they also reported that foreigners created the tyrannical factory-wide rules and sometimes violently enforced them. Workers reported that both supervisors and foreign managers pulled them by their plaits and slapped them.[77] Worse than this treatment, according to some, was the use of suspension and fines, because these measures cut into their already-low pay.[78]

BAT's large factory compounds became militarized spaces of imperial governance and conflict. BAT hired guards who stood in doorways and workshops throughout the premises, a common practice in Shanghai factories. As one of the city's largest employers, BAT helped set a standard for foreigner-labor relations.[79] Workers, however, struck against BAT more than any other company in China, waging fifty-six strikes between 1918 and 1940.[80] Workers' opposition was a nearly constant force pulling the company to reform its governance practices. In 1928, the company agreed in writing to workers' right to form a union and bargain collectively, seven years before workers gained this right in the United States. In addition, workers argued with some success for the right, at least on paper, for no dismissal without cause, compensation for injury on the job, and a six-week unpaid maternity leave. They also won some paternalist programs, including health benefits and a company-funded school for the children of workers at the Pudong plant.[81] These gains did not extend across all of BAT's factories and could remain subject to some supervisors' discretion, but they did change daily life in the factories.

When workers struck in great numbers in the 1920s, the rising spirit of nationalism and anti-imperialism also generated political pressure from Chinese authorities in Shanghai and beyond for BAT to respond, resulting in the creation of some union agreements and industry regulations. The 1920s strike wave against BAT was part of a larger uprising across numerous industries. Communist organizers appeared at BAT factories in 1921, focusing first on skilled mechanics. Though most workers did not become party members, they took advantage of new resources to address long-standing grievances.[82] In 1925, the May 30th Movement explicitly linked workers' concerns to students' and merchants' anti-imperial politics. At BAT's Shanghai factories, the strike lasted for four months, the longest work stoppage in Shanghai history. In 1926 and 1927, the new Guomindang municipal government in Shanghai supported the Three Armed Uprisings, which helped secure a written agreement between the company and the union, a significant win for workers.[83] In 1932, the Shanghai municipal government passed factory codes governing industry in the city.

BAT certainly did not simply give in to escalating pressure but fought back against strikers. First, the company hired more watchmen and police guards, making the factory compound a more militarized space. Sometimes this show of force convinced striking workers to back down but workers more often defended their right to strike with wooden clubs, rocks, and, on one dramatic day, weapons stolen from the local police station. During the era of Three Armed Uprisings, as BAT experienced a protracted work stoppage, British ships landed in Pudong and British soldiers "defended" the factory.[84] Chinese authorities protested these actions of a foreign military on Chinese soil, but their inability to stop BAT laid bare the global relations of power that underwrote the company's presence in China and made the corporation's status as a colonizing force undeniable.

BAT foreigners who had assumed Chinese workers were docile, then, found that workplace order evolved as a product of a complex imperial struggle rather than as an implementation of their modernist vision of universalizing industrial processes. At the very least, Chinese workers had successfully challenged ideas that they were submissive and worked "eighteen hours a day without the assistance of a labor union." Indeed, the tobacco industry secured its first union agreement in Shanghai almost a decade before tobacco workers in the US South. By 1928, James Joyner, a BAT salesman from eastern North Carolina, saw Chinese workers as model strikers. When he heard about the large Gastonia, North Carolina, textile strike, he remarked in a letter home that it seemed like Gastonia was taking a "leaf out of China's book."[85]

Factory Discipline and the Struggle over Workplace Governance in North Carolina

In contrast to dramatic and sometimes violent strikes in China, most of the US cigarette industry unionized in the 1930s without resorting to a work stoppage. No militia marched on the factory workers, no bricks flew, no batons fell. However, such struggles had been occurring in the Southern textile industry, as James Joyner noted about Gastonia. Numerous local strikes culminated in the general strike of 1934, when both South and North Carolina called out the National Guard to repress thousands of picketers, resulting in guards killing seven workers in Honea Path, South Carolina.[86] In response to this and other massive strikes, Congress passed the Wagner Act (National Labor Relations Act), which required employers to negotiate with unions selected by a majority of workers. The act went into effect the week that Southern tobacco workers first approached management for a contract, explaining the relative peace in the tobacco industry, and marking the beginning of a new moment in corporate-labor relations in the US.

As significant as was the divergence signaled by the Wagner Act, there are other insights to be gained by exploring the struggle for governance in the US tobacco industry in relationship to that in China. The bright leaf network failed to be a conduit for the universalizing of US industrial order. However, understanding BAT's corporate imperialism in China's factories helps reveal the multinational corporation's role in creating global labor forces by using markers of gender, race, and native place origin. In the US, the struggle between managers, union members, municipalities, and extralegal "gangs" created factory workforces and shaped the social order in factories in the context of Jim Crow segregation.

Perhaps the major difference in the daily operation of factory discipline between the US and China was that in the US, all factory floor supervisors were white men, without exception. BAT's Chinese floor supervisors were both men and women, came from a variety of native places, and tended to hire and supervise workers who shared their gender and native place. Chinese supervisors answered to white male foreigners, who also made their presence known to entry-level workers at the factory. In the US, then, factory hierarchy carried different potentials for creating workplace climate and discipline. As should be clear by now, Chinese workers did not find that commonalities with supervisors ameliorated the oppression of factory life. In the US, the fact that all floor supervisors were white men had two principal consequences: it gave opportunity and license to sexism and sexual harassment of all women, and it cut off promotional avenues for African Americans and women. This was no glass ceiling but a very opaque and unapologetic one.

When the ATC's successor companies moved cigarette manufacturing to the South, all followed the normative practice of leaving hiring and firing in the hands of foremen. Though African Americans experienced the worst abuses, changing the behaviors of foremen was one of the principal motivations for unionization for both white and black workers. As in China, Southern workers complained most about those elements of workplace discipline that regulated their bodies: the lack of breaks, the struggle over bathrooms, and sexual harassment. At least US workers did not complain about a lack of ability to nourish themselves during the workday, because factories began to build cafeterias on site.

Until World War II, workers applied directly to foremen for jobs, usually through an introduction by another worker, giving supervisors enormous discretionary power. Dora Scott Miller recalled that Liggett and Myers in Durham "didn't have a employment office . . . At the time when I was hired, somebody would take you up there and recommend you to the foreman." For Miller, that person was her cousin. Likewise, Martha Gena Harris obtained her job at the ATC factory in Reidsville because her cousin spoke to the foreman. Even the 1919 passage of child labor laws did not significantly diminish foremen's control over hiring. In response to the new laws, the companies created employment offices, but those offices did not hire but merely checked proof of age. At RJ Reynolds, foremen and managers alike did not trust the employment office to be able to select appropriate people for particular jobs. This authority extended to the right to fire. Miller recalled that, especially before unionization, "If they didn't like you, they'd fire you in a minute. . . . Sometimes he'd send you home for two or three days, and then sometimes he'd fire you. Weren't nobody to take up for you."[87]

Accounts abound of foremen cursing at workers, calling them names, hitting them, or getting into fistfights. Dora Scott Miller recounted that George Hill in the Liggett and Myers plant would "get on top of that machine and watch you see if you was workin' all right and holler down and curse . . . 'D . . . go to work!'" Foreman McKinsey in the RJ Reynolds stemming room had an informal bodyguard with him at all times who would join in if McKinsey picked a fight with a worker. Nannie Tilley reported that foremen at RJ Reynolds "used fighting and cursing as a means to increase production."[88]

The bathroom was a particular site of struggle, especially for African American workers. Bertie Pratt noted that at Durham's Liggett and Myers factory there were no scheduled breaks, and workers had to get someone to relieve them in order to leave their posts for the bathroom. Ruby Jones of RJ Reynolds reported, "When you'd go to the rest room, you was timed. And if you didn't come out at that time, you went to the desk." Foremen would "come in the dressing room, in the toilet where the women were ... [and] say, 'You've been in here long enough. If you ain't done, you won't get done.'" Men also reported being harassed to get out. At the ATC factory in Reidsville, a foreman punished a white worker for engaging in union organizing by not allowing him to use the restroom all day. At the same time, white women reported that there was a box of free cigarettes in their bathroom, suggesting that they could take enough time to smoke a cigarette before returning to their posts.[89]

Sexual harassment was a problem throughout the factory but especially in the stemmery, where white foremen apparently felt no hesitation at telling coercive sexual jokes and groping black women. Women knew that they had to navigate these encounters carefully or be fired. Dora Scott Miller reported that her foremen, George Hill, had "pets on the job—quite a few pets." Hill would "Give them a break, didn't work them hard like he did the rest of them." Anne Mack Barbee recalled that her foreman, Mr. Vickers, who "cussed all day," would also tell "old nasty jokes" and expect women to laugh along and flirt with him. This caused tension among the women. "I don't tell no nasty jokes with nobody. I said [to a friend], I'm a person. I said, well if you're scared you're going to get fired and let them man fumble your behind and say all that ... [trails off]." Robert Black remembered that at RJ Reynolds, "I've seen [foremen] just walk up and pat women on their fannies and they'd better not say anything." The structure of factory hierarchy gave more opportunity for sexual harassment of black women than white or Chinese women. The use of female supervisors in China did not prevent all sexual harassment but made it less central to the functional operation of factory discipline. In the US, white women did recall sexual harassment but the proximity of other white men in the cigarette department may have led foremen to be more careful. Certainly, the wider system of Jim Crow, particularly its sexual ideologies that cast white women as pure ladies and black women as sexually deviant, lent impunity to white foremen's harassment of black women.[90]

The multifaceted ways that Jim Crow segregation could inflect factory disciplinary structures is illustrated by the unionization of the Lucky Strike factory in Reidsville, North Carolina, where workers initiated a union drive with the Tobacco Workers International Union (TWIU) in 1933. Despite the ultimate success of their efforts, the racism of the white union local, the international union leadership, and the company's management worked together to control all workers at the factory, while reserving the worst treatment for African Americans. The company's differential repression of African American and white union members and the white local's remarkable relationship to the black local show that, as in China, legal and extralegal policing forces played a role in the functioning of the factory system.

African Americans in Reidsville's ATC factory began the union movement that would lead to a 1937 contract for the Reidsville, Richmond, and Durham plants. Five men and five women chartered TWIU Local 191 on December 3, 1933. (The TWIU, an American Federation of Labor affiliate, only had segregated locals.) The new union local immediately sent a delegation to the manager's office to demand chairs for women in the stemmery. White women in the cigarette department already had access to chairs as well as cleaner bathrooms, white uniforms, and free cigarettes. By demanding chairs for women, the black local registered an eminently practical as well as extremely symbolic protest in defense of black women's bodily health and dignity. Though the union was not recognized by the company, management granted the demand. Workers' success at obtaining the chairs became part of African American workers' pride: narrators still recounted this story seventy years later.[91]

Four months later, in March 1934, the white workers of the Reidsville plant formed their charter to create Local 192. After successfully organizing over six hundred dues-paying members into the union over the next six months, the secretary skipped town with the local's treasury. Hundreds dropped out of the union. Nevertheless, Local 192 rebuilt its membership and began coordinating with Durham and Richmond ATC workers. By early 1935, the company began efforts to undermine the union.

The ATC pursued a divide-and-conquer strategy with the white and black locals. Despite segregation, the black and white locals would have to work together in negotiating an agreement with the company. In February 1935, the company began firing members of Local 191 for union activity, but did not fire any white union workers.[92] This proved effective at undermining unionization because the white members of Local 192 and the white TWIU leadership were slow to come to black workers' defense, despite repeated requests for assistance. In March, Local 191 members engaged in a letter-writing campaign to TWIU president E. Lewis Evans requesting he send an organizer to assist. "I am swimming and don't want to drown," wrote Alice Williamson. "This seem to be a mighty hard pull without a organizer," wrote C. C. Caldwell. "We had several of our members to be discharge without a good cause we need a man to help to look after these things... stick up to your promas [sic] because I am sticking up to mine." Esther Jones wrote, "We are having much trouble here in the factory send us an organizer so we can stop it." Evans finally wrote back but refused to send assistance because members were behind in their dues.[93] While Local 191 tried frantically to bring in members and dues, the management continued to fire its members.

The leaders of white Local 192 were distressed by Local 191's declining membership but did not stand up to the company or the union on its behalf. Rather, they followed Evans's lead in believing the onus was on black workers to join the union and pay dues, despite the obvious mounting risks. In fact, the white membership base was growing and Local 192 increasingly saw the black local as a problem. In April 1935, the white local took the extraordinary step of trying to violently intimidate black workers into joining the union. The president of Local 192, Radford Powell, who later served as a representative in the state assembly, explained the vigilante action years later:

> The meanest thing we ever did was in trying to organize the black people, a new organizer came to help us. On his advice a large group of us took our cars one night and raced down through the black section of town trying to scare the people into joining the union. Some of them were sitting on their front porches with shot guns in their laps. Some of the boys painted skeletons on the doors of some of the houses. This was a terrible mistake for which we paid dearly and one that I will always be ashamed of. We then filed a partition [sic] for an election with the Labor Board.[94]

Membership in Local 191 plummeted to six; only its officers remained. Without mentioning the night ride, E. V. Boswell, financial secretary of the white union, wrote to Evans, bewildered about "our colored local here in Reidsville, No. 191. They seem to have landed at the bottom. All the best men of local #192 worked like h—! trying to bring this local up to date.... They will insult you now if you mention union to them."[95] On July 12, 1935, the company completed its campaign to destroy the union by firing Will Huff, the president and most important leader of Local 191.[96]

The vigilante action of Local 192 makes more sense when considering the role of the Ku Klux Klan in maintaining social order in the bright leaf belt. The Klan had an active history in Rockingham, Caswell, and Alamance Counties in the Reconstruction Era, especially in 1868 and 1869, when Klan members "rode the countryside at night pulling freedpeople from their homes, beating them, burning their dwellings, and threatening worse if they did not sign work contracts and stay away from the polls."[97] Many African Americans and one white state senator lost their lives to Klan violence. Though these events were over fifty years in the past, many whites remembered them with pride.[98]

In addition, the Klan revived in the 1920s in Rockingham County, as it did in many areas of the country, and established a consistent presence in Reidsville, including in Local 192. When queried about Powell's narrative of the event, two elderly former members of Local 191 separately insisted that Powell's *was* a Ku Klux Klan ride and that Local 192 had many Klan members in it. Local 192 and the Klan were "one and the same," said James Neal. "They were the segregationists."[99] Former Local 191 members recalled other Klan rides in the 1920s and '30s that were indistinguishable in form from Powell's union ride.

The claim that the white local engaged in such action to compel blacks to join the union may strain credibility, but evidence suggests that white members sincerely believed that African Americans could be motivated in such a way. Night riders sometimes presented themselves as enforcing moral codes, including punishing farmers who sold tobacco to the American Tobacco Company rather than to tobacco cooperatives or giving a warning to men who did not support their families.[100] Local 192 certainly had no practical motivation to bust the African American local, though that was the result. The vigilante tactic is evidence of just how profoundly the racialized violence of segregation pervaded the ATC workforce, the town, and the union.

With the black union completely destroyed, the company focused on undermining the white union. On June 14, 1935, Local 192 met with the company for the first time to ask it to sign a proposed agreement. The Wagner Bill, which required employers to bargain collectively with unions that had been selected by a majority of the bargaining unit, became law just three weeks later, on July 5, 1935. The company turned to town business leaders for assistance. Later that week, the *Reidsville Review* published a front-page story about a mass meeting of prominent businessmen who passed a resolution "endorsing the past and future policies of the American Tobacco Company." The businessmen detailed how easy it would be for the company to move its Reidsville production to Richmond or Louisville, and asked, "Must we let a small minority ruin and wreck our future well being?"[101] Powell recalled, "About that time the fire chief condemned the union hall. We moved to another and he condemned that one. We were planning to move again and he sent for me . . . [and said] I will condemn that as soon as you move."[102] In addition, white union members experienced daily altercations with angry citizens in town.

The company also drew on the police and judicial system in its fight against the union. In August, Local 192 official William Herrod and Mrs. George Belmany were arrested for adultery and fornication, a charge designed to humiliate them not only because it was a sex charge but because it was known as a charge reserved for controlling African Americans. "There has never [before] been a white couple tried for adultery and fornication here in Reidsville insofar as the court can tell," wrote E. V. Boswell to E. Lewis Evans. Herrod appeared publicly in court on September 1 and again on October 5. Boswell continued, "A man . . . warned Powell and I to watch our step, that we would be in trouble just like Bill if ever the opportunity presents itself, whether we have done anything wrong or not."[103] Using municipal resources, the company effectively informed the white union leaders that the forces of Jim Crow segregation would be turned against them. Meanwhile, the company stalled its legally required bargaining process with the union.

Only after the National Labor Relations Board heard a formal complaint against the company's refusal to adhere to the Wagner Act did the ATC locals from Reidsville, Durham, and Richmond obtain a contract on April 15, 1937.[104] The ATC had unsuccessfully argued that because there were two elected bargaining units, Locals 191 and 192, neither met the federal requirement of representing a majority of workers. In response to this, the two locals agreed that Local 192 would represent Local 191 in the bargaining process. In addition, through the long process of rebuilding Local 191 and gaining an agreement from the company, Radford Powell had attended Local 191's meetings, supposedly to help them organize, but in fact, the white local mediated the black local's relationship with both the TWIU hierarchy and the factory management. Union leader E. Lewis Evans explicitly told Powell that he was turning "the Colored people . . . over to you for supervision," and put Powell on payroll. He warned him that "the educated colored brethren" may be difficult to handle.[105] One result was that over the years, the company repeatedly drew 192 leadership into identification with the company to control the workers, beginning with black workers but extending to white workers as well.

Local 191's experiences demonstrate just how challenging it was to fight against a corporate power immersed in Jim Crow. Facing racism in the daily structure and life of the company, black workers had ample reason and considerable motivation to organize, but also faced greater opposition to their activism from the company and a white union that alternated between terrorizing and patronizing them. The Klan, as an illegal gang, became a volatile and unpredictable force in labor conflicts but ultimately worked to the company's favor. Indeed, from the perspective of factory discipline, the white union came to embody company power by adopting an intermediate position in its patronizing representation of the black union.

Conclusion

The story told here deviates markedly from the machine-driven tale of modernity, whereby an industry developed fully in the United States and then exported that development abroad. Rather, the bright leaf network connected cigarette factories in the US and China, especially through the constitution and governance of an increasingly global workforce. White male managers sifted labor flows into job classifications, drawing on gender, age, and race or native place as key techniques for labor differentiation. While the practice of hiring a workforce based on race/ethnicity and gender was commonplace, the bright leaf tobacco network intensified the principles of segmentation, making the cigarette factory compound more radically divided than the cigarette factories of the US North, the Nanyang Brothers' cigarette factories in China, and the textile factories of the US South. The bright leaf network also instituted parallel disciplinary techniques in both countries, using spatial segregation of workers, cigarette giveaways, and uniforms to assign differential status to job categories.

The disciplinary techniques that managers deployed to constitute and govern these labor forces were remarkably intimate and bodily. By requiring workers to sort themselves daily through different gates and doorways and wear particular uniform colors, management obliged workers to enact categorical differences. Cigarettes endowed some job classifications with the aura of the commodity. The particular hardships and humiliations of factory discipline also worked at the bodily level: struggles over lunch, the bathroom, pregnancy, and sexual harassment all speak to the vulnerability that workers brought to their lives in the corporation. Decades ago, the lawyer Abram Chayes suggested that all who have "sufficient intimacy" with the corporation should be considered members and have a voice in its governance. By this definition, the cigarette factory workers of China and North Carolina certainly should count as corporate members. Only through the unions, which the companies fought as long as they could, did workers gain some say in governance and begin to ameliorate the most extremely oppressive conditions that they faced.

Indeed, the stories told here suggest that the factories not only had their own internal systems of governance but that corporations functioned as important sites of governance in their societies, becoming a form of polity. In both the US and China, factory hierarchies entwined with local political hierarchies and legally ensconced inequities. The factories became fiefdoms that structured workers' economic and social lives at work but in certain ways outside the workforce as well. In China, the militarization of the factory made the imperial functions of workplace order clear, just as union activity took on anti-imperial significance. In the US, the Wagner Act enforced negotiations between the company and unions, saving late-organizing cigarette workers from the violent skirmishes with police and military forces endured by Chinese cigarette or US textile workers. The story of the first TWUI contract with the ATC underlines the significance of the Wagner Act but at the same time should alert us that the economic and political violence of Jim Crow segregation—including within union politics—constituted a disenfranchisement that not only ran parallel to imperialism but, via the bright leaf network, cannot be entirely disentangled from it.

The value that the workers contributed daily to the corporations' mission of making cigarettes fed the companies' growth in both the US and China. In the late 1910s and 1920s, Camel cigarettes in the US and Ruby Queen cigarettes in China simultaneously became massively popular, far beyond corporate expectations. These two brands, both containing bright leaf tobacco, signaled a new moment globally in brand innovation and the rise of the cigarette. Chapter 5 tells the story of their surprising triumph.

5

Of Camels and Ruby Queens

In the mid-1910s, it finally happened: cigarettes' popularity spiked in both the United States and China. This steep ascent in sales intensified in the early 1920s and continued for two more decades. The US and China went from lagging behind Europe and the Middle East in cigarette consumption to being the world's largest markets. But success did not come in the way that industry leaders had expected or planned. Instead of seeing a rise in popularity across many cigarette brands, in both the US and China the rise in cigarette consumption was attached to the triumph of a single brand: the Camel in the US and the Ruby Queen in China. The unanticipated social power of these two big cigarette brands was inextricable from the unprecedented rise in cigarette smoking. Camels and Ruby Queens became famous, recognized and known far beyond their consumer base and became structuring agents of everyday life.[1] By the early 1920s, Camel cigarettes were the most popular brand in the world, and Ruby Queens were second.[2]

The bright leaf network participated in both of these phenomena, making them part of a shared story. As we have seen, the US and Chinese industries were intricately connected through the network's flow of bright leaf tobacco, bright leaf seeds, manufactured cigarettes, and personnel from the US Upper South to China. Additionally, because the industry was expanding in the Upper South at the same time, the bright leaf network was an international conduit for knowledge about business and labor management. The close association between corporate development in the US and China certainly accounts for many of the preconditions for the rise in cigarette consumption, especially the development of functional bright leaf tobacco agriculture and cigarette manufacturing programs in both places.

The rise of the big brands does not fit well within the typical story of the West-to-East motion of modernity, however, because managers did not craft a new business strategy in the US and impose it on China. Rather, unrelated events led to the introduction of a new competitive situation in both places. In 1912, James A. Thomas, frustrated by the limitations of BAT's own sales force, launched a competitive experiment by turning over the marketing of two brands to high-ranking Chinese employees, Zheng Bozhao and Wu Tingsheng. In the United States, the Supreme Court dissolved the American Tobacco Company monopoly and distributed its assets, including brands, among four large successor companies, setting them into competition with each other. Each company scrambled to round out its offerings of tobacco products, which led to the release and promotion of new brands.

The newly competitive circumstances prompted marketers to intensify advertising and to focus unprecedented resources on particular brands. Zheng and Wu invested in Ruby Queen and Purple Mountain, respectively, by design of their competitive experiment. The ATC successor companies launched new brands with bursts of advertising—the Camel cigarette was one of them. Though Camels and Ruby Queens boomed simultaneously, then, they did so as a result of distinct business conditions, rather than one echoing a strategy set by the other.

Big cigarette brands like Ruby Queens and Camels achieved a scale of popularity heretofore not considered possible in the tobacco industry. Before this time, tobacco companies fielded a very large number of brands. In the US, other products, such as Pillsbury flour or Uncle Ben's rice, had captured a mass market with a single brand, but marketers perceived consumers to have few idiosyncratic taste preferences in basic wheat flour and rice.[3] The going wisdom was that the great variation in consumers' taste in tobacco products and purchasing capacity dictated a finely distributed array of brands. Companies typically wooed new consumers by releasing a brand offering an original flavor or a different grade and price. Though there were obvious trends in preference and some brands became especially popular in particular markets, it did not occur to tobacconists that vast numbers could be convinced to consume the same cigarette. When they discovered this, the cigarette industry moved to the cutting edge of corporate attempts to mine and shape the potentials of branding.

The boom of Camels and Ruby Queens accordingly prompted the reorganization of the industry around the model of big brands. By the 1920s in China, the Nanyang Brothers Tobacco Company focused on their Golden Dragon brand, while Huacheng Tobacco Company heavily promoted My Dear.[4] In the United States, the "big four" tobacco companies all produced big brands. RJ Reynolds began the rout with Camels, and the new American Tobacco Company answered with Lucky Strikes; Liggett and Myers reformulated Chesterfields; and Lorillard was last to successfully follow suit with Old Golds.[5] The triumph of big brands established a particular tobacco blend as the trend and seemed to spark an accompanying rise in cigarette smoking in general.

The wide travels and lively biographies of Ruby Queen and Camel cigarettes foreshadow the extensive capacities of the brand seen in late-twentieth- and twenty-first-century life.[6] A portion of the brands' salience certainly came directly from suggestions planted by corporate marketing, but their considerable power to organize social experience arose through cigarette and brand circulation in daily life.[7] Cigarette smoking was uniquely personal yet held possibilities for public ritual and the performance of affiliation; cigarette brands layered with these elements of the commodity to become powerful catalytic elements of new social connections and formations. Indeed, Camels and Ruby Queens became a notable part of public life, known far beyond those who actually smoked them. The Chinese word for a name brand, *mingpai*, literally means famous brand, conveying the star power that certain brands achieve. With and without direct corporate intention, famous brands shaped public relationships beyond commodity consumption in ways not fully controllable by marketers, including those in the corporation itself.

Our story begins with the simultaneous ascents of Ruby Queen and Camel cigarettes in the 1910s as a result of business practices that, in the wake of new and unexpected competition, created concentrated advertising and made possible a new scale of brand popularity. Famous brands circulated through many community contexts, some that were not defined by commodity consumption per se, including the use of the Ruby Queen brand in the May 30th Movement in China and the role of the Lucky Strike brand in the localized baseball culture in Reidsville, North Carolina. These two cases reveal dramatically different ways that the brand exceeded its purely promotional function and became a symbolic resource for people pursuing other goals.

In the May 30th Movement, activists skillfully rebranded Ruby Queen/Da Ying cigarettes as a symbol of imperialism. They also marshaled the power of the brand to cohere community identification but directed it toward commodity boycott and a critique of capitalism, rather than product promotion. In Reidsville, North Carolina, African American factory workers used the Lucky Strike brand to confer value on African American manhood through the creation of the Lucky Strike baseball team. Their use did not entail a rejection of the cigarette, the brand, or capitalism; rather, it staked a claim to these dominant forms of value that the corporation and town associated exclusively with whiteness. In these two divergent examples, famous brands were deployed in unexpected ways to engage with the corporation itself.

The Rise of Ruby Queens

In 1912, James A. Thomas, Henry and Hattie Gregory, Zheng Bozhao, and Wu Tingsheng arrived together at the Shanghai harbor from a tour of BAT's New York headquarters and the bright leaf belt of North Carolina; all were prepared to take part in a new moment in BAT's venture. We have already seen how Henry Gregory built up the agriculture department while Hattie Gregory created a Southern home in Shanghai, as well as how Thomas directed the building of new factories. At least as much significance attached to the experiment in sales upon which Wu and Zheng embarked. BAT had always relied on Chinese salesmen—in fact, it had depended on Wu and Zheng specifically since day one—but this collaboration put power to direct the corporation's sales strategy into the hands of the Chinese businessmen.[8]

In the first seven years of BAT-China, Thomas had patterned BAT's distribution system after the vertical sales management organization of the ATC monopoly in the US. Thomas organized the country into divisions headed by foreign managers and staffed by Chinese and foreign sales representatives. Chinese representatives identified and contracted with local Chinese dealers to carry BAT cigarettes exclusively—similar to the exclusive agreements that Duke imposed in the US in the 1890s. Sales representatives gathered data on sales and market conditions that executives studied when determining sales strategies at headquarters. In other words, Thomas instigated a well-defined hierarchical and scientific system of management designed to optimize feedback from the field while ensuring executive control. The dream of Western modernity was that such rational, bureaucratic practices could be scaled up and universalized, that they were the mechanism by which Western capitalism could transform more primitive societies and expand across the globe.[9] The problem was that Thomas's plan didn't work.

Frustrated with the degree of distribution he had achieved, Thomas set up a sales competition between Wu and Zheng, giving each of them responsibility for marketing a brand and the freedom to work outside BAT's established sales system. Thomas offered Wu the Ruby Queen brand but Wu felt its Chinese name, Da Ying (Great Britain), sounded proforeign and unpatriotic. He chose Purple Mountain (Zijin Shan) because it referenced a famous sacred site near Nanjing, which he felt would appeal widely.[10] Zheng accepted responsibility for the marketing of Ruby Queen/Da Ying, which was only appropriate since he had suggested the Chinese name during the anti-American goods movement seven years earlier. Each merchant needed a business form in order to market the brand. Wu became general manager of a new wholly BAT-owned subsidiary, the Union Commercial Tobacco Company. Zheng Bozhao used the long-established trading company, Yongtaizhan, where he had risen through the ranks and already served as general manager.[11] BAT was still a monopoly, but Thomas introduced an element of competition, thereby setting the conditions for the emergence of the big cigarette brand.

Wu's tactic was to move to Beijing and tap political connections in the hope of negotiating an official Chinese tobacco monopoly that would be entirely controlled by BAT-China. State tobacco monopolies existed in France, Italy, Spain, Austria, Turkey, and Japan, and additional countries were considering shifting to this model.[12] After three years of effort, Wu failed to achieve a formal link between BAT and the Chinese government. Wu did set up sales branches and utilized BAT's advertising, but he was unable to make great headway with Purple Mountain because of his inability to enlist a wide enough net of sales agents.[13]

Zheng (fig. 5.1), by contrast, succeeded beyond anyone's hopes. His spectacular success was due first and foremost to his access to an extensive merchant network based on native place ties.[14] Zheng hailed from Guangdong; merchants from this province spread through much of China. Zheng also offered dealers much better terms than the BAT sales management system had allowed, dis-

Fig. 5.1 Photo credit: Cheang Park Chew (Zheng Bozhao), from *A Pioneer Tobacco Merchant in China*. James A. Thomas. 1928. Courtesy of the publisher, Duke University Press.

pensing with BAT's exclusive agreements. BAT had set up financial requirements for its dealers, especially those who wished to obtain cigarettes on credit rather than purchasing them outright, which placed the risk on the retailer and protected the company. Zheng eased those requirements, allowing less wealthy retailers to become BAT dealers and rewarding success with ever-better terms.[15] Merchants readily signed up to carry the cigarettes. Zheng covered the same territory as BAT's own sales force but gained much better distribution through his retailers. Ruby Queen cigarettes flowed through the channels Zheng opened and achieved unprecedented sales.

Zheng's success was especially surprising because Ruby Queens were not a cheap cigarette but rather were priced in the middle range, out of reach for most poor people, including cash-strapped farmers. This cigarette was never intended for the masses. In the early 1920s, BAT salesman Frank H. Canaday noted that Ruby Queens were the most expensive cigarette on the market in many country market towns and only merchants and town officials smoked them. In some very small places, they were entirely unavailable.[16] Ruby Queens sold well in the treaty ports, but they made their mark because they attracted consumers of moderate income in hundreds of cities and towns across China.[17]

Part of the Ruby Queen phenomenon unquestionably came from the fact that Zheng's merchant networks spread advertising far and wide at the same time that they distributed the cigarettes. There was nothing unusual about Zheng's advertising practices or the advertisements themselves; what was new was the concentration of advertising resources on a single brand. BAT supplied Zheng with as much Ruby Queen advertising as he requested, made at the same printing plants as advertising for all of BAT's brands. While BAT's sales teams promoted several brands at once, Zheng's extensive distribution of one brand made Ruby Queen advertising seem to spring up everywhere. In particular, BAT's outdoor advertising methods combined with Zheng's reach to make the Ruby Queen brand famous far beyond its consumer base.

BAT's and Zheng's sales teams advertised in a few standard ways. When sales teams visited a locale to sell product to dealers and retailers, they also created hype by staging a parade, running a cigarette stand with prizes, giving away cigarette samples and calendar posters, and plastering advertising posters all over town. In addition, they left distributors and retailers with posters, calendars, handbills, and other advertising materials for display in shops. The brand itself was not new in China, and many other companies advertised in just this way.[18] BAT's advertising, however, was extremely high quality. According to Carl Crow, a Shanghai advertising agent, BAT made "labels and cartons and advertising material" on site in "the largest color-printing plant in the world and one of the finest."[19] James Hutchinson, a BAT employee from North Carolina, recalled that the Pudong plant was "as modern and well equipped as the leading ones at home."[20] The high quality of BAT's color advertising made posters eye-catching and desirable items unto themselves, quite apart from the commodity they promoted.

BAT's advertising crews tiled walls with multiple thirty-inch-by-forty-inch

Fig. 5.2 Wall with cigarette posters. Courtesy Joyner Library, East Carolina University.

posters that repeated the brand name and logo again and again, with the goal of putting up "as many posters as the space would allow" (fig. 5.2). Hutchinson recalled: "As soon as the coolies had made up their hot paste we set out and . . . plastered posters all over the town."[21] Of course, other companies did the same, especially patent medicine companies, but BAT posters became especially ubiquitous. Author Pearl Buck objected to the effect: "Huge cigarette-posters plaster[ed] those dark and ancient bricks . . . defiling their somber age with crass and glaring colors."[22] Hutchinson himself commented, "It was nasty work—deliberately defacing house walls with a lot of glaring, shouting, colored paper. I loathed it."[23] BAT and its subsidiaries postered so widely in China that by the 1920s, local governments tried various mechanisms for taxing them.

Posters on town walls made the brand prevalent in public spaces, but the special calendar poster migrated to people's homes. Calendar posters were, according to Hutchinson, "the big advertising smash each year."[24] Issued at the New Year, they featured a large colorful scene rendered by Chinese artists, with a calendar at the top or bottom margins or along the frame. Canaday noted, "The colored hangers and calendars are the most popular form of advertising in China. We use them in packet redemption schemes and all sorts of ways to push the business."[25] Such schemes offered customers prizes for redeeming their empty cigarette packs, with the highest number receiving the best prize. In the 1920s, Zheng's subsidiary Yongtaihe offered a wide range of prizes at cigarette stands, including soap, paper fans, mirrors, and enameled washbasins,[26] but Canaday explained, "It is almost a necessity up here to have hangers as the first prize of any prize scheme if it is to have any life to it."[27]

Indeed, calendar posters were so popular that they became commodities unto themselves and circulated the company's brand independently of BAT through picture hawkers. Canaday explained, "Some hawkers make a life business of peddling these pictures in the streets and the retail market value of a good one will go to 60 and 70 cents Mex."[28] Many companies issued calendar posters—BAT did not originate this scheme—but BAT's posters were very high quality and designed by prestigious Chinese artists[29] (fig. 5.3). Hutchinson described the careful process the company went through to ensure the high quality and success of the calendar posters:

> Leading popular Chinese artists, chiefly girl head specialists, were paid a good retaining fee to submit preliminary sketches nine months ahead of Chinese New Year.... These roughs were then sent out to all Division headquarters for the Chinese members of the staff to vote on their respective merits and check the titles to see that the characters carried no local double meanings.

The calendars allowed the company to measure how much "buzz," to use today's term, their advertising created. Hutchinson continued,

> The check on the calendar's value as an advertising piece was the price at which it sold on the market. Within ten days after distribution had started, picture hawkers all over China displayed them with their other wares on the main shopping streets. If the price rose to 11 or 12 cents each, the calendar was a success.[30]

Calendar posters were an especially effective way of spreading brand names into private homes through much of China, including rural areas. When sold by picture hawkers, the brand escaped the direct commercial interactions of the company and took on a social life of its own, to the further benefit of the company.

Fig. 5.3 British American Tobacco Company cigarette calendar poster. Courtesy Robert Brown Gallery.

Many more people viewed Ruby Queen posters or had a Ruby Queen calendar in their home than regularly smoked the cigarettes. Cochran argued that BAT achieved better information distribution than *anyone* in China, including any other company, missionary group, or governmental body. By the late 1910s, BAT's annual advertising budget was 1.8 million yuan, which was nearly half the amount of the national education budget of the Republican government.[31] One Chinese journalist in the 1930s remarked, "Many rural Chinese villages still don't know who in the world Sun Yat-sen is, but very few places have not known Ruby Queen (*Da Ying*) cigarettes."[32] The rise of Ruby Queens is significant not only because unprecedented numbers smoked the cigarette but also because it signaled a new era of the brand.

Zheng's success led to permanent changes in BAT's structure. Most notably, BAT put Zheng in charge of an entire sales structure that ran parallel to BAT's own. In 1923, BAT and Zheng formed a subsidiary called Yongtaihe that gave Zheng even more decision making power, since he owned 49 percent of Yongtaihe's stock and served as general manager (discussed in chapter 7).

Additionally, though BAT maintained a number of brands aimed at different markets, it shifted its sales and advertising strategies to develop big brands. BAT was especially interested in fielding a cheaper cigarette that might capture an even larger market. A BAT representative reflected on the changes in policy for Hankou in 1924. His unit now engaged in concentrated advertising of Hatamen, a cigarette that tasted like Ruby Queens (100 percent bright-leaf tobacco) but was made of lower grades of China-grown leaf. He wrote:

> Some years ago when some of the London and New York Directors were out here. . . . they decided that we could not [advertise in a concentrated way] for the reason that we had so many different brands to advertise. During the past three years we have proved that this decision was incorrect for we have advertised practically nothing but the 'Hatamen' packet, and from a Department that was selling practically everything under the sun this has become a one brand Department.[33]

In the wake of the Ruby Queens' success, BAT shifted its strategies well beyond the marketing of Ruby Queens themselves.

The rise of Ruby Queens also changed the Chinese industry as a whole because it prompted Chinese competitors to issue heavily advertised copy brands in an attempt to get in on this action. Nanyang Brothers Tobacco Company invested in its Golden Dragon brand and, by 1924, Huacheng Tobacco Company issued My Dear; both were copies of Ruby Queens in tobacco blend and price.[34] Cigarette advertising materials in large and midsize cities became even more ubiquitous. In treaty ports, and especially Shanghai, large cigarette advertisements covered the city so pervasively that one was seldom out of view of a cigarette ad. Large electric billboards crowded Nanjing Road and other commercial boulevards, and billboards spread through the countryside as well, lining main commercial roads and canals (fig. 5.4).[35] The big brands of cigarettes did not

Fig. 5.4 Ruby Queen billboard. Courtesy Harvard Yenching Library.

displace all other brands—companies still issued brands at different grades and prices—but it shifted corporate branding and advertising strategies.

Ruby Queen's ascendance came unintentionally from the historically specific conditions of doing business in China. Independently, a very different set of events transpired in the US that resulted in the emergence of the big brand, Camel cigarettes, the Ruby Queen of the west. These simultaneous developments transformed the global story of cigarettes and set the stage for them to become the signature products of the burgeoning, modern consumer culture.

The Rise of Camels

The rise of the big brand in the United States, like in China, followed the introduction of a new kind of competition within the monopoly, prompting a concentration of advertising and marketing on a single brand. The dissolution of the American Tobacco Company and the formation of the "big four" garnered wide criticism for keeping largely the same people in power.[36] The court required Duke to choose between heading the successor ATC and BAT, and he chose to stay with BAT. Executive positions in the successor companies, however, went to high-ranking employees in the monopoly. The new companies were so huge that smaller ones still did not have a chance. In fact, advertising budgets of the Big Four soon rocketed out of reach of most smaller companies; within a few years the cigarette industry was even more dominated by huge firms than it had been during the ATC monopoly. As a model for creating expansive competition in the industry, then, the dissolution plan failed miserably.[37] However, the competition that ensued among the Big Four did have a huge impact on the story of the brand.

The rise of Camel cigarettes is usually explained as the result of brilliant advertising tactics on the part of the entrepreneur, R. J. Reynolds—particularly one set of teaser ads. Published in local newspapers when Camels were being introduced in 1913 and 1914, the first ad simply pictured a camel with the word "Camel"; the second featured the same graphic with the caption "The Camels are Coming"; the third read, "The Camels are Here!" and introduced the cigarette. As great as this ad sequence may be, it was not out of step with advertising of the era. The company itself did not focus overmuch on the ad. Late in 1914, Reynolds reported that it had "tried out 'Camel' on the soft-pedal system in thousands of stores" and was about to begin a bigger campaign.[38] Instead of a brilliant single advertising scheme, a larger shift in business practices prompted the emergence of big brands.

The Big Four rushed to introduce and advertise new brands, not with the intention of creating a big brand but simply in order to round out their product offerings. When the Supreme Court divided the ATC, it did so by manufacturing plants. Cigarette brands made at a particular factory went with the factory, which meant that none of the four companies received a full menu of tobacco products. For example, the new ATC received cigarette and pipe tobacco factories in Richmond, New York City, Durham, and Reidsville, while RJ Reynolds received the large chewing tobacco factories in Winston-Salem. Consequently, the new ATC received brands that represented 46.3 percent of cigarette production and 42.1 percent of pipe tobacco production, while RJR received pipe tobacco brands that represented only 3.4 percent of national production and no cigarette brands.[39] Lorillard received 19 percent of cigarette production, but all of its brands were pricey Turkish brands, no cheaper ones. Though the new ATC was flush with cigarette and smoking tobacco brands, it had no Turkish–bright leaf blend cigarettes, a mid-price-range product with rising popularity. All of the four companies, then, introduced new brands in order to be competitive in the full range of tobacco products. They paid particular attention to pipe tobacco and cigarettes because these were rising in popularity. And when they launched their brands, they advertised them heavily.

The first big tobacco brand emerged in pipe tobacco, not cigarettes. In the monopoly dissolution, RJ Reynolds received a lot of chewing tobacco and just one well-established pipe tobacco brand, Prince Albert. The ATC monopoly had introduced Prince Albert pipe tobacco in 1907 as a high-priced shredded burley brand to avoid competition with Bull Durham and Duke's Mixture, popular granulated bright-leaf tobacco brands of different grades and prices.[40] The ATC monopoly was so determined to avoid competition that when the RJ Reynolds branch had proposed a new, granulated bright leaf smoking tobacco named The Advertiser, Duke approved the brand only on the ironic condition that Reynolds not advertise it. By the time of the ATC dissolution, Prince Albert had garnered a solid base but it still lagged far behind Bull Durham and Duke's Mixture, capturing only 3.4 percent of the total pipe tobacco market. RJ Reynolds poured cash into advertising Prince Albert, and the result was shocking: by 1913, this one brand had 36.7 percent of the high-priced plug-cut pipe tobacco market.[41]

The other three companies, meanwhile, took note and built up brands copying Prince Albert: the new ATC promoted Tuxedo; Liggett and Myers developed Velvet; and Lorillard worked up Stag. The US Commission of Corporations reported in 1915 that these four brands were "of the same class of smoking tobacco, and all of them have been extensively advertised since the dissolution. Apparently there has been during this period more competition for business in this type than in any other of the types of manufactured tobacco."[42] All of this advertising had an unintended result: it made burley *the* popular type of smoking tobacco. Consumers rushed to these four brands, abandoning Bull Durham and Duke's Mixture. Indeed, the *entire* increase in sales on smoking tobacco in 1912 and 1913 was due to these brands, while others declined.[43] Thus, the four burley brands demonstrated to the surprised tobacco executives that concentrated advertising could be very effective at creating trends. Still, not all of the company executives understood fully the potential for widespread brand affiliation until the same thing happened in cigarettes.

The Big Four tobacco companies responded to the uneven distribution of cigarette brands much as they had that of smoking tobacco: they introduced new brands to round out their offerings. There were three main types of popular cigarettes on the market: inexpensive bright leaf cigarettes, high-priced cigarettes of imported Turkish leaf, and Turkish–bright leaf blends.[44] The ATC monopoly had recently achieved its biggest success with Fatima, a mid-range Turkish–bright leaf blend. Fatima went to Liggett and Myers, so the new ATC introduced a copy of it named Omar, and Lorillard released a copy Turkish–bright leaf blend named Zubelda, as well as a cheaper brand that was likely all bright leaf.[45] None of the companies initially intended to create a big brand. They all understood the cigarette market as segmented by class and regional differences in taste: bright leaf cigarettes remained popular in the South, Turkish–bright leaf blends did best in the Northeast and Middle Atlantic, and all-Turkish cigarettes catered only to urban elites.[46] The companies all offered multiple cigarette brands to serve different markets.

The strongest showing by far in 1912, the year after dissolution, came from Fatima, Omar, and Zubelda. Having three virtually identical cigarettes on the market did not depress individual company sales; on the contrary, all three brands succeeded and defined a new trend in taste.[47] Percival Hill, president of the new ATC, remarked on the change in brand strategy, saying that before dissolution, the monopoly would not have released a brand similar to Fatima but would have introduced one that sold at a higher or lower price in order to appeal to a new group of customers.[48]

RJ Reynolds, having received no cigarette brands in the dissolution, introduced four in 1913: Red Kamel, a high-priced all-Turkish cigarette; Reyno, a cheap all–bright leaf cigarette; Osman, a Turkish–bright leaf blend; and Camel. Though Reynolds labeled Camel a "Turkish blend," it was not a direct copy of Fatima, Omar, Zubelda, and Osman. Instead, the company quietly included a dose of cheaper burley tobacco, the tobacco in Prince Albert, and lowered the amount of expensive Turkish. In addition, the company priced it a dramatic five cents lower than the competitions' Turkish blends, at ten cents for a pack of twenty cigarettes.[49] There were burley cigarettes on the market, but most cigarettes, and certainly the ones considered "better," contained either Turkish or bright leaf, both of which gave a lighter smoke. Reynolds floated all four brands on the "soft pedal system" for two years to see which gained traction in the market.

In response to repeat orders for Camels, Reynolds determined to pour massive advertising dollars into its promotion in an attempt to replicate the Prince

Fig. 5.5 From the collection of Stanford Research Into the Impact of Tobacco Advertising (tobacco.stanford.edu).

Albert success (fig. 5.5).[50] In December 2014, Reynolds announced an advertising blitz to distributers and retailers, writing, "Pin your ears back and get ready for the big breeze for we are starting something real." The company announced a national campaign to begin with a two-page spread in the *Saturday Evening Post*, the first time cigarette advertising appeared in that publication, and promised that "we will put a fortune" into a campaign "that will be as broad and convincing as that which has won fame for Prince Albert."[51] Camels did indeed respond exactly as Prince Albert had: the brand quickly made surprising gains over other cigarette brands and in 1917 captured 35 percent of the market.[52]

The other three companies scrambled to create copycat blends that would compete with Camels. In 1915, Liggett and Meyers modified one of its nine existing cigarette brands, Chesterfield, to resemble Camels; in 1916, the new ATC introduced Lucky Strike cigarettes. Lorillard released Tiger cigarettes in 1915, but this brand failed, leaving Lorillard out of the contest until the introduction of Old Gold cigarettes in 1926. Nevertheless, Reynolds had a head start. By 1923, Camels comprised 45 percent of the national cigarette output. It might have risen even higher, but Reynolds could not build production facilities fast enough to keep pace with orders.[53]

Though Camels remained the most popular cigarette in the world for several years, Lucky Strikes and Chesterfields also boomed, and the prevalence of cigarette smoking increased dramatically. By 1926, Lucky Strike, Camel, and Chesterfield held 85 percent of the cigarette market in the US. Camel's primacy lasted until 1928, when Lucky Strikes matched it in sales. In 1930, Lucky Strike pulled ahead, but through the 1930s, the three traded places as the top-selling cigarette, with Old Gold making a strong fourth-place showing. Each company invested large sums in advertising these single brands and attempted to recruit more types of consumers to them. As in China, the new brands were not only big in sales, they were also known far beyond those who purchased them. Camel, Lucky Strike, Chesterfield, and Old Gold became inescapable features of US life.

Cigarette advertising grew impossible to avoid. The bulk of the companies' budgets in the 1910s and '20s went to newspapers and magazines. The smallest town and college newspapers published black-and-white cigarette ads and nationally circulating magazines like the *Saturday Evening Post* carried high-quality colorful ads. As in China, sales representatives made face-to-face contact with retailers, took orders for cigarettes, and brought advertising for store displays. Camel sales agents, for example, carried colorful three-dimensional window displays that required assembly on site. By the 1930s, colorful ads served as decorations in poor people's homes, much as in China. Outdoor advertising also was a significant aspect of the budget, as it was in China, and billboards sprouted up in cities and towns and on highways all over the country (fig. 5.6). Baseball stadiums were especially significant early sites for cigarette billboards as compa-

Fig. 5.6 Lucky Strike billboard, Caswell County, North Carolina. Photo Marion Post Wolcott, courtesy Library of Congress.

nies vied to associate themselves with America's favorite game.[54] Certainly, not everyone smoked Camels, or any cigarette, but by the 1920s, virtually everyone knew Camels, Lucky Strikes, Chesterfields, and Old Golds.

Despite the corporations' success with big cigarette brands, the brands were not fully under their control. In both China and the US, famous brands became part of the symbolic repertoire of the general public. As different as they were, the use of brands in China's May 30th Movement and the Lucky Strike baseball team in North Carolina reveal how they could exceed their promotional function and structure a wide range of experiences and relationships.

Ruby Queen Cigarettes and the May 30th Movement

The anti-imperialist May 30th Movement reveals how BAT's famous cigarette brands could escape the company's control and be turned against its interests. Anti-imperialist protesters attacked BAT's corporate process at nearly every stage, from factory strikes, to disruption of distribution and sales, to consumer boycott, virtually shutting down BAT's business for several months and significantly affecting sales for two years. In addition, BAT's very success at spreading Ruby Queen/Da Ying and Hatamen brands made the brands ripe for rebranding with protesters' meanings. The foreign element of all cigarette brand stories—visible in the foreign names and lettering on cigarette packages—could shift quickly from exotic cachet to a sign of imperialism. Protesters especially put BAT on a defensive scramble to control the damage to Ruby Queen/Da Ying. In the process, the brand helped cohere protesters' sense of community and fuel the Chinese National Products Movement, clearly demonstrating that big brands could give shape to human experience beyond their product promotional purposes.

BAT became a focus of the May 30th Movement almost from the outset because the movement targeted British and Japanese companies. The movement had its origins in a coalition that cohered during a workers' strike the previous February against the Japanese Nagai Wata textile factory in Shanghai, when the company replaced forty male workers with lower-paid youth. Students and some forty other groups, including women's groups and merchants' associations, joined in support of the workers. The strike spread to other Nagai Wata factories, lasted a month and involved somewhere between 17,000 and 40,000 workers. In May, the same factory went on strike again and the strike spread in the same way. Though the strikers' immediate demand was for pay increases, students presented it as "a battle for national salvation" from "foreign exploitation."[55]

The labor struggle soon became a much wider movement. Police had arrested and refused to release six peaceful student protesters. May 30th was the day that the foreign court would try them. By late afternoon, several thousand people had converged on the police station in protest. The British commander of the Shanghai Municipal Police reportedly felt threatened by the large group and gave the order to fire, killing twelve and wounding dozens. The next day, Shanghai merchants did not open their shops in protest; by the following day the city's workers went on a general strike, including those at BAT factories. Activists immediately called for a boycott of all Japanese and British goods, and the movement quickly spread through student and other networks to Beijing and much of China.[56] BAT, being the largest employer in Shanghai and having advertising on every corner, posed an immediate, visible, and symbolically salient target.

Fifteen thousand workers at BAT's Pudong factory joined the strike and stayed out of work for four long months, returning only when the company promised to raise wages, improve treatment of workers, and not dismiss them without cause. The strike had the support of the new Shanghai General Labor Union, a Communist group with connections to organized gangs to which BAT supervisors belonged, though the political stances of the workers varied widely. Women in the packing department, especially, emerged as strike leaders and became part of strike propaganda teams that spoke and leafleted in Shanghai and the surrounding area.[57] BAT's factories were so hampered by strikes that the company sold bright leaf tobacco it could no longer store without risk of spoilage to the Chinese-owned Huacheng Tobacco Company.[58]

It is difficult to say how central the brand was to these strike activities, but the brand did pervade Pudong workers' experiences. Not only did workers produce and package Da Ying cigarettes and receive more or fewer of the cigarettes for their own use, but the factory itself became branded. In oral histories, workers referred to the factory as the Da Ying factory as a matter of course, making its British connections very clear and tying the factory itself with the brand.

Protesters stopped cigarette distribution at multiple points, with occasional opportunity for theatrical public demonstrations. They visited BAT's cigarette warehouses and sealed them shut in many provinces, including Jiangsu, Zhejiang, Jiangxi, Hubei, Henan, Anhui, Hunan, and Shandong.[59] BAT salesman Frank H. Canaday reported that "several hundred students have just gone through our 15-A depot at Tsiningfu [Jining], checked the stock (some 200 cases) and sealed the go-down [warehouse], warning the depot keeper not to deliver a single carton until the Shanghai case is satisfactorily settled."[60] When BAT depot keepers did not cooperate, activists burned the stock.[61] Protesters also issued public warnings to dealers to cease selling "enemy goods," specifically naming Hatamen and Da Ying. "If you persist in your mistake and do not correct your behavior, we have other ways to deal with you," read one threatening notice, "such as destroying your goods in public and issuing fines."[62] Indeed, protesters did issue and collect fines ranging from twenty to two hundred yuan and directed the funds to the ongoing strike effort in Shanghai.[63] Canaday reported that BAT's Beijing-area dealers observed the boycott, though he believed they had been "forced by agitators."[64] In one case, a "profiteer" bought twenty cases of Ruby Queen with plans to resell them. Students intercepted and fined him, but he enlisted Yongtaihe's lawyer to fight the protesters. Five thousand people gathered, "unanimously agreed" to burn the stock, and proceeded to do so.⁶⁵ The public destruction of branded cigarettes appropriated both the commodity and the power to revise its public significance.

Protesters also launched a consumer boycott of BAT that entailed rebranding Da Ying and Hatamen cigarettes. Canaday reported from Beijing in June that demonstrations were large and orderly and represented considerable class diversity. "The morning parade was made up largely of clerks, shop assistants, students, etc. They carried boycott banners and gave the usual yells. I did not see the afternoon demonstrations but understand that they were participated in mostly by coolies."⁶⁶ Protesters distributed leaflets that identified Ruby Queen and Hatamen as British and therefore as enemy products. One leaflet read, "Are not Hatamen cigarettes, which are even more noxious than opium, the products of a British company? . . . If you do not want to be colonial slaves, please do not smoke Hatamen cigarettes! Please rise up and join us in resisting British products!"⁶⁷ Activists also destroyed or wrote over posted advertisements and tried, with little success, to pressure newspapers to cease printing BAT advertisements; newspapers claimed they could not break the one-year contracts that they signed with BAT.⁶⁸

In the 1905 anti–American goods movement, BAT had rebranded Ruby Queen cigarettes by adding the Chinese name Da Ying (Great Britain). Now that British goods were under fire, the company tried to rebrand Ruby Queen cigarettes again, by claiming that they were American goods and by renaming them Red Tin Pack. Yongtaihe, the distributor, asserted that it was a Chinese company. Meanwhile, fresh Ruby Queen posters appeared newly stamped "made in America."⁶⁹ Hatamen, BAT insisted, was a Chinese cigarette, made in China of 100 percent Chinese tobacco. BAT tapped its political contacts for reinforcements. The American consul-general in Shanghai issued a statement, which BAT publicized through newspaper advertisements and handbills, asserting that "the 'Ruby Queen' brand of cigarettes were made of American tobacco and manufactured exclusively in the United States of America."⁷⁰ BAT also requested that the local government in Ningbo release a statement that Ruby Queen, Pinhead, and Lao Dao were made in the US and that Hatamen was a Chinese cigarette. BAT pressured its dealers in that area to sign on to this request as well.⁷¹ In very short order, then, leaflets and a variety of official statements got a rather garbled message out that people had no reason to boycott BAT's big brands.

BAT's strategy backfired as outraged protesters ramped up their attention to BAT's big brands. The *Shanghai Journal of Commerce* reported that "the letter from the American Consul-General at Shanghai, declaring that the cigarettes of a certain tobacco company are manufactured in America, has given rise to an opposition among people of all classes. Handbills are being circulated, advising the public not to allow themselves to be deceived."[72] Canaday reported that the Yongtaihe manager in Beijing, Mr. Cheng, "is putting up new 'Victory' and 'Ruby Queen' posters, specially stamped 'Made in America' but the newspapers have immediately had articles stating that this was a masquerade." He noted that one Beijing paper daily printed "facsimiles of BAT and Wing Tai Vo [Yongtaihe] brands and urged people not to buy them."[73] Indeed, leaflets distributed in Hangzhou by the Zhejian Law School May 30th Support Union declared, "The cigarettes of the British American Tobacco Co. Ltd. are British products! Pinhai [Pinhead] is not American but definitely British. Red Tin Pack is Da Ying and a British product!"[74]

BAT's attempts to cast Da Ying as American-made provoked particular ire. "If this brand is really made in the United States, why not call it 'Da Mei' [Great America] instead of Da Ying [Great Britain]?," asked one newspaper writer. "The word 'Ying' is not difficult to understand."[75] A leaflet distributed in Suzhou similarly responded, "They think that there are no intelligent people in our country and regard all of us as blind illiterates."[76] Yongtaihe also came under particular attack for claiming to be a Chinese company while declaring Ruby Queen to be American. "Before the boycott, this company was a British company," declared one newspaper sarcastically, "and after that it became Chinese or American. In the future, they can freely choose to be Japanese, French, Russian, or Italian.... Do not smoke their cigarettes and this company will quietly fail."[77] Protesters kept BAT on the defensive as the company struggled to regain control of the public discourse.

Activists also revised BAT advertisements or created advertisement-like posters to present new counterpromotional messages. Defacing BAT's advertisements became a common sport; to make it easy, activists in Suzhou issued stamps with radical slogans:

Oppose oppression by powerful countries!
Cancel the unfair treaties!
Permanently boycott Japanese products!
Never smoke pseudo-American-made British cigarettes![78]

One hand-drawn boycott poster urged people not to smoke Ruby Queen or Hatamen cigarettes. Two cigarette packs, displayed much as they were in advertisements, framed a graphic of a turtle smoking a cigarette.[79] Any Chinese viewer would understand the insult: the turtle was associated with female genitalia; this poster suggested that a person who smoked BAT cigarettes was a pussy. The consumer boycott of BAT appropriated capitalist tools to stigmatize the brands.

Protesters successfully destabilized BAT's brands and associated Ruby Queen and Hatamen with foreignness and imperialism. This was an ever-present danger for BAT, not only because it was indeed a foreign company but because all cigarette brands in China had a foreign, and usually a Western, association. That exotic element had long proved a selling point, much as with Egyptian cigarettes in the US. In China, English brand names and lettering on packing helped cigarettes to retain a foreign reputation for decades after their introduction. Indeed, in the 1920s, the Chinese-owned Huacheng Tobacco Company put English names and lettering on its packages, even as they advertised them as Chinese-made in order to appeal to the growing National Products Movement. In other words, cigarettes in China were not a blank slate, waiting for any brand identity to be imposed on them, but bore traces of the geopolitical history of their introduction into China.

Activists repeatedly interfered with BAT's cigarette branding by making those political traces more visible. Some protesters argued that it made no difference if BAT's products were manufactured in the US, or even China. Activist Chen Xuyuan argued that cigarettes "made in America" were no better than British cigarettes because of the US's Chinese Exclusion Policy, which he called a "burning shame" for Chinese people. The policy completely excluded workers and made exemptions for students and businessmen, but students attending colleges in the US faced intense scrutiny, humiliation, and harassment when attempting to enter the US. "Even if we have a valid passport, we encounter difficulties and it is not easy for us to enter," said Chen, "[but] those Americans coming to our country can freely enter." Chen acknowledged that Hatamen cigarettes were made in Pudong, but argued that "the capital, business decisions, and factory management are all under the control" of BAT. "Being manufactured in Pudong does not make cigarettes Chinese products."[80]

The May 30th Movement's rebranding of Ruby Queen and Hatamen as imperialist gave impetus to the National Products Movement and to Chinese cigarette brands. "Native cigarettes," wrote one protester in a public letter to dealers, "are high quality and in sufficient supply to guarantee profits. Why lose your dignity and aspirations to follow foreigners?"[81] That summer, more than two hundred new Chinese-owned cigarette companies sprang up in Shanghai in order to take advantage of the sudden unmet demand for China-made cigarettes.[82] Nanyang had for several years overtly advertised their cigarettes as native products; they increased such tactics after May 30th, despite having incurred criticism themselves for their ownership by overseas Chinese. They found that even their mildewed cigarettes sold and they expanded production as quickly as they could.[83] BAT tried stealthily to get in on the action as well. It introduced a number of brands with only Chinese names and characters, including Lao Dao, which means Old Knife, and Shuang Shih, which means Double Ten, referring to October 10, 1911, the date of the formation of the Chinese Republic, a brand that explicitly played to Chinese nationalists.

BAT tried hard over decades to shift its branding to prevent Chinese consumers from practicing politics through the brand, with decidedly mixed results. BAT eventually recovered from the May 30th protests and Ruby Queen cigarettes made a comeback. Nevertheless, anti-imperialist sentiment continued to be a significant factor in BAT's business in China. Ruby Queens, and cigarettes in general, retained their foreign connotation until after 1949, when the communist government nationalized BAT.

Lucky Strike Baseball Teams and Segregation

In some ways, African Americans in Reidsville, North Carolina, had a relation to the brand nearly opposite that of Chinese protesters: rather than contesting it, they embraced and claimed it as belonging to them by taking it as the name of their premier baseball team. Yet African Americans were also performing a kind of rebranding. The ATC attached the Lucky Strike brand to whiteness at the factory, in Reidsville, and in national advertising. By laying claim to the Lucky Strike brand and associating it with black masculinity, African Americans in Reidsville countered aspects of white supremacy's value structure, even as they accepted and even furthered cigarette promotion.

Lucky Strike cigarettes transformed the town of Reidsville because the company put the first Lucky Strike factory there. As the cigarette brand rose in popularity, jobs became plentiful, migrants arrived from the countryside, and businesses boomed. The factory painted its smokestack with the words LUCKY

STRIKE, making it resemble a lit cigarette. Lucky Strike was the new modern cigarette, yet it was uniquely Reidsville's. Workers and residents referred to the local ATC as the Lucky Strike factory and there emerged a Lucky Strike Glee Club, a Lucky Strike Orchestra, a Lucky Strike bowling team, and Reidsville itself became the Lucky Strike Town. It was patriotic to smoke Lucky Strikes in town and few dared to speak against the company. "Everyone in the city believes in Lucky Strikes," reported the local newspaper. The brand was proudly shared.[84]

When the ATC management started an all-white male Lucky Strike baseball team, the team augmented white male workers' privilege at the factory and reinforced white ownership over brand, cigarette, company, and town. The team fit well into the company's other uses of the brand to organize workplace hierarchies along race and gender lines, including uniforms for white women in brand colors and free cigarettes to white workers. Team membership carried workplace privileges, including augmented pay and reduced hours. The team also advertised the brand and associated it with whiteness. When the Lucky Strikes traveled for games, the players wore branded jerseys, and smoked free Lucky Strike cigarettes while in the dugout.

The ATC followed an emerging Southern practice when it created a baseball team as part of its paternalist program for labor control. In the 1910s, company-sponsored teams sprouted up all over the South, particularly representing textile mills but also tobacco factories. Companies responded to the astounding popularity of baseball nationwide. Regularly called "America's game" or the "national pastime," baseball was the most popular sport in the country, and the best major league baseball players were national heroes.[85] By the 1920s, leagues of semiprofessional company-sponsored teams spread across the small-town South, and fans turned out in droves. Baseball often drew crowds of spectators in the thousands, even in towns the size of Reidsville. Herbert "Rabbit" Leary, centerfielder for the Lucky Strikes in the late 1930s, recalled that "Reidsville really supported their teams. It was a baseball crazy town."[86] By the mid-1930s, the aggregate investment in baseball generated a lively minor league system with some teams feeding their best players into the majors.[87]

Team and factory management enmeshed, with a job serving as partial payment for playing on the team. As company teams proliferated, young white men who could knock a ball over the fence found their opportunities for employment almost magically enhanced. Pat Griffin explained, "In 1929 I was playing for Leaksville. I pitched and won two games against the Luckies. Then the manager of the Luckies asked me to come play for them. At the same time he hired me to work for American Tobacco Company." The job was the more lucrative part of the semiprofessional package, worth more to most players than a position on a fully professional team that offered only seasonal employment. Sherman Hoggard gave up a position on the professional Winston-Salem Twins to play for the Luckies in 1929. He explained, "My friend Barry Worsham, knew the Luckies' manager, Hap Perry, who was at the game. They sent me in as a pinch hitter in the ninth inning and I hit a home run and we won. Mr. Perry gave me a place on the team and a job at American Tobacco Company.... To keep my job I had to play baseball."[88] Howard Briggs applied for a job at the American Tobacco Company in 1932, hoping his experience on professional baseball teams would help him obtain it. He remembered, "My older brother Lefty told me that American Tobacco was hiring, so I ended up working 30-some hours unloading boxes and playing semi-pro ball."[89]

Payment on the Lucky Strikes was minimal and irregular and came mostly from ticket sales and the generosity of fans. "People paid about 50 cents to get in each game," recalled Hoggard. "At the end of the year the team would split the profits. One year I made $92!" Pat Foy Brady remembered that when his father played for the Luckies, "The players didn't get much money ... so there was this tradition that, when a guy had a really good game, you passed the hat." His father hit three home runs over the fence in one game, Brady recalled, and fans gave him "about 10 or 15 dollars that day." Though the cash for their play clearly mattered to players, the job provided the consistent pay envelope that allowed them to settle down in Reidsville and raise families. Indeed, Griffin and Hoggard played for three years and Briggs played for seven, but all three kept their jobs at the factory for over forty years.[90] White men's privileged access to jobs based on their eligibility for the baseball team reinforced factory job segmentation that reserved the best jobs for white men.

Playing for the Lucky Strikes conferred honor and special treatment at the factory and in town. During the baseball season, players practiced thirty minutes after work on nongame days and left work early three or four times per week for games.[91] Briggs recalled, "My boss [foreman] was also the president of the team, so he set it all up. I would ... sneak out early in the afternoon and catch the team bus if we were playing on the road."[92] Augmented payment and a shorter day: there were no more coveted perks than these in a factory town. In addition, players were town heroes. The *Reidsville Review* covered each game in great detail and town businesses, including the factory, closed early when the team had an afternoon home game. White and black residents alike attended the games, though African Americans were confined to the bleachers for white baseball games.[93]

The Lucky Strike team provided unique regional advertising for the brand. Though textile and tobacco companies all over the South funded baseball teams, Reidsville's was the only company team named after a cigarette big brand. Other tobacco companies fielded teams named after smoking tobacco big brands, such as Reynolds's Prince Alberts or the Durham Bulls. These teams emerged in the 1910s, when smoking tobacco was much more popular than cigarettes. Indeed, the Reidsville factory's first team was named the Red Js, after a smoking tobacco of the same name, and its stadium retained that name even after the team became the Lucky Strikes.[94] Textile mill baseball team names referenced their mill (Proximity after Proximity Mill) or a play on their town name (the Snow Hill Billies), or an aspect of textile production (the Weavers, the Spinners), or the kind of textile product they made (the Towelers, the Blanketeers), but not a brand name. Likewise, tobacco preparation factories had team names like the Tobacconists, the Leafs, and the Auctioneers.[95]

With the Lucky Strike brand name emblazoned on their chests, players carried and smoked free cigarettes at the games, becoming sponsored athletes who performed product placements. In these ways, the ATC management marshaled the value of the brand in promoting social investments in the segregated company and town while also advertising cigarettes. Furthermore, as workers and town residents attended the games, read about the team in the papers, and encountered players in town and at the factory, they made the Lucky Strike brand a part of their daily personal lives.

The racial politics of the Lucky Strike team echoed national marketing strategies. Cigarette companies promoted the famous brands within a "national market," which meant that they targeted white consumers exclusively. African Americans may have purchased their goods—some companies depended on their dollars to survive—but business practices pointedly developed around segregation logics that meant that African Americans would not be visibly recognized as consumers in national print culture and advertising. Brands that gained national attention in this climate belonged in an intangible but intentional way to white consumers and carried an aura of whiteness.

Nationally circulating brands like Aunt Jemima's Pancake Mix or Uncle Ben's Rice carried the mark of this process explicitly. These brands offered consumers a feeling of mastery in their homes while positioning white consumers as modern in relationship to primitive, servile African Americans.[96] African Americans certainly could purchase these brands if they had the cash and desire to do so, though African American sociologist Paul K. Edwards found in his 1932 study of African American consumers, perhaps unsurprisingly, that the Aunt Jemima brand was widely criticized and despised. African American women, in particular, objected to the brand's use of a head rag and old-fashioned clothing; that is, they objected to the positioning of the figure as primitive in opposition to the modern white consumer.[97]

The failure to market cigarettes to African Americans was not an oversight. In 1928, renowned white public relations guru Edward Bernays recommended to the American Tobacco Company that it run a Lucky Strike advertising campaign in black newspapers around the country. Bernays had received an office visit from Claude A. Barnett, an African American marketer with aspirations to build a marketing firm that would represent the "Negro market."[98] Barnett pitched an idea for a newspaper article or series about the Richmond ATC factory, focused on African American workers. This article, complete with pictures of the leaf department's drying and stemming rooms and personal stories of workers, would highlight the importance of the ATC in hiring African Americans and, of course, would be paired with Lucky Strike advertisements. Very similar kinds of articles about factories focused on white workers ran regularly in white newspapers nationally. Bernays thought it was a great idea. He wrote to ATC president George W. Hill's assistant, giving the idea his personal stamp of approval. Hill, without explanation, dismissed it out of hand.[99] That same year Bernays recommended a very large number of other campaigns, some of them quite unconventional; virtually all of them gained approval.[100]

Indeed, so intent was the American Tobacco Company on not associating its product with African American consumers or workers that its promotional materials about its factories obscured the fact that blacks worked there. These materials were not advertisements per se, but rather informational materials provided to stockholders, company managers, and interested parties such as government inspectors and newspaper reporters. One ATC promotional pamphlet, for example, pictured cigarette making and packing rooms filled with orderly white workers at machines. The picture of the stemming room, where African American women typically worked, showed only a white-shirted white manager holding a bunch of tobacco leaves as though he alone were about to stem them.[101]

When African Americans in Reidsville fielded their own all-black baseball team named the Lucky Strikes, they intervened in the whitewashing of US life by appropriating and reassigning the value of the brand. In other words, they challenged the brand's use as a technology of white supremacy, while refraining from countering the company directly. Black businesses in town joined together to sponsor the team, providing players with uniforms and partially covering travel expenses. The team played its home games in Reidsville's Red J stadium, creating its schedule after the white team had claimed its home game dates there.[102]

The Lucky Strike team's appropriation of the white team name followed a nationwide customary practice and asserted representation of the town. For example, the white Charlotte Hornets represented the town of Charlotte, North Carolina; the Charlotte Black Hornets *also* represented Charlotte on the diamond. The Atlanta Crackers claimed an explicitly white name; the Atlanta Black Crackers' name betrayed the team's sense of humor. The Washington Black Senators' name carried a hint of critique, as the players were the only black senators in town. In white discourse, only the white team remained racially unmarked, casting the black team in a secondary, derivative position. When African Americans discussed the teams, however, they navigated around stigma with care. For example, Lucky Strike pitcher William Davis said, "The factory did have a team, but it didn't have a black team. We had one of our own." He paused to choose his words. "They were the Luckies—the white team. You see, we couldn't—we Luckies, they Luckies. We had to be the 'Black' Luckies."[103] The parallel naming practice echoed the community-building strategy of African Americans under segregation as they built businesses, churches, and schools: it accepted, on the surface, a secondary status, but also made a claim on parity and representation.

By appropriating the cigarette brand, the Lucky Strikes created social relations around baseball that celebrated African American masculinity, the particular sensibility of which was captured in watercolor portraits made of each player in 1937. Davis kept his portrait for over sixty years displayed in a place of honor in his den. The portrait abandons sports photography's promise to capture action or masculine fitness and conveys instead an inner emotional experience. Davis is pictured from the knees up, wearing his uniform and glove, with his arms at rest. The viewer is positioned as though looking up at Davis's face, which is bathed in light; a small border of white around the face further centers our attention. His face is serene, relaxed, almost ethereal. His eyes look straight ahead, focusing in the distance. The blue uniform has a large round patch on the left breast that reads "Lucky Strike." The artist has used pen and ink to outline some aspects of the figure, including the words "Lucky Strike" and the borders of the figure. "Wm 'Hoss' Davis, 1937" is written in the lower corner (fig. 5.7). The artist slipped between the stereotypes of the black working class — physically threatening, amoral, or ridiculous — and the middle-class tradition of uplift, which relied on realist photography to bear witness to accomplishment. In doing so, the artist answered New Negro theorist Alain Locke, who called for black art to render the "internal world of the Negro mind and spirit."[104] In the baseball portrait, Davis embodies a glowing and manly dignity.

As an appropriation of the brand, the Lucky Strike team cut two ways. On the one hand, African Americans redirected the brand's value to black men and African American culture — precisely what the ATC worked hard to avoid. At the same time, the company got free advertising. Though black businesses had put up the funds, the Lucky Strike team did not advertise black products or services but those of the very company that had excluded black employees from the company team. Tellingly, the local ATC management provided the team just one perk it accorded the white team: free cigarettes to carry to the games. While the ATC would not sponsor the black team or market its cigarettes in syndicated black newspapers, it abetted the local promotion of cigarettes by the team within segregated spaces. African Americans' use of the brand set in motion multiple and contradictory events and affects.

The rise of the Lucky Strike brand, then, became entwined with struggles over Jim Crow segregation at the workplace and in the town of Reidsville as both white and black people used the brand as a resource for the creation of social value. The ATC management guarded the commodity's association with whites and whiteness, but African Americans created brand stories that supported their own cultural formations and wove the Lucky Strike brand into a larger oppositional culture that simmered within segregation.

Fig. 5.7 Portrait of William "Hoss" Davis, 1937, artist unknown. Courtesy William Davis.

Conclusion

The rise of the big cigarette brands, Camels and Ruby Queens, was a turning point in the global cigarette industry. The famous brand developed accidentally, rather than by express design, in both the US and China in response to the introduction of competition into a monopoly and a resulting concentration of advertising resources into single brands. Companies learned that single brands had capacities to woo much larger markets than they anticipated. Famous brands became a notable part of public life, functioning in mutable ways as part of a shared symbolic language even for those who did not smoke. The resulting shift in the industry in both places was dramatic and swift, and soon other companies introduced copycat big brands. By defining a trend among cigarette smokers, rather than dispersing the market into segments, the famous brands also seemed to encourage the trend of cigarette smoking.

This story deviates from the prevailing explanation for the spike in cigarette smoking rates in the 1920s. For decades, historians have held that soldiers picked up the practice in World War I and brought it home to civilians. This explanation holds merit but it does not fully explain the embrace of big brands that is so definitive of the era. US companies provided a range of pipe smoking tobacco and cigarette brands to military outlets in World War I, not just the Turkish–bright leaf–burley blends.[105] In addition, World War I had a negligible impact on daily life in China and cannot explain the concurrent rise in smoking there. This chapter's understanding the simultaneous rise of Camels and Ruby Queens has required a closer look into the social history of business practices.

For the story of the global cigarette, the rise of Ruby Queen and Camel cigarettes is significant in a number of ways. First, the simultaneous but independent boom in sales challenges notions of modernity that assume a West-to-East flow of business development. In both places, company leaders discovered the potential of the single brand accidentally, when a new kind of competition prompted more concentrated attention to a single brand than in the past. Furthermore, BAT had not succeeded in gaining sufficient distribution of Ruby Queens; only after it abandoned its rationalized, modern distribution system and empowered Zheng Bozhao to implement his own business practices did it achieve anything approaching the sales it had hoped for outside treaty ports. In the US, the rise of Camels depended on concentrated advertising; in China the dynamic was different, relying both on a new distribution system and advertising.

Finally, this chapter demonstrates that the big cigarette brands were not entirely blank vehicles for advertising messages, but rather some of the relations of their production stuck to them. In China, virtually all cigarette brands carried English lettering because of their introduction as foreign products. While the foreign aura of cigarettes had appeal in China as in the US, BAT ran into trouble as anti-imperialist sentiment rose. Giving Ruby Queen cigarettes the name Da Ying was a solution in 1905, but a problem in 1925. When Da Ying factory workers went on strike in conjunction with consumer boycotts, protesters and strikers succeeded in disrupting and co-opting the famous brand. In the US, there was no visible mark on the packages that linked the famous brands with whiteness, but the concerted, conscious, and multifaceted efforts of the companies to define the brands as white sent a clear message to African Americans. The African American–funded Lucky Strike baseball team did not so much break the spell of commodity fetishism as hone in on the magic. As an example of the ways brands become embodied ways of reframing relationships to a wider social world, the baseball players' actions are poignant testimony to an alternate African American value system.

The triumph of the famous brands reveals much about the way cigarette companies sped to the cutting edge of the new science of branding, but it reveals less about the shifting nature of the popularity of cigarettes themselves. What were cigarettes coming to be in public life? A crucial international public scene of the cigarette during the interwar era was the cabaret that played jazz music for dance-mad men and women. It would be difficult to overstate just how popular and widespread the jazz dance craze became in large cities like Shanghai, Paris, Berlin, and New York. The cigarette became a constant companion in the jazz clubs, and the association between the two changed cigarettes and jazz alike. Chapter 6 explores this phenomenon.

6

The Intimate Dance of Jazz and Cigarettes

By the 1920s, cigarettes were booming in both the US and China, just in time to coincide with the international jazz dance craze. From Charlotte to Shanghai, the commodities of jazz and cigarettes joined forces by both accident and design. In the US, people danced to jazz music in mostly segregated venues that included jook joints, warehouses or armories, big-city cabarets, little hole-in-the-wall clubs, vaudeville shows, high school gymnasiums, hotels, dance halls, and restaurants. In Shanghai, entrepreneurs rushed to build new clubs, cabarets, and hotel ballrooms for the burgeoning throngs of dance-crazy foreigners, among whom were many BAT employees. Situated in the foreign settlements, new cabarets epitomized the colonial privileges of the treaty port, now more elaborate than ever. The cabarets provided posh surroundings, live bands, and young taxi dance girls available for hire. Cabarets began as foreign-only places, but by the late 1920s an increasing number catered to elite Chinese customers, though typically still in segregated contexts.[1] Offering cigarettes for sale at cabarets and other jazz venues was an innovation that probably seemed obvious to cigarette salesmen and venue owners alike. Linked by atmosphere, rhythm, and bodily intimacy, the experiences of jazz and cigarettes entwined.[2]

Tobacco companies seized upon this synergy and made it part of their explicit marketing strategies. In particular, they sponsored jazz radio shows in both the US and Shanghai, tying cigarette branding to the aura of jazz and selecting the jazz that filled the airwaves, thereby spreading the association to consumers beyond those who attended the clubs. In Shanghai, BAT-sponsored shows played American jazz records nightly while advertising high-grade cigarette brands to a mostly foreign clientele. In the US, cigarette companies sponsored jazz shows on radio's new national networks. Shows such as "The Lucky Strike Hit Parade" and "The Camel Caravan" became a dominant way that jazz hit the airwaves, giving cigarette companies a prominent role in shaping the jazz that most people heard.[3] Driven by marketing concerns, cigarette companies carefully shaped radio jazz to have "hot" elements but to avoid the suggestion of blackness. Accordingly, with just two exceptions, they hired and gave national exposure only to white bands. Companies interjected advertisements for cigarettes throughout the jazz shows, further linking the two commodities and endowing them with shared affective powers.

By the 1920s, jazz was an ever-changing, globally circulating commodity. "Jazz" referred to popular dance music that included a wide variety of styles. Though jazz's syncopation emerged within African American musical communities and people generally understood it to have black origins, US immigrant and white musicians also played and sang jazz songs from its earliest days, and musicians the world over played jazz and left their own marks on the form. The artistic innovations of black musicians and dancers cannot be extricated from jazz's status as a globally circulating capitalist commodity; touring, recording, and distribution opportunities developed unequally within racially shaped markets in which merit was rarely a singular or primary consideration. In the US, white bands claimed the best gigs and most of the recording deals, giving them ever-wider exposure, while black bands generally traveled more and further to make money. In Shanghai, in contrast, African American bands were at a premium as the "originators" of jazz, but also in the competition were Filipino, white American, Japanese, Russian, and Chinese musicians.[4] Race and nation, in the composition of the bands and the music alike, was ever present in jazz.

Jazz music everywhere was for dancing.[5] One-step, two-step, fox-trot, lindy-hop, and eventually the jitterbug. It was not until cool jazz and the bebop era, when musicians intentionally interrupted the dance-ability of jazz, that fans sat down to listen. In the interwar era, jazz dances circulated the globe along with hit songs. Touring bands, especially in Asia, often included dancers who performed and taught the steps to eager audiences. In Shanghai, cabarets supplied dance hostesses from China, Japan, Russia, Korea, France, and other places so that the largely male business community could dance to the music. Even radio announcers presented jazz as dance music and assumed that many listeners were dancing to the music in their homes. Jazz was not simply aural, but was movement and intimacy, a highly participatory bodily experience.

Cigarettes became an integral part of the public scene of jazz dance and accrued their particular emotive power as the two commodities circulated internationally. Cigarettes' seemingly natural fit as punctuation to a night of dancing left a lasting global mark on the commodity's life. Far more than could advertisements, the association with jazz imbued cigarettes with sexiness and cool. Cigarettes and jazz were central among several commodities attached to an emerging globally circulating but highly mutable set of trends that took root in urban nightlife. Often labeled "modern," these trends included sexual expressiveness, pleasure seeking, and the breaking of convention. The controversial figure of a fashion-savvy, sexually expressive "modern girl" emerged within and became a symbol of this broader international phenomenon.[6]

The flow of jazz and cigarettes through urban leisure contexts in major world cities did not Americanize the world, however, because in every location these commodities took on distinct cultural and political valences. Even in Shanghai, the synergy between jazz and cigarettes generated a certain set of associations for BAT's foreign employees enjoying the imperial culture of foreign-only cabarets, but quite different ones in Chinese-owned cabarets catering to Chinese clients. Likewise, the confluence of jazz and cigarettes created a particular ambiance on national radio in the US, but carried very different potentials for African American workers attending jazz dances in cigarette factory towns. By examining cigarettes in the context of jazz, it is possible to discern the production of cigarettes' cultural appeal at points where the blunt tools of corporate marketing nudged the dynamic and unpredictable unfolding of cultural community.

Accordingly, this chapter explores how the synergy between jazz and cigarettes developed in cabarets and radio shows in both the US and China. But corporate employees not only helped produce this phenomenon, they also participated in it. Foreign businessmen like the young men in the bright leaf tobacco network were prime customers for Shanghai's cabarets. The same foreign representative might make a sales visit to the cabaret in the afternoon and return in the evening for dancing, punctuated by a few smokes. Likewise, BAT's high-ranking Chinese businessmen were exactly the kind of clientele that some cabarets sought by the late 1920s. Shanghai's cabarets were too expensive for Chinese factory workers, but African American cigarette factory workers in North

Carolina turned Jim Crow to their advantage, for a change, by booking world-class jazz acts into small Southern towns. In these ways, the confluence of jazz and cigarettes also became sites of intimate encounter that variously produced and contested corporate hierarchies.

Jazz and Cigarettes in Shanghai Cabarets

Western businessmen built Shanghai's first cabarets soon after World War I, and the dance palaces quickly proliferated, stimulating and profiting from foreigners' desire for opulent places to celebrate the fruits of Western imperialism.[7] Shanghai's seven thousand foreign companies, BAT prominent among them, imported a steady stream of young male representatives who required entertainment.[8] Cabaret owners hired live jazz bands and dance hostesses, or taxi dancers, who danced with the overwhelmingly male clientele for a fee. An English-language guidebook conveyed the centrality of cabarets to Shanghai's burgeoning nightlife:

> Shanghai has its own distinctive night life, and what a life! Dog races and cabarets, hai-alai [sic] and cabarets, formal tea and dinner dances and cabarets, the sophisticated and cosmopolitan French club and cabarets, ... prize fights and cabarets, amateur dramatics and cabarets, treatres [sic], movies and cabarets, and cabarets—everywhere, in both extremities of Frenchtown, uptown and downtown in the International Settlement, in Hongkew [Hongkou], and out of bounds in Chinese territory, are cabarets. Hundreds of 'em![9]

The segregated nature of the international settlements is captured by the guidebook's reference to the majority of the city as a "Chinese territory" that was "out of bounds." BAT employee Irwin Smith expressed the same sentiment when he remarked, without irony, that Shanghai "was just like any other foreign city except the Chinese were there."[10] Cabarets initially epitomized treaty port imperialism.

Because of its lucrative cabaret scene, Shanghai became a very important place for jazz music in Asia. The music scene was profoundly international: musicians from the US, the Philippines, China, Russia, and Japan all competed for gigs. The most successful obtained extended contracts in Shanghai, but many

Fig. 6.1 The Candidrone dancefloor in Shanghai. Used by permission of the University of Missouri-Kansas City Libraries, Dr. Kenneth J. LaBudde Department of Special Collections.

traveled a circuit that included Singapore, Hong Kong, Manila, Calcutta, and Shanghai.[11] Like musicians, taxi dancers converged in Shanghai from near and distant places including China, the Philippines, Russia, Korea, Greece, and Japan. The jazz scene in Shanghai was so established that Japanese musicians went to Shanghai to learn "authentic" or "American" jazz.[12] At the same time, the international mix of musicians and the need to cater to a Chinese clientele by the 1930s led to innovative new sounds.[13]

Jazz music and cigarette smoking created a distinctive space and atmosphere in the cabarets. Cabaret floors in both Shanghai and the US reserved a large central area for dance, with a periphery of tables for clients who wished to smoke cigarettes, drink alcohol, and talk.[14] At the Canidrome, for example, a very large space in front of the bandstand encouraged dancing couples to fill the center of the room, circled by a narrow band of small, intimate tables (fig. 6.1). At some clubs, dance hostesses occupied one side of the dance floor while clients eyed them from the other before inviting one to dance or join their table. Sound and smoke combined to define the ambiance. BAT employee James Hutchinson noted that when he entered the Cercle Sportif Français in 1930, the music was "deafening" and the cigarette smoke so thick it took his eyes several minutes to adjust so he could discern faces.[15] Cigarettes and jazz thus functioned symbiotically to shape the interior of cabarets.

Cigarettes and jazz shared a distinctive capacity to structure the cabaret experience. Unlike other commodities present at the cabaret, such as cosmetics or fashionable clothing, cigarettes and jazz dance had a temporal quality. Both could be used to mark time: one more cigarette, one more dance. Both also could motivate the body through space: to the center of the dance floor or to the periphery to a table. As jazz dance charged the body in movement, cigarettes animated the body at rest. Cigarette smoking and jazz dance thus were the medium through which the body became part of, even constituted by, the cabaret.

Furthermore, the activities of jazz music and cigarette smoking carried distinct powers to transform bodily and emotive sensations in ways that not only facilitated eroticism but came widely to serve as signs of the erotic. Both cigarette smoking and jazz dance allowed physical intimacy between customers at a cabaret. The one-step, the fox-trot, the Lindy hop: the point for fans of this era of jazz was an expressive capacity to embody the music in dance. "Listening" was a full-body experience. Spinning together on the cabarets' large dance floors allowed strangers to touch each other's bodies in ways otherwise forbidden. Dancing to jazz music combined precision and skill in executing set dance steps with personal improvisation and expression. "Hot" syncopated rhythms favored dances that moved the hips and could require considerable physicality, in contrast to nineteenth-century European ballroom dances, which held the body more stiffly erect.[16] Likewise, physically lighting one's own or another's cigarette, sharing cigarettes, and breathing the smoke of another's cigarette all engaged the mouth, hand, and breath in a sensuous and intimate activity. Jazz dancing and smoking carried expressive capacities that were at once shared and distinguished by personal signatures. The space of cabarets became widely celebrated and decried for this eroticism, a defining element of the jazz craze.

The eroticism of the cabaret experience carried a racially transgressive element for many foreigners in Shanghai. Consider this scene of intoxication from an English-language guidebook to Shanghai in the 1930s, under the heading "Dancing and Music":

The throb of the jungle tom-tom; the symphony of lust; the music of a hundred orchestras; the shuffling of feet; the swaying of bodies; the rhythm of abandon; the hot smoke of desire—desire under the floodlights; it's all fun; it's life.... There's nothing puritanical about Shanghai.[17]

The "throb of the jungle tom-tom" efficiently references notions of primitive African rhythm at the heart of the racialization of jazz. Pointedly containing "nothing Puritanical," jazz according to this view was sexually passionate, primitive, and black, in contrast to rational, civilized whiteness. Jazz's racial status was ambiguous in Shanghai, as elsewhere. Many white American musicians played in Shanghai cabarets (and resented the competition of Filipinos and Russians). Paul Whiteman achieved early and uproarious fame in Japan and elsewhere in Asia. US recording companies discriminated against African Americans, a fact with international consequences because jazz recordings were a key way that jazz first circulated in Asia.[18] Still, notions that African Americans offered a more authentic jazz sound clearly circulated among the foreign elite in Asian ports of the 1920s.[19]

Many Chinese people in Shanghai also came to see jazz as originally or in essence black. In 1933, celebrated fiction author Mu Shiying reflected these sensibilities in his story titled after a popular brand of cigarettes, "Craven 'A.'" The story was narrated by a lawyer, Yuan Yecun, and described his infatuation with Yu Huixian, a sexually adventurous taxi dancer who favored Craven "A" cigarettes. Mu described the cabaret as a primitive space:

> Around wildfires of the neon lights were sitting a tribe of earthy people....
> Clapping hands, blowing trumpets, yelling, as if in fear that wild animals were coming out of the forest to attack. Under the Japanese-style paper lanterns were a group of people cut off from civilization, throwing themselves into the feeling of the barbaric music.

Yuan became captivated by Yu. He bought her cigarettes, which they smoked together. Later, he watched her dancing with another client to a rhumba, "swaying her head and her shoulders like a pendulum . . . her hair spreading outward like a parasol. . . . I was . . . looking at this black African woman."[20] Mu cast the taxi-dancer as a tragic figure exploited by men; jazz and cigarettes in his story combine to create a heady, primitive atmosphere. Thus, the transgressive eroticism of the cabaret experience engaged and fueled the racialization of jazz, making possible new investments in racism through a wide range of controversial pleasures.[21]

As "Craven 'A'" indicated, cigarettes only added to the sexual charge of the cabaret atmosphere. Taxi dancers very typically smoked cigarettes and drank alcohol with their clients in the cabarets. An anticigarette campaign in the turn-of-the-century US and England had linked cigarettes to alcohol as intoxicating substances and associated both with moral lassitude and physical degeneration, notions foreigners brought with them to China. Considered a masculine prerogative in the US and Britain, women's cigarette smoking held particularly transgressive and sexual connotations. For Chinese people, no analogous proscription against women's smoking existed; women typically smoked pipes or cigarettes as they wished within the household. Smoking in a public place, however, was as new and transgressive for women as the public leisure palaces themselves.[22] Thus, the Shanghai guidebook writer's phrase the "hot smoke of desire" uses the word "hot," associated both with jazz rhythm ("the rhythm of abandon") and an exhaled breath, to merge the effects of jazz and cigarettes with "desire."

Jazz and Cigarettes in US Cabarets

The international cabaret dance craze helps explain the dramatic rise of women's cigarette smoking in the interwar era.[23] As chapter 1 argued, the first public scene of the bright leaf cigarette had been the British gentlemen's club, a homosocial institution that was alive and well in Shanghai's International Settlement. In the US, where such clubs were not as prevalent, elite men and women tended to divide after a social dinner: men to the smoking lounge and women to the parlor. In the twentieth century, however, heterosocial forms of amusement were on the rise internationally, dancing venues being especially participatory and potentially romantic.[24] In both the US and China, cabarets became notable places for the appearance of the "modern girl," who danced, wore daring new fashions and hairstyles, and smoked cigarettes.[25] The cabarets, then, became a new public scene for the cigarette, one that redefined its reputation in an intimate, embodied relationship to jazz dance.

It would be difficult to exaggerate the importance of dance to the cabaret experience. In the 1910s, the dance craze took off, as the interracial pairing of James Reese Europe's band with white professional dancers Irene and Vernon Castle popularized the fox-trot and other ragtime dances and working- and middle-class people caught dance fever.[26] By the 1920s, people from Reidsville to New York to Shanghai danced variously in public dance halls, large cabarets, smaller clubs, school gymnasiums, tobacco warehouses, armories, and living rooms. "Jazz" became a catch-all term to refer to popular songs, especially ragtime and Tin Pan Alley songs, that circulated widely through sheet music, traveling bands, and the new technology of records. Not all of these songs would sound like jazz as we know it today, but the dance craze was rooted in syncopated music that invited bodies to move in new ways.

Because of jazz's roots in African American culture, jazz dance was highly controversial among whites. Many equated the jazz dance craze with danger, sexuality, racial intermixing, and the loss of respectability.[27] At the same time, jazz dance became an expression of newness, youth, and expressive vitality. Ironically, nineteenth-century blackface minstrel shows parodying "primitive" black music set the stage for twentieth-century black music and dance to be seen as quintessentially modern and liberating. Nineteenth-century white blackface performers presented music that they declared to be black and performed blackness as primitive, sexual, violent, and irrational in order to project a fantasy of "civilized" whiteness.[28] Having displaced the "low" onto a racial fantasy, however, the minstrel tradition helped white people later hear jazz music as liberated from white convention and Victorian constraint. The "low" was precisely what appealed. Though white musicians performed jazz from the outset, making the music more acceptable to middle-class white audiences, the popular association with primitive blackness remained; the tension between renunciation and rebellious embrace remained critical to the emotive power of the music for white audiences.[29]

In New York, Chicago, St. Louis, New Orleans, and other cities, African Americans built performance spaces and created jazz scenes, but they also toured from the outset, both in the US and abroad. In the segregated US, most white hotel and club owners hired white bands for the better-paying gigs. African American entrepreneurs had difficulty raising the capital to build high-end clubs and their African American clientele, for the most part, could not afford to pay top dollar. This sent the vast majority of musicians on the road for lower-paying gigs, playing for both white and black audiences.

The New York City cabaret scene was as segregated as Shanghai's. Harlem's famous Cotton Club, like the Canidrome, catered to white patrons only. African American musicians and dancers performed exoticism for white audiences who were slumming for the evening. Duke Ellington and His Orchestra, for example, long played the Cotton Club with a stage set called "The Plantation"; professional dancers' costumes and movements catered to white fantasies of slavery. Josephine Baker became famous in the US and Paris for her primitivist dances, and perhaps especially her banana skirt costume. The Savoy, on the other hand, was a large and famous black-owned cabaret that admitted black patrons. The constraints on the style of music played there were fewer. While white-owned clubs gained the most fame, small black clubs proliferated. *Variety* magazine reported in 1929 that there were eleven large cabarets that catered to whites only in Harlem, but over five hundred "colored cabarets, of lower ranks."[30]

As in Shanghai, cigarettes and jazz developed a synergy in these clubs. Also as in Shanghai, the layout of the clubs favored a large central area for dancing surrounded by small tables for resting, drinking, and smoking. "Cigarette girl" became a job classification in the US as clubs hired young women to sell from table to table (in China, boys performed this work). In the black journal, *The Messenger*, a fan of opera described the experience of going to basement black music clubs of 1924: "[One was] plunged deep beneath the ground, free from ventilation, where one's clothes become thoroughly saturated with tobacco smoke and where no complaints can be made against this generally recognized impossible music."[31]

Cigarette Corporations Produce Jazz Radio Shows

In both the China and the US, cigarette companies soon sought to reinforce the lucrative association between cigarettes and jazz in the cabarets, and they seized upon the new medium of radio. Radio was so new that stations scrambled to fill airtime. In both countries, corporate sponsors not only supplied advertisements to be played on air, they produced the entire content of their shows. In the US, sponsors bought time in one- to two-hour segments and relied on marketing firms to create the shows.[32] This gave cigarette corporations a significant creative role in the kind and quantity of jazz that hit the airwaves, as well as in how that jazz would be contextualized and connected to cigarettes by announcers.

In Shanghai, BAT sponsored a recorded jazz radio show that was broadcast several nights each week. The company relied on James Hutchinson, originally from North Carolina, to curate the programs:

> An English advertising house opened a broadcasting station. The company decided to try it out on a high-grade cigarette and I spent the best part of a week selecting records from the vast stocks imported by the four leading music houses. Then the company presented me with a small receiving set to check up on our tri-weekly programs, and from six in the evening until eleven I listened to . . . American jazz.[33]

Given that the company designated a "high-grade" cigarette as sponsor, the intended audience was elite foreigners. No doubt, the announcer made sure that listeners heard the brand name of the sponsoring cigarette between each and every number.

BAT-China likely took inspiration from the widely known recent success of cigarette-sponsored radio shows in the US. In 1928, the American Tobacco Company created a sensation by sponsoring the *Lucky Strike Radio Hour*, which became so popular that it shaped the sound of jazz on the radio. Other companies followed suit with a barrage of shows, including the *Old Gold Hour*, the *Camel Caravan*, the *Chesterfield Show*, and the *Raleigh-Kool Program*.[34] A key difference between shows in China and the US is that US regulations stipulated that radio shows could not use music recordings but had to use live performers. Recording companies, threatened by the new medium, had pushed for a 1922 licensing rule against radio play of phonograph records that lasted until the mid-1930s.[35] For this reason, cigarette-sponsored shows became a primary means of exposure for bands in the US. The cigarette shows did not simply transmit jazz, they also shaped and standardized a jazz product, making jazz part of the branding of cigarettes.

The rule against playing recorded music on the radio had the unintended effect of empowering wealthy corporations to shape the kind of jazz heard on national network radio, squeezing out the jazz aficionados involved in local radio production in some urban areas. The ATC employed two public relations firms, Lord and Thomas, and Ivy Lee, to develop and manage the show each week. Lord and Thomas hired a huge forty-piece house band, B. A. Rolfe's Lucky Strike Orchestra, to anchor the show. The orchestra created the big sound that people associated with live cabarets. The PR firm also booked guest bands—sometimes very famous ones—nearly every week, sometimes playing in the studio and sometimes broadcasting by live feed from a New York cabaret. Only large corporations would be able to compete with the ATC's new show.

The *Lucky Strike Radio Hour* billed itself as providing music expressly for dancing at home or a party. As a radio reviewer for the *Forum* wrote in 1932, "Tune in to [the *Lucky Strike Radio Hour*] if you like to dance."[36] At that time, radio technology produced a louder and clearer sound than phonograph recordings, so dancers scheduled private parties to coincide with the shows.[37] Radio jazz parties became so prevalent during the Depression that some musicians reported a resulting decline in live gigs.[38]

The *Lucky Strike Radio Hour* announcer explicitly linked cigarette smoking to the rhythms of a dance party. In 1929, for example, the announcer linked to the current marketing slogan by saying, "Should you sit out a dance, may I suggest, reach for a Lucky instead of a sweet."[39] Dancers were not likely to forget that Lucky Strike cigarettes brought them this music, since the announcer repeated the Lucky Strike brand name with each mention of the orchestra. In addition, advertisements for Lucky Strikes filled fully four-and-a-half minutes of airtime each hour, inserted in ten- to thirty-second sound bites between virtually every song.[40] Besides distributing jazz, then, the shows functioned as an hour-long commercial for Lucky Strikes, enhancing the synergy between the two commodities.

The *Lucky Strike Radio Hour* entailed the risk of associating the brand not just with the exciting ambiance of jazz but with blackness. The ATC thus crafted a standardized jazz product designed to please the most and offend the fewest. To do so, the company, guided by its PR firm, pursued three strategies. First, it created an executive group comprising a diverse range of white male corporate experts, but no musicians, who met weekly to monitor the show's production and content. In the executive group were ATC president George W. Hill, Edward Bernays from Lord and Thomas, a PR representative from Ivy Lee, and an executive from the NBC radio network. Second, Bernays developed the "Lucky Strike Radio Hour Formula," a set of guidelines for selecting music in order to shape a distinctive sound for the show. Third, the company established an unwritten rule that the show would feature only white musicians. It proved relatively easy to exclude blacks from the show, but much more difficult to maintain the emotive energy of jazz while fully purging blackness and its myriad associations from the show's sound.

The executive group met every Saturday morning at ATC headquarters for "dress rehearsals" of the *Lucky Strike Radio Hour*. There, the Lucky Strike Orchestra would play the numbers slated for that evening's show. Lore abounds about these meetings, including that Hill made ATC secretaries dance to the numbers to ensure their rhythmic danceability, and that he brought his nearly deaf aunt, who would tap out the beat with a pencil. If she couldn't hear the beat, it wasn't loud enough.[41] Even the lore does not find a place for actual musicians in making the weekly musical decisions. The real purpose of the "rehearsals" was to standardize the jazz product and avoid a sound that would associate the show with jazz's stigma.

Bernays circulated the principles for shaping the *Lucky Strike Radio Hour* sound, as played by B. A. Rolfe's Lucky Strike Orchestra, to the executive group. The "Dance Formula," he wrote, is "Breast of Chicken a la Rolfe, served hot without any dressing." Then he interpreted the statement:

> "Breast of chicken a la Rolfe": this means choruses and nothing but choruses. "Served hot": This means the characteristic Lucky Strike Dance Orchestra tempo, which is lively, rather fast and is an essential characteristic of the program. "Without any dressing": This means without any frills and furbelows in the matter of arrangements.[42]

While Bernays never explicitly mentioned race in this memo, the formula sought to include but limit aspects of the music that had particular associations with blackness. His rather cute metaphor, "breast of chicken a la Rolfe," associated white meat, and the upper body, with choruses rather than verses. African American dance seemed ribald and primitive to some Euro-Americans because it began at the hips, suggesting the lower body and sexuality, rather than at the shoulders and the upper body, like European dance. By playing choruses of popular songs, the executive group chose their up-tempo payoff parts. Certainly, something was lost by dropping more musically complex verses that built emotive energy in a song, but choruses retained the element of rhythm that got people on their feet.

Likewise, the formula forbade certain kinds of creative instrumentation: "no frills or furbelows." Bernays elaborated that "there shall be no extravagant, bizarre, involved arrangements, no 'pigs squealing under the fence.'"[43] This was a delicate balance to strike. One reviewer of the show praised the "amazing number of unusual instrumental combinations and effects achieved... [including] a musical saw... and a man who can play a bass tuba... [like] a cornet. When the bass tuba artist gets into action on a lively air, he... sounds for all the world as if a huge elephant were cavorting about." Yet, the reviewer approved, there was no "ultra jazz so commonly effected by others."[44] In the Hit Parade, the formula required the orchestra to play a song exactly as it was already known to the public, "without variation, interpolations, or new ideas."[45] In this way, the executive group reined in creative musicality and improvisation in order to create a standardized music product. "Ultra jazz" and squealing pigs did not necessarily reference only African American jazz, but black jazz bands, particularly those that did not package their music for white audiences, did often emphasize hot rhythm and musicianship.

Though the executive group wished to limit creativity in the music, it also explicitly stated that the "hot" aspect of the music should define its signature sound. The formula for *Your Hit Parade*, ATC's successor to the *Lucky Strike Radio Hour*, specified, "pick hot numbers full of rhythm, full of shoulder shake, full of dance."[46] By the 1920s, some whites used "hot" to refer to highly rhythmic kinds of jazz, and associated hotness with their fantasies of blackness, including sexual promiscuity, violence, and general depravity.[47] While many used hotness to refer primarily to rhythm, the executive group associated it with tempo and volume. Each Lucky Strike advertising sound bite on the *Radio Hour* was to be followed by numbers that were "particularly lively and snappy."[48] *Musical Digest* noted disapprovingly that the ATC "sponsored radio programs in which fast and loud playing bands, and rapid-fire speakers, have predominated."[49] *Variety* called *Your Hit Parade* a "big, brassy, breathless show" and declared it "the noisiest show on the air."[50] Thus, the executive group did not wish entirely to drain the music of hotness—the racialized element—but rather to contain creativity and render the product predictable.

Another key technique that the executive group used to preserve hotness while neutralizing risk was to mix old and new songs in its playlist. The slogan of the *Lucky Strike Radio Hour* in 1928 was "The Songs that Made Broadway Broadway." Bernays emphasized that the road to safe musical choices was the well-traveled one: "Not songs that are *making* Broadway Broadway," he reminded the executive group, "but the songs that *made* Broadway Broadway. People like to hear things their ears are attuned to, not new numbers. Songs that have so rung in the public ear that they mean something, recall something, start with a background of pleasant familiarity."[51] Bernays wrote a review of a March 1929 show for the ATC reflecting these standards: "From a musical standpoint, I think the Lucky Strike Radio Hour on Saturday was great.... It showed that the old and the new will mix; it had enough variation in tempo to please everybody from the most conservative to the jazziest dancers."[52]

The executive group paired the explicit *Lucky Strike Radio Hour* formula with a tacit policy of excluding black bands from the show; this policy built on discrimination already well established in live gigs and recordings. White hotels and ballrooms hired white bands to entertain their clientele; far less common was the high-end venue, such as Harlem's Cotton Club, that hired black bands to play to white audiences. The nascent recording industry was, if anything, worse than the live scene for black musicians. Corporate giants Victor and Columbia refused to record African Americans and ignored African American consumers until the 1920s, when each carried a ghettoized list of "race records" for the African American market. Black-owned Black Swan Records formed in 1921 in response to this discrimination, but was hampered by financial challenges and ceased to record in 1923.[53]

The medium of radio at first held great promise in this bleak landscape, as newly formed local stations scrambled to fill hours of airtime each day. Remote hookups made cabaret broadcasting possible by the mid-1920s, which became a way for black bands that had good cabaret gigs to get on local radio.[54] In New York City, WHN had a leasing agreement with Loews Inc., the vaudeville empire that controlled many of the city's clubs. In 1924, WHN began cabaret broadcasting from Loews venues using Western Union telegraph lines. By 1925, the station had remote hookups in more than thirty New York clubs, including the Cotton Club, and regularly broadcast both white and black jazz bands for local radio audiences.[55] It is not clear how much, if anything, stations like WHN paid the bands, but the practice was widely seen as an inexpensive way to fill airtime, particularly late at night. In this way, hot jazz performed by black musicians circulated on local radio in large urban centers.

New York City's more respectable WEAF broadcast the *Lucky Strike Radio Hour* and became the basis for the NBC Red network. With one notable exception, when the *Lucky Strike Radio Hour* used a live feed, it broadcast from an elite hotel playing its usual kind of white band. Such hotels already exerted a conservative force on the kind of jazz that musicians played, but the *Lucky Strike Radio Hour*, and the plethora of cigarette jazz shows that followed, put

Fig. 6.2 Benny Goodman and his orchestra became the *Camel Caravan* house band in 1936. From Tobacco Truth Industry Documents (https://www.industrydocumentslibrary.ucsf.edu/tobacco/).

these shows on the emerging national networks. In 1928, the word "national" in National Broadcasting Company was aspirational; the vast majority of radio listeners that year listened to local stations exclusively. By 1931, however, NBC boasted seventy-six stations and listed an annual profit of more than $2.3 million; by the mid-1930s, radio became a truly nationwide urban medium, though its reach into rural areas was still spotty. By capturing the networked jazz show, cigarette companies exerted considerable control in shaping nationally circulating jazz.[56] In 1936, the *Camel Caravan* hired Benny Goodman's band as a house band, one of the few famous groups that had black members (fig. 6.2).[57] None of the cigarette companies, however, ever hired a black band as its house band, nor did black bands receive invitations to perform as guest bands on the shows.

An exception proves this unwritten rule. The *Lucky Strike Radio Hour* made radio history when the executive group hired the famous Cab Calloway and his band as guests on December 29, 1931, broadcast via a live feed from the Cotton Club. Calloway often played at the Cotton Club when regulars Duke Ellington and His Orchestra were on tour. The African American *Atlanta Daily World* celebrated the appearance saying, "His Highness of Hi-De-Ho . . . will be the first colored orchestra to be featured on an important commercial broadcast."[58]

Calloway was one of the most distinctive African American bandleaders in the country. Known for his big personality, dramatic performance style, and hot jazz, Calloway would later popularize the zoot suit and publish a glossary of "hep" jazz slang. The *Lucky Strike Radio Hour* paid him an impressive $1,500, the same rate that went to white guest bands.[59] His appearance is evidence that the executive group was serious about its desire for hot music.

The executive group must have been disturbed by the public response to the performance because the experiment was never repeated. Indeed, the ATC's failure to hire the even more famous Duke Ellington on one of his regular nights at the Cotton Club—a move that would use already-established contacts—suggests that the Calloway experiment was deemed a failure for the *Lucky Strike Radio Hour*. The executive group hired a black band just one more time: Marion Hardy's band played on the *Lucky Strike Radio Hour* from an eighteen-passenger airplane flying over Manhattan on May 19, 1932, a stunt that flaunted two new technologies: the airplane and the remote radio pickup. However, Hardy was an unknown and his race was never mentioned on the air. The *Atlanta Daily World* informed its readers after the fact, saying, "[You] probably did not know you listened to a Negro band."[60] The newness and dangers of air travel likely diminished competition for the show from white bands. Hardy's performance reveals that blacks could play on the *Lucky Strike Radio Hour* as long as no one knew they were black and the gig endangered their lives.

All this in the service of selling cigarettes. Cigarette company marketers clearly hoped that cigarettes would glean an element of racialized cool from the public scene of jazz without accruing the stigma of blackness. But if the cigarette changed due to this synergy, so did the sound and economics of jazz music, especially in the US, as cigarette companies shaped the kind of jazz that played on national radio and the kind of musicians—white—who played it.

As much as companies tried to capture and control the synergy between jazz and cigarettes, the power of the international jazz dance craze was that it carried no single meaning or message. One's relationship to it depended on who and where one was. As organizations, the cigarette corporations took particular shape through daily encounter, partly through this assemblage that they helped to make. The remainder of this chapter considers how foreign cigarette company employees in China, African American musicians in China, and African American cigarette company employees took up residence in the corporation in relationship to the intimate dance of jazz and cigarettes.

Jazz and Cigarettes for the Foreigners of BAT-China

The cabarets became a key transformative site for BAT employees as they gained new identities as corporate agents in the larger world of global capitalism. Indeed, the cabarets' initial purpose had been precisely imperial; BAT employees joined many other representatives of foreign businesses in rooting their corporate culture in the cabarets. Irwin Smith gained his footing in BAT on the cabaret dance floor. He recalled attending the cabarets with coworkers:

> Well, you know Mr. RH Gregory, he was out there. He always thought a lot of the young fellows.... He was a fine old gentleman and there wasn't anything stuck up about him. He'd take you to the cabarets. He'd buy the tickets and put them on the table and say, "Boys, let's have some fun," and you'd dance all you wanted to. He'd sit there, he wouldn't do any dancing, but he'd sit there and have a lot of fun. He and this Joe Honeycutt used to go cabareting quite a bit.[61]

By hosting the young employees and buying their dance tickets, Gregory made the evenings into corporate events that tied the cabaret's pleasures to the privileges of employment with BAT and fostered a cohesive business culture.

Young BAT foreigners built identifications with a global corporate imaginary in part by dancing at Shanghai venues with taxi dancers from a variety of countries.[62] Hutchinson recalled:

> Along the dimly lighted criss-crossing streets behind the Astor House were dozens of international dance halls and cafés, some with Chinese dance partners, some with Japanese, some with Russian and some with a mixture of the three, including sensual mixed-breeds.[63]

In Hutchinson's erotic imagination, "mixed-breeds" were more "sensual," a view rooted in the socially constructed fears and attractions of racial crossing. The nationality and race of dancers gave a cabaret its character and became a commodity for clients' consumption. Canaday recalled that in 1923 he "went to some dance and cabaret place and danced with some Russian, French, and Greek girls there till about 2am." He also recalled another night when they hopped to various places, drinking beer and "kidding the girls—Koreans—Russians—Lord knows what."⁶⁴ Some taxi dancers enhanced their cabaret pay with sex work after hours.

Canaday's phrase "Lord knows what" to refer to the range of nationalities of taxi dancers aptly expressed the imperial privilege he experienced at the cabaret. The economy of the cabaret prompted and drew on globalized movements of women, erased their stories, and offered them as a leisure commodity for the pleasure of foreign businessmen so that nationality became a kind of flavor in a smorgasbord of global options: Russian, French, Chinese, Korean, Greek, "Lord knows what." Thus did the privileges of the cabaret for BAT employees blend an ability to know and have with an erasure and an unknowing so characteristic of imperialism in general. Foreign BAT employees' public intimacy with dance hostesses gave an erotic charge to their new position as representatives of a transnational corporation in the global city of Shanghai, as they enacted unequal relations on the dance floor.

Like other cabaret clients, BAT employees smoked cigarettes between dances, but in their case they consumed the products they promoted during the day, provided gratis from the company. Smoking the company product could produce a proprietary identification with the company and the product, despite the low-level positions that most of them held. Though this typically went without explicit comment, Canaday noted that when BAT rented a cabaret for a party, his "main amusement [came] from watching [his coworker] Stanley Grey's efforts to thwart the 'squeeze' of too many cigarettes and cigars by the Chinese boys employed to sell the company's products from table to table."⁶⁵ Chinese workers typically increased their meager wages through the moral economy of the "squeeze," a surcharge of cash or products. Grey must have been unsuccessful if his efforts were entertaining enough to distract Canaday from other pleasures. For Chinese workers, cigarettes and cigars were valuable items that they could smoke themselves, use as gifts or barter, or sell for cash.

BAT employees tended to hop from club to club in ways that emphasized male homosocial bonding, strengthening ties within the company and with other foreigners. Frank Canaday recorded a night's festivities in his diary. He began the evening with drinks at the American Club (men only) with a few friends. There they met a few more men and proceeded by car to the Western Tavern in Hongkou, where they heard an "American jazz band" and two more men joined them. "Our party of eight was now having too good a time looking and talking to give any time to the Russian girls across the room who seemed interested. None of us deserted the table to dance." Canaday knew that the expected behavior was male-female dance, and he took pleasure in his group's refusal of that script. At 1:30 AM, all eight took a car to Del Monte. He used a feminine image to convey his group's appearance and spirit of togetherness: "There two more men joined us and we began to look like the front row of a glee club as we occupied a row of tables across one end of the dance floor." Soon a portion of this crowd ("six men still 'with us' when we counted noses in the car") went to Mumm's Café, ate ham and eggs and "took our turns dancing with a Russian girl." At 4:00 AM, Canaday finally went home to get "a few hours sleep before office time."[66]

From the men's-only club to "taking our turns" with a professional dancer, the practice of club hopping opened up a space for public intimacies between men to be as or more intense than the male-female couples on the dance floor. Canaday did not mention it, but Chinese men undoubtedly offered the group cigarettes for sale whenever they were seated at cabaret tables and members were almost certainly smoking as they lounged, talked, and listened to jazz music. In this way, the synergy between jazz and cigarettes in the spaces of cabarets supported a kind of normative corporate bonding that carried possibilities for homoerotic experience in excess of the imperial script.

For BAT representative James N. Joyner of North Carolina, a relationship begun at a Beijing cabaret with Liu Lilin, a Chinese taxi dancer, turned into an ongoing relationship. Joyner arrived in China in 1912 and remained there, unmarried, for twenty-four years. Unlike most foreigners, Joyner did not entirely expunge his sexual relationships from his records.[67] He recorded that the company kept branch offices supplied with condoms (most of the time) and that he contracted a sexually transmitted disease in 1926.[68] He met Liu in 1932, when he was about forty years old and she was thirty-one, certainly at the upper end of the age range of dance hostesses. She had left her husband at age thirty because, as she put it, "he has a bad heart and mortgaged my land."[69] She had a son whom she supported with her earnings, but he did not live with her because "surroundings are bad for him." Joyner enjoyed Liu's company and over two years slowly paid for more of her expenses, even while she kept working at the dance hall, an arrangement that was quite typical.[70] When Beijing closed its cabarets in 1934, Liu became Joyner's full-time mistress.

Joyner expressed his affection for Liu through letters and gifts, including black and colored gauze dresses, red shoes, and a subscription to *Cosmopoli-*

tan Magazine.⁷¹ "I am glad that you are no longer engaged in cabaret work," he wrote, "as such work might at times prove disagreeable," he wrote her.⁷² Joyner brought Liu to his home in Jiujiang for the seven months preceding his return to the United States. When his work postponed her move to Jiujiang, he quoted a romantic song lyric, "My 'journey' still has two long months to go before it ends 'when lovers meet again."⁷³ She later recalled, "When we were at Peiping [Beiping] our love was great. . . . When we were at Kiukiang [Jiujiang], our love was again glorious."⁷⁴

When Joyner left the country, he moved Liu to a boarding house in Shanghai and gave her money for a year's support. He told her if another man came along she should "stick to him" or she should go back to her husband. Joyner gave careful thought to his role in Liu's life and believed that setting her up for a year was an honorable response. He wrote to a Tang Foshu, a BAT coworker:

> I am writing this well into the Pacific headed back home. With regard to my little personal affair, thanks to your kind assistance, everything went smoothly upon my arrival in Shanghai, the young lady was comfortably settled in the hotel when I arrived and later we made arrangement for her removal to a permanent rooming place. With regard to the future — "what man know" — I have the personal satisfaction, in a matter of this kind, to realize that I didn't "duck," but "played the game."⁷⁵

And indeed, absolutely nothing would be expected of Joyner by anyone, except perhaps Liu herself.

As a retired dance hall hostess with an ex-husband, Liu's economic future was bleak, a fact that she did not hesitate to convey to Joyner. She sent him numerous letters, drawing upon different women to take down her words. As to his suggestion she find another man, she wrote, "My dearest, you know I am new to this place and have no relatives; how am I to look for a good man? I am now thirty five years of age and I have only one man whom I really love and that man is you." She speculated about why he had failed to write her—had she spent too much on nice clothing? Had he taken a lover in the United States?—and asked repeatedly when he would return to her. She tried expressing her grief: "I often feel unhappy and wept for my bad luck. . . . I wish I am relieved of all this worry and be suddenly overtaken by Death or a sudden shell to strike me dead." She wished he would send for her: "Do you like me to come to America to see you, if you do, you will remit to me some money and I will come to America immediately." Short of that, however, she hoped he would send money: "I am getting old, have no money and do not know how to earn money. I do not know how I can carry on. My dearest, I hope you will again help me and put me on my feet again."[76] Certainly, her economic distress was genuine. Joyner kept her letters, but left no record of a reply. They offer a rare glimpse of the ways that BAT's foreign business culture's reliance on the imperial social relations of cabarets could play out in dance hostesses lives.

Liu's experience was far from the stereotype of the carefree Chinese modern girl, who attended the cabarets as a consumer, wore the distinctive Chinese dress called a qipao, had bobbed hair, and smoked cigarettes. However, elite dance hostesses in clubs catering to Chinese clientele did sometimes approximate the modern girl image. Some Chinese taxi dancers achieved a starlike status, becoming well known and celebrated in the tabloid press for their beauty, style, and dancing ability. This fame was two-sided: taxi dancers were censured in much of Chinese society as lightning rods for debate about the relationship between foreign companies, changing Chinese cultures, and emerging national sensibilities.[77] Thus, the cabarets were not static places of fixed social relations but shifted with ownership, employees, and clientele. Jazz and cigarettes, however, remained.

African American Jazz Musicians in Shanghai

Though entrepreneurs built Shanghai's cabarets explicitly and exclusively to serve the foreign elite, after only a few years the economy of the cabaret scene shifted, creating new room for Chinese entrepreneurs and, incidentally, African American musicians. The cabaret scene was so hot in the immediate post–World War I era that it quickly became overbuilt and, by the mid-1920s, some owners wished to sell. Two Chinese entrepreneurs, known by their surnames Dong and Feng, became major players in the Shanghai cabaret business. They soon bought several cabarets, including the large and famous Canidrome Ballroom.[78] The Dong-Feng Company contracted with scores of African American musicians and bands to work in Shanghai.[79] Apparently, many foreign-owned cabarets had worked with the booking agency Hamilton House, which booked white performers. S. James Staley, a white musician, claimed that Hamilton

House was "in charge of most of the theatrical, music and entertainment bookings throughout the Far East." Bandleader Whitey Smith recalled that Hongkong and Shanghai Hotels, Ltd controlled top-level hotel booking throughout China. No African American musician mentions these agencies.[80] At the same time, some foreign and Chinese cabaret owners wooed an elite Chinese clientele.

Chicago piano player Teddy Weatherford became a booking agent for Dong and Feng by the early 1930s, building on a pipeline between the African American jazz scene and Shanghai that had been years in the making. African Americans played in Shanghai by 1920. In 1924, drummer Jack Carter traveled to the Philippines, where he joined the Manila Hotel Band, a gig that lasted until the summer of 1925, when the entire band relocated to Shanghai. Billing themselves as the New York Syncopators, the band signed a nine-month contract directly with the Hotel Parisien, soon renamed the Plaza Hotel. In May 1926, Carter returned to the US and hired a new band, including Weatherford on piano and Valeida Snow on trumpet; Snow also sang and danced. In 1927, Snow returned to the US to hire more personnel and in 1928, the band took a tour through Asia. Weatherford, too, returned to hire his own band, and eventually worked with Dong and Feng.[81] When Carter's orchestra performed in Singapore in 1928, the *Straits Times* celebrated that a "real 'colored cabaret' [has] reached Singapore at last."[82]

African American musicians noted that the Shanghai cabaret scene was different in both economy and affect from US elite cabarets or hotels. Buck Clayton recounted the businessmen treating them to a banquet:

> We were met by the bosses of the Canidrome Ballroom, Mr. Tung and Mr. Vong [Dong and Feng]. They were two very rich Chinese, one fat and the other skinny, but both were very nice people. They greeted us with a huge welcoming committee and soon we were at a banquet that was really something else. I didn't know what I was eating and I couldn't eat with chopsticks but I was happy.[83]

Trombone player Happy Johnson recounted that Dong and Feng treated them to "a real Chinese fashionable dinner of about 75 courses. Gee, I never saw so many different foods in all of my life." Dong and Feng might have viewed African

Americans as an inferior race, but by extending business courtesies to the musicians they stood in stark contrast to whites in the US. Under Jim Crow, African Americans were denied courtesies and titles, even in business exchanges. Two years before Jack Carter went to China, for example, a white hotel proprietor in Florida hired Carter's band to play in his white-only ballroom, but assured the white public that he did his best to "segregate the men while they were with me and to make [them] feel like servants." In Shanghai, in contrast, jazz musicians were treated like "the Emperor of China."[84]

Black musicians often described their work in Shanghai as free from racism because they could get high-class gigs and enjoyed wide popularity. Happy Johnson reported, "This was the break of our lives for over here is a real golden opportunity for our profesh. China is the last resort for the colored musicians. We are the talk of the Far East and the Orient, and it seems too good to be true, and it is true." "There is absolutely no color line over there," Johnson told a reporter when stateside to hire more musicians, "all Negroes need to do is produce." Buck Clayton later recalled, "I still say today that the two years I spent in China were the happiest two years of my life. . . . My life seemed to begin in Shanghai. We were recognized for a change and treated with so much respect." Perhaps the strongest testimony comes from musicians who chose to stay in East Asia. Reginald Jones, bass player in Clayton's band, married a Filipina woman and settled in Shanghai. Ernest (Slick) Clark, trombone player, went to Shanghai in 1935 and stayed. Weatherford married an Indian woman and lived in Shanghai and Calcutta until his death. Irene West, manager of the MacKay Twins in Asia, wrote in the *Baltimore Afro-American*, "Do you wonder why so many of your race, who make good abroad, never wish to return to this Jim-crow America! Would you? The Chinese, Filipinos and Indians look upon the colored American with admiration and respect. He is somebody in the lands of color."[85]

Though African American musicians found relief from US racism, Shanghai's foreign settlements were imperial, segregated spaces that imposed their own racist practices. African Americans found themselves barred from entering most public places unless they were working. US marines stationed in Shanghai, US businessmen, and white US musicians all resented the prestige that African American musicians enjoyed. Marines picked fights with bands on more than one occasion by hurling racial slurs. In one case, Clayton lost his gig at the Canidrome because he became embroiled in a fight on the cabaret floor with a marine who objected to his presence on the stage with a white female dance troupe from California. US Southern businessmen, possibly some from BAT, then called for Clayton's band to be fired; Dong and Feng complied. Clayton suspected that white musicians had staged the entire conflict in order to get his band ousted.[86] If Clayton was correct, racial solidarity among US residents of Shanghai reasserted white supremacy in treaty port relations. Shanghai's cabaret economy thus brought particular potential pleasures and dangers for African American musicians. They did not escape the imperial context but took up an ambivalent place within it.

As for African American musicians' experience of jazz dance and cigarettes, the evidence is spotty but compelling. Clearly, as producers of the music, African Americans had a different relationship to the space and to dance than clients, just as BAT employees had a particular experience of cigarettes. Many bands at this time incorporated dance steps into their performance and appreciated and responded to enthusiastic or skilled dancers on the floor, fostering a creative relationship between dancer and musician. Like customers, they formed relationships with taxi dancers, but these relationships carried a unique range of potentials. For some, the ready and inexpensive sex with taxi dancers was an exciting privilege and they responded much as did young white men who worked for BAT. Langston Hughes traveled in Shanghai at this time and recalled Irene West asking him to speak to the young MacKay Twins, a tap dance duo from Los Angeles's Central Avenue, about slowing down in their sexual exploits. According to Hughes, the young men "were both feeling their oats—and sowing them. Between the White Russian women and the Japanese girls, the boys almost never got back to their hotel at night."[87] Buck Clayton recalled his entire band going as a group in rickshaws to the doctor to receive treatment for sexually transmitted diseases they contracted in the treaty port. African American musicians could tap imperial pleasures, despite the fact that those pleasures had not been assembled with them in mind.

At the same time, something happened at the cabarets that made dating relationships possible between African American musicians and taxi dancers. Hughes noted that he went with the band for breakfast at Teddy Weatherford's home in the Chinese part of Shanghai. The location itself was significant: virtually all BAT employees lived in the international settlements and some reported fear of traveling through the Chinese city. African Americans, in contrast, encountered discrimination in the settlements and typically rented places to live in Chinese neighborhoods. At Weatherford's apartment, the band members' wives and girlfriends made a Southern-style breakfast; the diverse group of Russian and Japanese girlfriends or wives followed the instructions of the African American wife who had traveled with one band member from Harlem.[88] The group enjoyed a meal similar to those served at the Gregory home, but with very different relations of production and consumption. African American musicians and taxi-dancers shared a common workplace, which might have enabled them to socialize on more equal terms, leading Hughes to refer to the taxi-dancers as girlfriends rather than as mistresses.

African American musicians do not discuss cigarettes in their archived remembrances, but cigarettes' powers are on display in a posed photograph of Clayton with a rickshaw driver on a street in the International Settlement of Shanghai (fig. 6.3). The rickshaw picture was a genre among colonial travelers to China, and many BAT employees sent such a picture home to family members. It established the Westerner as imperial and leisured in relation to the menial, "primitive" labor of the rickshaw driver.[89] In Clayton's picture he is standing in the foreground, dressed in a fine white suit and is conspicuously holding a cigarette and white derby hat. In contrast, the rickshaw driver is off to the side, dressed in simple pants and shirt, hat, and bamboo shoes. His shirt is open to the waist, revealing his body and indicating his physical labor.

Clayton explained that his own clothes were a benefit of the favorable exchange rate experienced by all Westerners, and acquired specifically for jazz performance:

> As soon as we hit Shanghai we were smothered with tailors who made suits, hand-made suits, for such ridiculous prices that we were ordering them like we were millionaires.... I wouldn't be exaggerating if I said that I believe we had more uniforms than Duke Ellington. We had tuxedos of different colors, we had full dress suits both in black and in gray colors, we had many white suits of different materials.[90]

The cigarette and the suit serve as props from the cabaret that allow Clayton to extend his claim on celebrity onto the public street. The white suit, bought for performance, and the cigarette's central placement mark his special status. The cigarette, held before his chest, helps create a contrast with the Chinese rick-

Fig. 6.3 Buck Clayton and rickshaw driver in the International Settlement. Used by permission of the University of Missouri-Kansas City Libraries, Dr. Kenneth J. LaBudde Department of Special Collections.

shaw runner. While Clayton is at leisure and holding a cigarette, the rickshaw runners' hands are occupied by holding the rickshaw handles. At the same time, Clayton is posed in the International Settlement, the part of the city where he was most likely to face racial discrimination. Clayton's status as jazz musician is rendered here as a resolute claim to imperial power, reaping and representing the benefits of Western privilege in Shanghai. The musicians could end up on either end of power relations, victims of imperial racism in the foreign settlements or privileged as relatively well-off foreigners.

Jazz and Cigarettes in Reidsville and Durham, North Carolina

Just as discrimination in US network radio play, recording deals, and high-end gigs pushed African American musicians to take opportunities in China, so did it prompt even internationally famous musicians to tour the African American South. Cigarette factory workers in Reidsville and Durham, North Carolina, had the money to host them, even during the Depression. Cigarettes were not hit as hard by the Depression as other businesses. Factories ran on a reduced schedule, but many people kept their jobs. As spectacular entertainers, African American musicians offered audience members a way to be hip, modern, and exceptional: African Americans in Reidsville and Durham were listening live to the best and most modern music in the world, and it belonged to them in a special way.

Jazz dances brought globally circulating music and international styles and dances into the context of local Jim Crow segregation practices. While African American musicians in Shanghai could capitalize on the contradictions of colonial modernity, there was no confusion about their status in the US South under Jim Crow. No matter how accomplished and famous the musician, they faced the same rules and risks of segregation as did the cigarette factory workers who went to dance to their music. At the same time, African American musicians could offer models of exuberant modernity in opposition to Jim Crow's daily stigmatizing of blacks as primitive and backward.

The story of jazz dance in Southern cigarette factory towns counters the classic jazz origins story. Historians of jazz acknowledge the roots of jazz in the rural South but locate the creative centers of its development in New Orleans and Northern cities. They link music to the story line of the Great Migration, which celebrates African Americans' "modernization" as they move from the primitive rural South to the modern urban North. Jazz histories invoke tours in the South only to document segregation's indignities. With rare exceptions, the creative element in black music is not attributed to these repeated, almost constant tours, but to the Northern clubs. Meanwhile, small-town Southern African America is portrayed as isolated.[91]

African American cigarette factory workers would have been surprised to hear that they were isolated, particularly when it came to music. In their seventies and eighties when I interviewed them, they enthusiastically recounted the great performers who came through town in the swing era: Duke Ellington, Cab Calloway, Jimmie Lunceford, Louis Armstrong, Pearl Bailey, Billie Holiday, and many others. These were big events that drew people from the surrounding area to the local armory or to a tobacco warehouse. Shows were as frequent as every two weeks. In Durham, the local black newspaper, the *Carolina Times*, advertised the shows. In Reidsville, advertising was by flier in black business districts. James Neal recalled:

> They would attach [fliers] to the telegram poles, or in your restaurants or your cafes they would place them in . . . the front window, so as you pass by you'd see this placard there . . . "Dance, such and such a night." And it would tell you who was going to be there and everything. . . . It might last three hours, four hours, the band get going pretty good they'd start jamming for a long time. They'd give you enough for your money, you know.[92]

Because of the shows, Reidsville and Durham became hubs in North Carolina's black culture.[93]

Bandleaders like Ellington, Calloway, and Lunceford brought full bands to Reidsville's armory and put on showy dance-oriented performances, linking Reidsville to an international circuit of contemporary sounds, dances, and style. Musicians also often danced, sometimes even while continuing to play their instruments. Duke Ellington's trumpeter, Ray "Floorshow" Nance was known for his spectacular dancing. The players in Jimmie Lunceford's band were widely known as the supreme performers of show-band novelties. Arranger Eddie Durham recalled, "They would come out and play a dance routine. The Shim Sham Shimmy was popular then and six of the guys would come down and dance to it—like a tap dance, crossing their feet and sliding." Lunceford's band also did impersonations of other bands, glee-club style songs, and occasionally the four trumpet players would throw their instruments high into the air, catch them, and hit the next note.[94] Performers dressed in elaborate formal attire: flashy suits and sparkling gowns. Ellington was known for always dressing "very sharp and fly," and other musicians had to travel to China to match his extensive wardrobe.[95]

Performers created big stage personas, larger-than-life embodiments of style and coolness. Calloway's unconventional style went over better with African Americans in Reidsville than it seemed to with the *Lucky Strike Radio Hour* executive group. In Ellington's words, Calloway was "the most dynamic personality ever to front a band. He established characters who existed in the realm of dreams, characters who attained their altitude on a curl of smoke."[96] Shows were total experiences, meant to capture the imaginations of audience members and dancers. This was as true in New York City and Shanghai as in Reidsville and Durham, but carried particular potentials in the Southern cigarette towns.

Segregation heightened the access that the black audience in the South had to performers. Reidsville had two hotels in the 1930s, the Piedmont and the Belvedere, but both refused to serve African Americans. Performers stayed in fans' homes, which opened up opportunity for after-parties. These should be seen as creative spaces. Duke Ellington recalled that he composed "In a Sentimental Mood" at such a party:

> It was one of those spontaneous things. It was after a dance at Durham, North Carolina, and they gave me a private party. Something was wrong with it. Two girls weren't speaking to each other. One girl had cut in on the other girl's guy and the other girl kept saying, "Of all the people in the world! That *she* should take my man!" I was sitting at the piano, one girl on each side, and I'm trying to patch 'em up, see? And I said, "Let's do a song" and that was the outcome, and when I finished they kissed and made up.[97]

Though Ellington commanded reverence from New York to Paris, the conditions of segregation brought him into intimate social contexts in the cigarette towns. Black residents could lay special claim to jazz musicians as "our own," just as they claimed black drugstores, schools, newspapers, and baseball teams as "ours" under Jim Crow. Despite his wealth, Ellington remained connected to the daily lives of Southern working-class African Americans.

The timing of the dances heightened the distinctiveness of this segregated experience and linked it to factory life for African American workers. Bands often stopped in Reidsville after playing Saturday nights in Atlanta or Washington. However, it was illegal to dance on Sundays in North Carolina. Shows typically started, as William Davis put it, "after Sunday," at one minute after midnight on Monday. African American Reidsville would dance all night and then limp off to work, exhausted. Mondays were long days, said Davis, but they were carried through by the enjoyment that they had had.[98] When they arrived at their jobs, where whites had better positions and higher pay, employers' notions of value and worth could be countered, at least for a while.

Indeed, at the dances segregation worked to privilege African American participation. Whites could attend, but they had to stand in a small area to the side, behind a rope. They could not dance, but could only look on. Neal recalled that only a handful of whites attended — perhaps twenty — in a crowd of hundreds of African Americans.[99] Every African American person I talked to in Reidsville remembered the dances at the armory, but I could not find one white person who attended or even remembered them. When I asked Marion Snow about them, she patently told me that "there never were dances at the armory." Indeed, it is quite likely that, as a young white woman, she would have had no way to hear of the dances, despite the smallness of the town. Advertised in the black business district, discussed within black circles, and beginning in the middle of the night, the dances could quite easily escape her notice. Snow spent her young womanhood in a town where shows featuring arguably the best and most "modern" music of the era occurred regularly, yet she told me, "There never was much to do in Reidsville for fun."[100]

The dances were formal affairs in the swing era, and built on a connection between dancers and the band as well as opportunities for intimacy between dancers. James Neal recalled that people dressed up for the dances, matching the formality of the performers. "At that period of time, some of the males would wear full tie, I mean regular suit and tie, and the ladies would go in there with their night dress on, you know, beautiful dresses on." Bands played a variety of kinds of songs and dancers responded. Ellington once said that swing music was "the musician and his audience talking things over."[101] Neal saw things much the same way, saying, "So you're dancing out there, one-step, two-step, jitterbug, ballroom, waltz, or whatever, whatever mood that they had set from the stage up there, threw the mood back into the dance hall. So you go with the flow there. It was very beautiful." Wynton Marsalis recalled that "Duke Ellington realized more than any other musician the value of dance rhythms and the value of the combination of jazz and dance. . . . He understood the importance of romance in body movement — the romantic aspect of body movement — to jazz music."[102] Neal particularly enjoyed the slower numbers that offered opportunity for romance:

They all had sentimental pieces, you know, because you with the one you wanted to be with.... You're out there that night to be with them. The female would have her head, like the song says, "lay your head on my shoulder," she'd have her head on your shoulder, she was oblivious to everything around her, whisper sweet nothings into each other's ear, you know. Beautiful days, back then.[103]

For Neal, the dances were about intimacy and grace, about going with the flow.

Ruth Davis, however, went in for the more athletic dancing and reveled in the joy of movement. "I ain't the slow dragging type," she said. She remembered many dances, including truckin', the Big Apple, and the Lindy hop. "And you add all those dances together, got the combinations, and you come out to the jitterbug." Davis danced at home with her twin brother until they perfected their moves: "Well, my brother and I we practiced all the time, and we've always had something to make music, you know. Old Victrola, it started out.... Yeah, we used to love to dance." However, Mr. Davis could not dance. William Davis admitted, "I had sisters who danced. My brothers, we were slow about dance." Mrs. Davis elaborated, "William would get on one foot and stay on it [for I don't know] how long. I had to get him off of that foot!" Mrs. Davis solved this problem by dancing with other men. She recalled learning the sexually suggestive Lindy Hop with her brother. "He threw me all between his legs and grab me all around his back. The first time he did that I got up and beat him all upside the head. I didn't know he was going to do that, you know! We hadn't done that part!" Mrs. Davis maintained her sexual boundaries and only had one problem with a man who held her too close; she stayed away from him in the future and did not let it cramp her style. "We used to go to all of the dances and I don't ever remember going to a dance and just sitting down. I never was what you call a wallflower because I loved to dance and it seemed like the guys loved to dance with me. He didn't care.... I like to move. I danced all the time. But I didn't let nobody throw me between the legs but my brother."[104] People in Reidsville had a multiplicity of experiences at the dances, but shared the events as a whole as one of the finest aspects of their local culture.

Musicians and dancers at the armory may have been segregated, but they created a space that author Ralph Ellison called "mockingly creative." Ellison as a teenager went to dances in Oklahoma City and is one of the only commentators to dwell on the meanings of these Southern tours of black jazz musicians:

And then Ellington and the great orchestra came to town; came with their uniforms, their sophistication, their skills; their golden horns, their flights of controlled and disciplined fantasy; came with their art, their special sound; came with Ivy Anderson and Ethel Waters singing and dazzling the eye with their high-brown beauty and with the richness and bright feminine flair of their costumes, their promising manners. They were news from the great wide world, an example and a goal; and I wish that all those who write so knowledgeably of Negro boys having no masculine figures with whom to identify would consider the long national and international career of Ellington and his band, the thousands of one-night stands played in the black communities of this nation. Where in the white community, in *any* white community, could there have been found images, examples such as these? Who were so worldly, who so elegant, who so mockingly creative? Who so skilled at their given trade and who treated the social limitations paced in their paths with greater disdain?[105]

It is difficult to maintain a view of the black South as "isolated" or "backward" when considering this history. The jazz dances did not transcend the conditions of Jim Crow—not by any means. Instead, they built on its contradictions and provided a context for African Americans, most of whom worked in some capacity for the cigarette corporations and were positioned under Jim Crow as primitive, to create themselves as modern subjects in a powerful way. Such events linked cigarette towns to the wider world and fostered a sense of elsewhere, of possibility.

Conclusion

Examining the public scene of cigarettes in the context of the cabarets and the international jazz dance craze enables a glimpse of some of the power that cigarettes accrued. Within the space of the clubs, cigarettes and jazz dance became linked, participatory, and embodied experiences. Jazz music reverberated with its racialized reputation, infusing the "modern" experience of smoking with a promise of transgression. The heterosocial nature of the clubs, furthermore, supported an emerging trend of women's smoking. The controversial figure of the modern girl was less a real person than a social and political category that emerged only after "girls" emerged as a global labor category and transnational corporations spread advertising far and wide and fueled transformations in urban leisure cultures. Internationally, the figure of the modern girl danced and held a cigarette. Cigarettes' new international aura of modernity, then, had a racialized and gendered element. Jazz left a mark on cigarettes.

The cigarette-sponsored radio shows make clear that cigarettes also left a mark on jazz music. In the US, cigarette company–sponsored radio shows became a very important "whitening" force in the national shaping and distribution of jazz music. Radio shows also spread the synergy between jazz and cigarettes into the homes of urbanites, and eventually even rural residents with their weekly radio dance parties.

Corporations set the conditions for the emergence of these cultural and economic phenomena, but corporate life also took shape within this cultural milieu. The synergy between cigarettes and jazz became part of a diverse but not random set of social formations. For BAT-China, the imperial access to a multiracial candy box of women taxi dancers proved to be a critical experience in the socialization of BAT's young foreign male staff. Little wonder that within Chinese culture, the economic status of taxi dancers stratified, creating an elite class of stars, while dancers themselves became intensely controversial figures, epitomizing Chinese "modernity" for some and imperial exploitation for others. Meanwhile, African American cigarette factory workers in the US South made segregation work in their favor by hosting epic dances with the most revered jazz musicians in the world while they smoked the most popular cigarette in the world. These teeming scenes of life all built personal investments in smoking; they no doubt carried more power than any advertisement.

As Ruby Queen cigarettes boomed in China, and the cigarette became part of a heady new heterosocial leisure culture, BAT also reconstituted its corporate structure. BAT's great success had shifted the terrain and given Zheng Bozhao power to redefine his place as a part and partner to BAT. At the same time, anti-imperialist sentiment throughout the world was on the rise, requiring BAT to legitimate its business in new terms. Chapter 7 explores how BAT remade itself in the interwar period in response to Chinese power both within and outside of the company.

7

Where the Races Meet

In November 1922, Zheng Guanzhu and Frank H. Canaday attended a Harvard-Yale football game together, an event that changed Canaday's life. Zheng was touring BAT's facilities in England and the United States, and had met Canaday just the day before in New York City, at Canaday's job interview with BAT. An enterprising Harvard graduate, Canaday came to his interview with two tickets in his pocket to the upcoming game. At the close of the interview, he offered the extra ticket in turn to each of the two Americans present and, after they declined, he turned with some hesitation and offered the ticket to Zheng, though he had no idea of Zheng's identity or importance. Zheng Guanzhu was the eldest son of Zheng Bozhao, the most important Chinese entrepreneur in BAT, whose selling organization was responsible for the triumph of Ruby Queen Cigarettes. The two young men attended the game together and Canaday received a job offer with BAT-China the following week. Two months later, he arrived by steamship in Shanghai.[1]

From its inception, BAT-China was made through cross-racial encounters, but those relationships took on new possibilities in the interwar period as Chinese businessmen gained more power in BAT. In 1921, Zheng Bozhao and BAT had established a subsidiary corporation named Yongtaihe. Though BAT owned a controlling interest in the new company, Zheng Bozhao retained considerable power because he owned 49 percent of the stock, chaired the board of directors, and served as general manager. The board of directors was split evenly between Zheng's men (Zheng himself, his brother-in-law Huang Yicong, and his eldest son Zheng Guanzhu) and BAT foreigners (Arthur Bassett, Joseph Daniel, and William Morris). The incorporation of Yongtaihe signaled how important Zheng's selling network had become to BAT and how much negotiating power Zheng had accrued. BAT honored this new phase of its collaboration with Zheng Bozhao by sending Zheng Guanzhu and Huang Yicong on a tour of BAT's New York and London facilities, much as it had done for the elder Zheng and Wu Tingsheng a decade earlier.[2] Another new day for BAT-China had begun.

Canaday became BAT's only entry-level foreign employee assigned to Yongtaihe, a position that enabled a more collaborative relationship with Chinese businessmen than foreigners experienced in the BAT parent company. Whereas BAT's main corporate hierarchy ensured that foreign employees answered only to foreign superiors, Canaday had Chinese executives as well as divisional and territory sales managers above him. BAT's system of "messes" provided traveling foreign sales agents respite from Chinese lodging, food, and culture, but Canaday often stayed at the homes of Chinese dealers. Any foreign BAT employee could find himself in an all-Chinese context, but this was typical for Canaday. In these ways, Canaday's job demonstrates that the development of Yongtaihe also required and produced new kinds of foreign corporate employees.[3]

Canaday's career in Yongtaihe offers a window into BAT's reconstitution in the interwar era and the new hybrid forms of business that emerged. BAT's purpose in forming Yongtaihe was to retain Zheng Bozhao's expertise, rather than risk having him start his own business; that is, BAT wished to induce enduring but controllable forms of Chinese commerce that would remain pervious to the continued expansion of US and British corporate capitalism.[4] Yongtaihe emerged as a hybrid of foreign and Chinese business forms and practices in which neither BAT nor Zheng could claim total power.[5]

The contrast between Chinese and African American careers in cigarette corporations is stark. Though the bright leaf network built the industry in both China and the US South, there was no parallel role in the US for African American managers and executives; indeed, the bright leaf network functioned specifically to completely close down such opportunities. Chinese and African American workers and servants filled analogous lower-level positions in the corporate hierarchy in both places, but BAT-China could not function without Chinese managers' knowledge and access to Chinese networks.[6] BAT's dependence was not temporary; in fact, Chinese businessmen increased their power in the company over time.

This chapter explores the development of new hybrid forms of business in Yongtaihe and the differences they made in the way that corporate imperialism functioned. Did rising Chinese power within BAT mitigate corporate imperialism's tendency to reap and sow uneven power gradations? During just this time, President Woodrow Wilson and others touted the idea of mutually beneficial financial stabilization and growth and distanced the US from explicit imperial aims.[7] Did Canaday's cross-racial encounters in Yongtaihe signal new possibilities for more egalitarian economic exchange? Ultimately, BAT maintained principles of racial separation in the hybrid, making it an organized mixture of Chinese and foreign forms, but not a meld that might challenge imperial hierarchies. Corporate imperialism entered a new era that accommodated powerful Chinese businessmen, but still generated uneven, racialized power within the corporation itself and in the wider society. Rising Chinese power within BAT was matched by a widespread increase in Chinese anti-imperialist nationalism. Challenged from within and without, BAT generated new corporate imaginaries in order to defend its actions in foreign relations debates.

The existence and shape of Canaday's position reveals the nature of Yongtaihe as a new corporate hybrid. How well did Canaday adapt himself to Yongtaihe's hybrid business practices and culture? This question reverses the typical gaze: for over a hundred years, people have queried whether and how well Chinese people could adapt themselves to Western modernity, with the presumption that any "Westerner" was already modern. But Canaday had to find his way within a distinctly Chinese version of "modernity" in Yongtaihe—that is, within a hybrid of Chinese and foreign capitalist forms—and in relationship to rising anti-imperialist nationalism. His struggles to do so reveal a great deal about the capacities and limitations of BAT.

Canaday's effort to adapt to Yongtaihe offers a worm's eye view of corporate change, but James A. Thomas, longtime head of BAT-China, waged an analogous struggle with Yongtaihe as BAT's most vocal interlocutor, engaged in highly public debates about corporations' role in foreign relations. In fact, Yongtaihe served as a symbol of BAT-China's simultaneous success and failure. Because of Yongtaihe, BAT sold cigarettes—lots of cigarettes. But BAT-China's dependence on Yongtaihe ran counter to the project and principles of modernity: that the superior, rational business structures of the West could be scaled up without significant change and transposed around the world.[8] The belief was that the rationality of capitalism inhered in bureaucratic hierarchies and bookkeeping methods and that these would rationalize even the most primitive places as surely as a cigarette machine would make cigarettes wherever it was installed. Differences of race, region, gender, and nation were to be tamed and rendered irrelevant or instrumentalized (as in the segmented factory workplace), transforming crops and workers at all levels into parts of the rational corporate machine. One point of this book is that it never really worked that way (even cigarette machines seemed to require more guns to operate in China than in the US). But Yongtaihe posed the biggest threat to the mythos of modernity that justified BAT's very presence in China. Thomas's sales hierarchy, based on the American Tobacco Company's in the U.S., had failed and only Zheng Bozhao's expertise had saved Thomas from embarrassment. The chapter ends with Thomas's attempts to reconcile these elements, rescue modernity, and defend BAT's business practices amid public anti-imperial critique.

Yongtaihe's Power and the Creation of Canaday's Position

When Canaday arrived in Shanghai, the foreign executives of BAT were annoyed, which made for a distressing welcome. As instructed by the New York office, Canaday went to BAT headquarters on his first morning in Shanghai and asked for Arthur Bassett, the principal foreign "advisor" to Yongtaihe. Bassett had hosted Zheng Guanzhu and Huang Yicong on their tour and had likely returned to Shanghai not long before Canaday arrived. Bassett informed Canaday that the company had "not yet decided upon [his] route of induction into the company," and to check back the following day. Canaday received the same message the next day, and the next. After five long days of anxious waiting, "Bassett sent me in to make the acquaintance of the Director of the main BAT company, a man of tough spirit whose welcome consisted mainly of the admonishment, 'If you don't work, you'll get fired!' I was a little taken aback by this harsh greeting at such long distance from home." Canaday later learned that this "had resulted from irritation with the New York office for sending out any further Western personnel at all," as BAT was in the process of replacing foreigners with Chinese sales managers.[9]

In contrast, Zheng Bozhao immediately hosted a Yongtaihe banquet in Canaday's honor. Though Canaday enjoyed the evening, he did not yet understand that he would be working for the subsidiary or even that there was a subsidiary. Years later, he described the evening:

> Taking a ricksha to [Yongtaihe headquarters on] Kiukiang [Jiujiang] Road, I joined my hosts of [Yongtaihe], and in a motor-car with the six top Chinese of the company, sped across the city to the huge Hing Fa Liu [Xing Hua Lou] Restaurant, where I had my introduction to the Chinese cuisine of Shanghai. ... The evening provided me also with my first experience of Chinese courtesies and of Chinese formalities traditionally a part of the Chinese male dinner parties which I was to encounter in all my business contacts throughout the land.[10]

I know of no other BAT foreigner who experienced a Chinese welcoming banquet upon arrival. Though Canaday did not recognize its significance at the time, this meal with the top men of Yongtaihe inducted him into the subsidiary.

How did Canaday's job come to exist in the first place if BAT-China had decided not to train any more foreigners for entry-level sales positions? The answer lies in the power that Zheng and Yongtaihe had accrued within BAT; the evolving role of foreign advisors in the subsidiary structure; and tensions between London, New York, and Shanghai headquarters that Yongtaihe could exploit. More than anything, the fact of Canaday's position reveals that Chinese businessmen had become more than coproducers of the corporation: they had gained the power to demand changes and direct the future.

Zheng's dramatic success at marketing Ruby Queen cigarettes made him ever more important to BAT's overall performance. BAT responded by giving Zheng more autonomy and control of additional brands. In 1919, BAT and Zheng negotiated a new contract that gave Zheng responsibility for Vanity Fair cigarettes and authorized him to expand his sales organization into a structure parallel to BAT's own sales force. James A. Thomas orchestrated the contract from London, as the director in charge of China, in cooperation with the new general manager of BAT-China, Thomas Cobbs, from Virginia. The contract entitled Zheng to assign dealers who reported directly to him and in turn recruited their own subordinate distributors. It also empowered Zheng to work directly with the company's twelve regional sales offices and warehouses in order to coordinate distribution and advertising. The company would make "advertising appropriations as it may from time to time think fit." The greater responsibility and protection for Zheng—the company agreed to cover 75 percent of any losses that he incurred—was matched by close supervision and control over aspects of the business.[11]

Even before the subsidiary formed, then, BAT began fielding two sales operations—promoting nearly identical competing brands—throughout much of the Chinese market. Both sales systems employed Chinese salesmen; the difference was the structure of the hierarchy. Following the ATC's system in the US, BAT's system divided China into divisions and territories, placed foreigners in supervisory positions, and hired Chinese salesmen to cover those territories. Chinese sales representatives enlisted local dealers and retailers, who had to meet BAT's standard requirements, and reported on market conditions, including the activities of competing brands, via a set form. Managers in Shanghai scrutinized these forms to set the most rational strategies and direct resources in the most reasonable directions. This system did not work well in China. Zheng Bozhao's structure tapped and further developed a network of merchants from his Guangdong native place origin and modified requirements for dealers and retailers depending on circumstance.[12]

The formation of the Yongtaihe subsidiary increased Zheng's share of the company's profits. His staff had already been responsible for 25 percent of BAT's sales.[13] Within a few years, Yongtaihe promoted seven cigarette brands and sold one-third of BAT's cigarette traffic. Eventually, Yongtaihe would sell 50–60 percent of BAT's cigarettes and entirely replace BAT's own sales system in the highly profitable sales division surrounding Shanghai.[14] BAT also began working with Zheng on creating a real estate holding company to own the company's properties outside treaty ports that the company had purchased (illegally) through Chinese "dummies."[15] In all, Zheng played a unique, indispensible, and ever-expanding role in BAT's success.

The role of foreign advisors in Zheng's business had evolved with his growing power. The 1919 contract stipulated that the company would "appoint a foreign Associate to assist and advise you in connection with the marketing of 'Ruby Queen' and 'Vanity Fair' Cigarettes." BAT explicitly retained the right to "nominate your foreign Associate," and the associate received his salary from BAT, not Zheng.[16] BAT appointed Arthur Bassett to this role, a job he must have performed ably, for he remained the principal advisor with the 1922 formation of the subsidiary.[17] In addition, three BAT foreigners sat on Yongtaihe's board of directors and participated in determining the subsidiary's development.

In the event that Yongtaihe gained permission to open advertising, distribution, accounting, or production departments, the subsidiary would be obligated to accept an additional foreign advisor for each. Advertising was the first department that Zheng hoped to open.[18] It made sense for Yongtaihe and BAT to have liaisons who understood departmental operations in each company and coordinated between them, but there were plenty of Chinese employees who could perform this work. BAT almost certainly stipulated that foreign advisors hold these roles in order to ensure identification with and loyalty to the foreign leadership of BAT. Once hired, Canaday was intended to advance quickly to be the foreign advisor for Yongtaihe's new advertising department, a situation that Canaday only slowly came to understand.

Canaday's hire appears to have been the result of Zheng Guanzhu pressing Yongtaihe's interests by playing one portion of BAT's foreign leadership off against the others. Zheng, Arthur Bassett, and the New York headquarters of BAT hired Canaday without approval from London or Shanghai. It is difficult otherwise to explain BAT-China's foreign executives' annoyance and lack of preparation at Canaday's arrival. Tension between the nodes in BAT's governance hierarchy helps explain why New York might act without proper consultation. From BAT's inception, US directors and managers held more power throughout the corporation than the British, especially in BAT-China. But the Supreme Court's 1911 dissolution of the American Tobacco Company required the ATC to sell its controlling share of BAT's stock, which made the Imperial Tobacco Company of Britain the major shareholder. Nevertheless, Americans retained influence and scores of positions through World War I because Britain's earlier and more extensive role in the war meant that US management, production, and distribution capacities were essential.[19]

After the war, however, the London directors began to push the company toward British control. This was not easy because New York headquarters continued to manage the essential US export trade in bright leaf tobacco and cigarettes. In addition, the large BAT-China branch had always had an especially strong American character, with American employees throughout. In 1920, BAT asked James A. Thomas to resign from the board of directors. Also in 1920, the British government (possibly at BAT's request) required that British companies in China have British directors.[20] This forced the many American directors of BAT-China to resign from their positions, though some stayed on as managers. Shortly thereafter, BAT-China determined that foreign sales and factory managers would be incrementally replaced by Chinese managers, another policy that affected American more than British representatives.[21] The stream of new US employees slowed and became more confined to the agriculture department. These policies may explain why Bassett and the New York office allowed Yongtaihe to hire an entry-level, American sales representative, despite the explicit new policy against such hires.

It is easy to see why Canaday's hire would be to Yongtaihe's benefit. If another foreigner would be required by BAT for the promised advertising department, Yongtaihe had an interest in securing the hire as soon as possible and in guiding the selection process, something that had not happened before. From BAT-China's perspective, it made far more sense to appoint an experienced foreign employee to Yongtaihe than to bring in someone new. Many recruits could not adapt to the requirements of life in China or the demands of the job.[22] In addition, BAT-China would have more control with a loyal employee who had demonstrated accord with the parent company's viewpoints. Of course, for the very same reason, Yongtaihe might wish to hire someone without such a history.

To understand why Canaday got the job over other applicants, we need to appreciate the significance of the Harvard-Yale football game. Only after being in China for several weeks did Canaday discover exactly who Zheng Guanzhu was and that the football ticket, definitively, had won him the job. He explained to family back home:

> My little stroke of hospitality to the Chinese boy at the Harvard-Yale game seems to have had exceptional results out here. His father has probably made the largest fortune of any Chinese in Shanghai and enjoys unlimited confidence on the part of the BAT Co. He is head, in fact, of a subsidiary organization covering all China, Chinese throughout.... There have previously been just three white men attached to this Chinese organization in an advisory and liaison capacity and I am now in training as a fourth.[23]

As Canaday's phrase "stroke of hospitality" implies, there was some luck at work. Though Canaday offered the ticket hoping it would cast him in a good light, he did not have the cultural context to understand why such a gift would communicate especially powerfully to Zheng Guanzhu.

In China, bringing such a gift to a job interview would be very legible as an expression of respect and deference. Rules of etiquette and gift giving built and sustained social and business networks. Honoring one's business partners or superiors with a compliment or gift acknowledged and conferred status while raising one's own. Business etiquette required myriad polite courtesies, compliments, and gifts as well as the avoidance of conflict or embarrassment.[24] In this light, the football ticket would not seem like a bribe but like an appropriate honor that recognized Zheng's considerable status and reflected well on Canaday's character as well.

In addition, Canaday's invitation to Zheng to attend a public social event may have conveyed deviation from the social codes — and foreign racism — that governed treaty port life. Though Zheng held considerable power, wealth, and opportunity as a part of his father's subsidiary, he worked in a context in which even Chinese elites endured regular insult. In 1922, some leisure spaces in the foreign settlements, such as the racetrack, welcomed an elite Chinese clientele, but gentlemen's clubs and many cabarets and hotels prohibited them. Most notorious was centrally located Jessup Park, which denied all Chinese people entry, except for amahs caring for foreign children. By racially sorting the right to enter, foreigners made Jessup Park into a visible scene of taxonomic violence against all Chinese people; the park's exclusions became a resonant cross-class Chinese symbol of the injustice of extraterritoriality as a whole.[25] The contrast between Shanghai as a site of segregation and insult for Chinese businessmen and a playground with legal impunity for foreigners contributed to a racially charged business atmosphere. Canaday's honorable gift of hosting Zheng at the Harvard-Yale game may have signaled to Zheng that Canaday had an ability to interact with Chinese colleagues on more respectful grounds than some BAT foreigners.

The Zheng family was not anti-imperialist — it benefitted lavishly from its relationship with BAT — but there are indications that the family viewed foreign control with caution, at least in relation to its own autonomy. Though Zheng worked closely with foreign companies, sold a commodity associated with the West, and invested in motion picture theaters, a culture industry associated with the West, he never adopted Western dress.[26] By the 1920s, the Western suit had become standard fare for Chinese businessmen in treaty ports, particularly those who worked with Western companies. Indeed, some hotels allowed Chinese men to be seated in their dining rooms only if they were in Western dress.[27] Zheng wore the Chinese changpao, a long high-collared gown worn over pants. He also lived in a Shanghai-style attached house rather than a Western-style home, and his company's headquarters was not located among the Western skyscrapers downtown.[28] The Zheng family also worked steadily to increase Yongtaihe's autonomy within BAT. Zheng Guanzhu's selection of Canaday for the advisor position may have built on a family strategy within BAT's corporate structure. The conditions of Canaday's hire reveal that Yongtaihe, through both cooperation and struggle with BAT, was emerging as a hybrid of foreign and Chinese business forms, interests, and cultural values.

Once Canaday arrived, Yongtaihe and BAT wrangled over who would train him. BAT wanted the training to occur within its parent structure, much as all foreigners had previously been trained. Yongtaihe, however, explicitly wanted someone who had not gone through, in Canaday's words, the "old orthodox field training within the old company" but would be trained by Yongtaihe, "uncontaminated by any loyalties or practices that he might acquire in the parent organization of BAT."[29] Yongtaihe knew that training included a host of experiences that socialized recruits into a particular company culture and taught them normative practices and skills. In other words, Yongtaihe wanted to control the environment and atmosphere within which Canaday developed and took up his position as a corporate representative.

Yongtaihe won this battle and trained Canaday but BAT successfully interjected a few demands. First, Canaday's training had the same two-stage structure as foreigners' training in BAT: Canaday accompanied a Chinese sales team on a number of day trips in and around Shanghai, and he subsequently accompanied a Chinese traveling sales team on a several-week trip outside of the treaty ports.[30] Second, BAT insisted on oversight near Shanghai. BAT sent the director's secretary, Mr. Wu, to observe Canaday's first several days of training; Yongtaihe put their "number one salesman, Ssu [Si]," in charge.[31] By assigning such highly placed men to supervise conventional sales tasks—meeting with Yongtaihe dealers, distributing handbill advertisements in teahouses—BAT and Yongtaihe signaled the importance of their collaboration in Canaday's training. Finally, Yongtaihe adopted BAT's policy of sending foreigners in pairs on sales trips when the company assigned the more seasoned C. L. Conrady to accompany Canaday on his first extended sales trip. After this training, however, Canaday would be the only foreigner on Yongtaihe sales trips.[32]

By sending Conrady and Canaday with the Chinese sales team, BAT ensured that Canaday's first trip would expose him more to BAT's foreign business culture than to Yongtaihe's Chinese business culture. Their trip mostly followed the Tianjin-Pukou railway. Canaday and Conrady sought out foreign accommodations whenever possible, including from foreign hotels and BAT messes in territory and divisional headquarters. From Jiujiang, Jiangxi, Canaday wrote to his brother, "We usually have managed to find a nucleus of foreigners most everywhere we have been so far and have had mostly our own kind of food." Canaday and Conrady even chose to eat with other foreigners when it meant missing banquets with local cigarette dealers, despite their business function.[33] When the group's route approached a BAT tobacco buying station, the two foreigners "knew we should have a welcome from the [three] Virginians in charge of our company's leaf-collecting station, [so] we could look forward to the evening with assurance and pleasure."[34] When the team neared Qingdao, Canaday and Conrady split off to stay at Stein's Hotel, where they met a foreigner with the BAT parent company and enjoyed German food.[35] Conrady initiated Canaday into BAT's foreign business culture as a buffer against Chinese culture.

Canaday also participated in BAT's business culture whenever he was in Shanghai or Beijing. During his first two weeks in China, Canaday stayed at the Astor House, the hotel that had housed BAT foreigners for two decades, after which he moved to a room in the home of a Southern family in the French Settlement. Canaday explained to his mother: "I can live in familiar surroundings and have a part of each day with American people. It gives me a chance to get used to the Chinese swarm gradually."[36] Describing Chinese people as a "swarm" reflected the US racist discourse against Chinese workers and neighborhoods that had justified Chinese exclusion.[37]

BAT's foreign business culture intersected with the larger imperial treaty port culture, as chapter 6 delineated, and immersed Canaday in an elite and segregated foreign world. Canaday joined the Powhattan Club, a gentlemen's club specifically for BAT foreigners, and the American Club.[38] These exclusively foreign male clubs offered dining, cocktails, lounge spaces, libraries, and special events, such as dances. Chinese men were not allowed unless they were working.[39] Canaday later reflected, "I think what upset me principally here when I first arrived was the impression that I was going to have to spend all my spare time around a club table drinking whiskey sodas."[40] Though this would not be Canaday's only experience, he returned to the club table whenever he was in a treaty port or Beijing.

While cross-racial encounter was pervasive in the BAT parent company, it was less scripted and controlled by foreigners in Yongtaihe. Canaday arrived with the same presumptions of Western superiority as other foreigners but had more opportunity to interact with Chinese colleagues and supervisors as full economic subjects. How would he face these challenges and evolve as a businessman? How would he respond to the May 30th Movement? Could he adapt to China's own "modern" business context?

Canaday's Culture Shock

Like most foreign recruits, Canaday arrived in China with personal hopes of advancement rooted in masculine individualism. With a degree from Harvard, Canaday had a much more elite pedigree than his US colleagues at BAT-China, including his bosses, but he had struggled to launch his career. After his graduation, Canaday had joined the College Training Corps with the intention of serving in World War I, though the war ended before he saw action. He landed a sales job in New York City, but when the firm failed, he moved back to square one. Canaday's older brother owned the United States Advertising Corporation in Toledo, Ohio, but Canaday felt that taking the offered position there "would be accompanied by a certain sense of retreat and defeat." Overshadowed by his brother's success, Canaday wanted to make it on his own, as the ideal of the self-made man demanded. A position with BAT-China seemed to be the transformative job that would give him "a worthwhile experience out of life."[41]

China seemed promising to many young men from the US precisely because they saw the country as primitive, a frontier in need of a cowboy. Canaday was thrilled that his job offered adventurous challenge and rapid advancement. He wrote home, "It's a real man's job—the kind men dream about—full of interest and excitement and change of scene, with enough discomfort and uncertainty to give it zest, and enough difficulties to bring all one's faculties into play." Canaday worried that he lacked the "hunting and sporting life as a youngster" that he felt would have prepared him for his job as salesman, though he had never perceived this as a deficit in his sales positions in the US.[42] Though rural China had been farmed for hundreds of years, foreigners compared it not to the agricultural US but to the North American frontier. James Hutchinson recalled that, upon his arrival in Shanghai, James A. Thomas gave him a pep talk that "paint[ed] a picture of the young pioneer who by grit and dogged determination wins his way to fame and fortune. I tightened my lips and swelled inside like a pouter pigeon."[43] The mythos of modernity contained a promise of individual heroism for the young foreigners of BAT who would bring a modern commodity to naïve, less-evolved China.

Virtually all foreign representatives had a rude awakening when these lofty expectations collided with daily life. Big corporations like BAT could seem incompatible with liberal individualism because they required men to subsume their will and creativity within a large hierarchy. By the time Canaday went to China, men worried that corporate jobs would feminize and render them dependent.[44] In turn, managers posed Euro-American corporate hierarchy as voluntary, that is, as a contract that individuals made in exchange for the competitive chance of advancement and greater independence. Still, the tension between individualistic expectations and the Euro-American corporate structure could be difficult to resolve.

If anything, one would expect the conflict between Canaday's masculine individualism and the corporate structure to be even more strained in the subsidiary. Yongtaihe's business practices reflected both Western management hierarchical structures that would be familiar to Canaday and Confucian-based ordering that would be new and likely opaque to him. While Canaday wanted to escape his family-based business connections, Yongtaihe was both literally and figuratively based on family as a hierarchy. Yongtaihe took some of its structure from Confucian ideas of filial duty and a notion of hierarchy as natural rather than a voluntary contract.[45] The Chinese members of Yongtaihe's board of directors shared familial as well as financial bonds. In a very common Chinese configuration, Zheng Bozhao shared leadership with his brother-in-law and eldest son. Zheng Bozhao referred to himself as Yongtaihe's "family head" and, by hiring primarily through ties based on native place, positioned his large sales force as fictive kin. Zheng's trading company had been strongly rooted in Zheng's native place in Guangdong, a management structure that persisted in the subsidiary.[46] This native place network was not simply a cultural overlay but served management functions, including creating pathways into the company and a cohesive business hierarchy and culture.

Of course, it is important not to essentialize or overdichotomize the role of Western individualism and Confucian filial loyalty in BAT and Yongtaihe. Though the fundamental philosophical assumptions underlying the sense of the individual and the system of law were very different, family and individual experiences were legible and important aspects of business in both Euro-American and Chinese systems.[47] Indeed, BAT-China's reliance on the bright leaf network made fictive kin part of BAT's corporate structure, softening its individualism. Southerners entered the company and found a home within it through the bright leaf network, much as merchants entered Yongtaihe through a Guangdong network that secured their loyalty. In addition, self-interest played a role in both profit-seeking companies. The difference is not one of human essence but of the organizing story that shaped perceptions, experience, and decision-making.

After arriving in China, Canaday found that his experiences repeatedly challenged his presumption that he was a modern, masculine individual in contrast to Chinese effeminate primitives who could not perform technical work. Writing during his training trip, he tried to re-create his surprise in a letter to his brother:

> After thinking of the Chinese workman for years as a soft-slippered laundryman it is a shock in a way to see him doing all the things that white men do at home. Imagine yourself suddenly in a land where a Chinese drives your motor car, where a Chinese is engineer and another fireman on the locomotive which pulls your train, where he is at the wheel of the steam-boat that carries you up through the most difficult stretch of navigation in the world, where he collects your tickets, takes your dictation, runs your office, collects your accounts, contracts for the building of your house, tailors your clothes, makes your furniture, sells your products, acts as a clerk in your bank, is in charge of your telegraph station, is your telephone operator.[48]

This remarkable list reveals just how unprepared Canaday was for China, precisely because of the ways that Chinese men were racialized in the US after four decades of Chinese exclusion. Canaday had absorbed an entrenched sinophobic discourse that held Chinese men to be failed as men: incapable of professional or technical labor and devoid of masculine individualism. When Canaday saw Chinese men confined to very few job categories in the United States, he saw inability rather than discrimination.

As Canaday traveled and trained with Yongtaihe, he found he had to rework his sense of racial difference. This was an experiential as much as an intellectual process. Adapting in China required that Canaday remake himself in relationship to his new companions. We call this process culture shock because of the visceral displacement that is created until the strange can be made more familiar and the sense of self adjusts. In particular, since Chinese men did "all the things that white men do at home," Canaday's series of revelations about Chinese men also destabilized Canaday's sense of his own role in China. Why was he needed? Who was he in relationship to these Chinese men who seemed so startlingly like white men? In other words, Canaday began to actively, and sometimes self-consciously, take up his position in direct relationship to his shifting racialization of Chinese people, especially his colleagues in Yongtaihe.

Canaday and other Western employees experienced particular dissonance on sales trips because they were surprised to discover that Chinese men performed all of the skilled labor while they completed menial tasks. "We were called salesmen," James Hutchinson recalled, "But, actually, we did no selling. The large majority of the foreigners in the company spoke no Chinese; interpreters and dealers took care of that end."[49] In addition to lacking language skills, foreigners did not have the cultural position or knowledge to navigate sales transactions. Instead, foreigners assisted unskilled Chinese laborers— whom they called "coolies"—with advertising tasks by distributing leaflets and cigarette samples, running a public cigarette stand, staging a parade, or putting up advertising posters. "I was doing work fit for the intelligence of a ten year old child," Hutchinson complained.[50] The job did not match foreigners' expectations of what modern global corporate representatives did in China, generating frustration and embarrassment.

While they adjusted to their unexpected roles in China, foreign employees worked out their dissonance in varying ways. Hutchinson and another BAT foreigner, for example, went in the evenings to "a Chinese girl house" and ordered Tsingtao beer in the courtyard. "Drinking toasts to each other, we ran down the company, particularly the Shanghai office and its off-hand ways and hidden motives, [and] the Chinese and their rotten habits. . . . [Then] we divided the empty bottles . . . and with savage satisfaction, crashed them one by one against the flower design painted on the wooden screen." Completing the inversion of his expectations, Hutchinson played out "savage" impulses against the symbol of "civilized" life, the Chinese flowered screen. Hutchinson emphasized that he smashed beer bottles specifically because of his assigned work. He wrote, "I was all in and depressed. So this is what I had come to China for. . . . For ten days I spent the whole of each spoiling walls [by putting up posters] and giving out sample cigarettes and handbills. There were nights when I smashed beer bottles with great gusto and satisfaction." Hutchinson slowly adjusted. Though he remained "not quite satisfied," he determined that a job in China was worth the "dirty work" he was asked to perform.[51]

Canaday did not smash beer bottles but rather used his pocket Kodak camera to work out the dissonance he experienced between his expected and actual role. Canaday took many candid and scenic photographs in China, but his posed shots are especially revealing of his mindset. Canaday occasionally fashioned scenarios for the camera that involved Chinese people, thereby discovering a creative way of interacting with his surroundings. These posed shots were far from a reflection of reality, but rather were Canaday's attempts to interact with and comment on his experiences. Though self-conscious, like all photographs they may communicate beyond the photographer's intentions. When he returned to the US, Canaday mounted these photographs in a scrapbook and added captions that offer further clues to how he wished others to view these highly intentional shots.

The dissonance between Canaday's expected and actual role became visible in a photograph that he took in October of his first year. The sales team was traveling by mule cart to Yucheng, a market town in Shandong Province. Since mules move slowly, "I walked most of the way," Canaday recorded in his diary, "giving out samples to countrymen and trying to talk with some of them with more or less success."[52] Canaday gave a cigarette sample to one farmer and then stepped back and took a photograph of him holding it (fig. 7.1).

This contrived photograph reveals nothing about Yongtaihe's selling or Chinese farmers' smoking practices, but it does capture Western fantasies of modernity. The photograph certainly distorts reality if taken as a record of Yongtaihe's sales techniques. Following the method common to BAT-China and Yongtaihe, the sales team had designed their trip to intersect with harvest festivals and other marketing days when farmers came to town from nearby small villages and the surrounding countryside; there was just such an upcoming festival in Yucheng. There, the team would resupply Yongtaihe's established retailers, put up advertising posters, run a cigarette stand, stage a parade, and hand out samples.[53] The sales team did not attempt to meet farmers in their fields; such a practice would be ridiculously inefficient. Canaday was simply entertaining himself when he took the farmer's photograph.

Though he had to bend events to do so, Canaday staged the iconic scenario at the heart of modernity's mythos: the moment when a Westerner introduced

Fig. 7.1 Photo by Frank H. Canaday. Courtesy Harvard Yenching Library.

a Chinese primitive to the Western commodity. When Canaday returned to the United States, he mounted this photograph in a scrapbook and gave it a caption that explicitly framed the picture as the moment of encounter in just this way: "A farmer doing a little fall plowing with a one-handled plow. Also trying a cigarette for the first time. He hasn't any idea of what is happening ti [sic] him." Canaday's caption drew viewers' attention first to the farmer's "one-handled plow," which would appear primitive in the US. Canaday's caption also declared that the farmer's cigarette was his first. Cigarettes had been available in the Yucheng area for well over a decade, however, so it is certainly possible that the farmer had smoked one — or many — previously. By concluding that the farmer "hasn't any idea of what is happening ti [sic] him," Canaday also implied that the photograph marked the moment that the farmer first contacted the forces of modernity and international capitalism, embodied in the commodity, and entered history.[54]

Ironically, when Canaday staged this photograph, he still did not entirely understand his own position within Yongtaihe. It would be another month before he fully understood that he was to manage a new advertising department. Like the farmer, Canaday participated in large processes of international capitalism, only some of which were fully known to him. By staging and photographing this mythic scene of a first encounter with a Western product, Canaday cast himself as the modern global businessman, the role he had expected to play in China.

While Canaday constructed a figure of the naïve Chinese consumer in this photograph, in others he worked out his relationship to Chinese laborers, or "coolies." In two pictures, taken the same day and location, Canaday convinced laborers to let him pose doing their work and enlisted someone to photograph him. In the first, Canaday carried a pole with a bulky load hanging from each end (fig. 7.2). In the second, Canaday posed on the same path with a harness used to pull a boat secured across his torso, its lines leading back to the boat (fig. 7.3). The "coolie" is behind him waiting to take his job back. These pictures are jokes, evidenced by Canaday's broad grin in both. Considering that Canaday was working with laborers on sales trips, these staged scenes draw their comic power from status anxiety and the transgression of category crossing. Posing as a "coolie" was meant to provoke laughter — clearly he could not do this labor — that would reinforce and naturalize the difference between himself and Chinese laborers. As chapter 4 showed, many people in the US believed that Chinese bodies differed inherently from white American men's bodies, allowing Chinese men to work exceptionally long hours at physically demanding jobs while requiring little food or rest. White men, in contrast, supposedly had lower physical ability but higher intellect and cultivated habits. Canaday's photographs staged larger imperial narratives in order to address his dislocation and reestablish his difference from, and superiority to, Chinese men.

Canaday's Development within the Chinese Subsidiary

Canaday did work hard to adapt to his surroundings, however, even as he worked out his discomforts. Within Yongtaihe, he had few foreign colleagues and much

Fig. 7.2 Photo by Frank H. Canaday. Courtesy Harvard Yenching Library.

opportunity to build business connections with Chinese people. After his training period, Canaday ceased avoiding Chinese business culture. When he traveled through Shandong, for example, he stayed for over a week at the large compound of a cigarette dealer named Mr. Wang, despite being in the vicinity of a BAT mess.[55] He described Wang as "a man of cultivated habits who commands respect and friendly feeling."[56] While a guest at Wang's home, Canaday was the

Fig. 7.3 Photo by Frank H. Canaday. Courtesy Harvard Yenching Library.

only foreigner present at nightly dinners and a trip to the theater. Canaday wrote home that "my days with the Chinese are full of the most rigorous social training.... I am kept rigidly on the alert all my waking hours to omit none of the innumerable forms of common Chinese courtesy."[57] Canaday also embraced his work with a language tutor, as did many BAT employees.

Specifically, Canaday was mastering rules of etiquette necessary for business meetings and banquets. Any business meeting in China began with tea and either a pipe or a cigarette. Canaday reported home that

> everything must be offered—and accepted—with both hands—even a cup of tea, or a cigarette. Everything proffered must be refused once or twice before it is accepted. And consequently, in offering anything the first and second refusal must never be noticed.... When the host [pours tea] himself one must rise

and bow and protest and place his right hand in a certain way alongside his cup signifying thanks.

Banquets additionally required discerning hierarchies of guests and the host. Canaday learned that guests vied to allow others to enter the dining area before them: "I've seen the dispute last for minutes on end. There is a proper place to stand on entering a room. The chairs in the room have a certain order of precedence."[58] In May, Canaday believed he had mastered Chinese manners pretty well; by October he had learned enough to be aware that "there are innumerable ways to miss the chance to be courteous that I *know*, and Heaven knows how many I am ignorant of."[59]

Yongtaihe offered Canaday a unique opportunity to interact with Chinese businessmen as colleagues and supervisors—that is, as full economic subjects—because he answered to both foreign and Chinese superiors. He wrote home that his "number one" (using the Chinese term for boss) was Arthur Bassett, but he also interacted with Chinese men as collaborators and authorities. For example, Canaday worked with H. T. Chen, the Jining territory manager, to develop a promotion plan for Ruby Queen cigarettes in Shandong Province, as the brand had lost sales in the last year or so. Canaday praised Chen to Bassett, writing, "Mr. Chen who has just taken over the organization of this territory for us is a man that I can work with very effectively and smoothly. He is interested and thorough, speaks English very well, and has friends in the right quarters to get things done." Chen and Canaday sent their proposal to Division Manager C. Y. Yi, requesting his revisions, and the proposal circulated to Zheng Bozhao and Huang Yicong as well as to Bassett and the BAT parent company for executive approval.[60] The same day he sent in the proposal, Canaday wrote home that the country he was encountering was "not the China that we know at home even when we study about it carefully."[61] His idea of China destabilized, Canaday held mixed and sometimes contrary investments in racial difference and cross-racial recognition.

Canaday took three extraordinary posed photographs of Chinese men smoking cigarettes that suggest pleasurable collaboration more than do his other posed photographs. These photographs are quite unique, first and foremost, because the subjects are smiling. Like people in the US at this time, Chinese people did not typically smile for photographs. In addition, the body

Fig. 7.4 Photo by Frank H. Canaday. Courtesy Harvard Yenching Library.

language in the photographs suggests an intimacy and comfort between photographer and subject: these are not wooden smiles or stiff postures. Why were these men interacting with Canaday in this way?

Canaday took these photographs one day in 1924 at a Tianjin market where Yongtaihe was selling cigarettes. Though Canaday does not identify his subjects by name, they were almost certainly Yongtaihe salesmen or cigarette retailers who were known to him. All three shots are clearly posed with the central focus on the cigarette, like advertisements. In figures 7.4 and 7.5, the men hold the cigarette in the Western style, between the first two fingers; this is notable be-

Fig. 7.5 Photo by Frank H. Canaday. Canaday's caption: "Always happy regardless of weather if he has a 'yuan juar' from the big factory. Behind him on the wall is a poster advertising 'Shuang Shih' or double ten cigarette, put out by the BAT to commemorate the tenth day of the tenth month, the national holiday dating from the founding of the Chinese Republic, so-called." Courtesy Harvard Yenching Library.

cause Chinese people tended to hold cigarettes between first finger and thumb, so that the palm would slide under the chin when inhaling, though advertisements often featured Chinese figures holding the cigarette between the first two fingers. In figure 7.4, the subject is standing with the cigarette very near his face, head tilted slightly back with a smile of satisfaction on his face. Canaday later captioned this photograph "Ah!" In figure 7.5, the subject stands with a large umbrella, the cigarette held near his head, and again, a large smile on his face. This man stands in front of advertising posters for Shuang Shih cigarettes, a Yongtaihe brand that attempted to cash in on Chinese nationalism, as it was named after October 10, the date of the founding of the Chinese Republic in 1912. In figure 7.6 the same man is sitting, still smiling, with a cigarette coming out of his mouth, his hands in his lap.

I puzzled over these photographs for a long time before ascertaining their likely conditions of production. My first clue was that Canaday had been looking for a commercial use for his camera: he had the idea to offer consumers who purchased a carton of Victory cigarettes a photo portrait of themselves. Canaday wrote copy for a possible poster to be translated into Chinese: "Return Home With Victory Buy 1 Carton Excellent Cigarettes—Get Your Own Photograph Free—A Good Picture of a Good man who smokes a Good Cigarette."[62] There is no evidence that such a scheme was approved, but it is possible that Canaday was on the lookout for other ways to use his camera in service of cigarette promotion.

The second clue is a Ruby Queen cigarette advertisement that appeared in newspapers and that Yongtaihe distributed as a handbill at markets, likely including the one at Tianjin (fig. 7.7). Twenty-two circles containing head-and-shoulder drawings of smokers surround a prominent pack of Ruby Queen cigarettes. The nineteen men and three women are presented to represent a variety of types, with a caption that reads, "Da Ying Cigarettes: the most popular in the country." Significantly, each person is smiling, to accompany the other caption that reads, "Anyone will feel cheerful after trying Da Ying." One man has his head tilted back, much like the subject in figure 7.4. The men all have the cigarette sticking directly out of their mouths with hands out of view, like figure 7.6; the women hold the cigarette between the first two fingers. The methods of holding the cigarette near, or in, the smiling faces, and even the tilt of the head, work to associate cigarettes with satisfaction in a visual vocabulary duplicated by the subjects in Canaday's photographs.

Another happy convert with a
B.A.T. cigarette.

Fig. 7.6 Courtesy Harvard Yenching Library.

Canaday and his subjects, then, appear to be enacting the advertising slogan "Anyone will feel cheerful after trying Da Ying." Perhaps Canaday had an idea to use these or similar photographs in advertising materials, though I know of no such materials. Alternatively, the men may have been passing the time. Either way, like figure 7.1, the shots distort history if taken as reflections of Chinese smoking habits. Rather, they are the product of the particular commercial and personal interests that Canaday and his subjects brought to the photographic event. What is most striking about these photographs is precisely how collabo-

Fig. 7.7 Handbill advertisement, Courtesy Harvard Yenching Library.

rative they seem: the men photographed seem genuinely amused and theatrical, especially in figures 7.4 and 7.5. The relationship between photographer and subjects seems relaxed and playful. Unlike the photograph-jokes in which Canaday performs "coolie" labor, here the joke is shared. At least on some days, Canaday was finding his way in Yongtaihe.

Canaday rose rapidly, as intended, and soon "[looked] after the Yongtaihe-

Corporation Advertising for the whole of China."[63] His job was living up to his hope for both adventure and individual advancement. He also wrote daily with a self-conscious plan to publish accounts of his China encounters in the future. Canaday spent 1924 setting up "his" advertising department, though it turned out not to be as independent of BAT's oversight as promised.[64] Canaday confronted a number of challenges, including political disruptions due to competing warlords, but he continued to find the work satisfying and productive. In 1925, Yongtaihe sent him to Beijing for a six-month intensive language program in preparation for more authority in the subsidiary. He wrote home, "Conditions in China.... couldn't be much worse and we did the largest month's business in our history in December. Here in Peking... our own companies now have 94 percent of the business in this vicinity."[65] To Canaday, the cigarette business seemed immune to the mounting political turmoil in China. "Everything hinders our business activities abominably and yet we go merrily on, building new factories and making millions of profits for somebody."[66] He wrote to a friend that he was considering settling in China permanently. "Don't tell my mother," he added.[67] At just this optimistic moment, he received a letter from his brother Ward reiterating the offer of a position in his advertising company in Toledo.

Canaday turned down his brother's offer in terms that revealed his own developing hybridity within the subsidiary. Canaday relished pointing out that Ward had failed to treat him "as a serious factor in business," and instead saw him "merely as a young brother to be tucked in somewhere and given a chance. ... you've read enough Emerson also to know why I can't 'come in' with you on that basis." Canaday was referring to Ralph Waldo Emerson's essay, "Self Reliance," masculine individualism's most famous treatise. However, he immediately continued by praising the quite different Chinese concept of "face," saying, "The Chinese try to handle all of their affairs so that no one in the deal 'loses face,' i.e. every One's [sic] personality is respected and his self-esteem is preserved."[68] He concluded by professing loyalty to Bassett and bragging about how his salary was promised to rise.

By linking self-reliance and face, Canaday posed an intriguing unification of nearly opposite concepts. For Emerson, the masculine individual must deny community. "Whoso would be a man," he wrote, "must be a nonconformist." It is easy to see why Canaday's conflict with his brother might draw him to Emerson's phrase, "I shun father and mother and wife and brother when my genius calls me," but such an idea would be offensive, if legible at all, in Confucian-influenced China, where filial duty ordered social life. Father, mother, wife, and brother were each due very particular forms of honor and respect. "Insist on yourself," insisted Emerson, "[and] disdain ... to conciliate."[69] Confucian-influenced business culture, in contrast, highly valued conciliation. One tempered one's genius and avoided truths or events that might be confrontational or embarrassing. One gained standing or face not by promoting oneself but by honoring others, for example, through a compliment or a gift or good manners.[70]

Of course, Canaday's understanding of both concepts may have been sketchy; nevertheless, the hybrid ideal he described of maintaining autonomy and dignity in business does make sense given his experience. To make it work, he united the concepts through their common elevation of masculine honor and imported the twentieth-century concepts of personality and self-esteem, which allow for a more positive sense of the individual within a group than does Emerson's radical individualism. Canaday's hybrid ideal bridged his upbringing and his present context; if nothing else, it indicates that he had found a satisfying place for himself in Yongtaihe.

Anti-Imperialist Nationalism and Beijing's Foreign Diplomat Culture

The May 30th Movement interrupted Canaday's progress, however, and ultimately led to him to abandon his remarkable job in China. He was studying Chinese in Beijing when protest erupted. Protesters charged that Yongtaihe was a partly foreign-owned company and successfully pressured merchants to drop its goods and officials to obstruct its activities. In Beijing, officials shut down the Yongtaihe branch office and arrested its Chinese manager, a sudden enforcement of a 1916 municipal law that prohibited foreign businesses from operating in the city. Canaday warned BAT that to acquiesce would "be a dangerous acknowledgment which might serve to identify us in all China as other than a purely Chinese company."[71] Within a couple weeks, Yongtaihe negotiated a compromise that reopened the Beijing office as a distributor, but not a branch, and allowed Yongtaihe to present itself as a Chinese company, revealing how central this subterfuge was to BAT's motivation in creating the subsidiary.[72] Canaday also tried to resolve the continually developing obstructions to cigarette distribution in the Beijing area. At the end of this frustrating period, Canaday broke his contract and returned to the US.

BAT-China and Yongtaihe would eventually recover from the May 30th Movement, but Canaday could no longer envision his future in China. Shortly before he left, he wrote home, casting nationalist China as a sleeping giant that was waking:

> I seem always to end up with empty hands. It is the same now—all the adventure and successes of this China career, apparently to come to an end in a month or a year with chaos and the wreck of good prospects. For if China rouses up and shakes herself, business twenty years in the making falls to the dust and with it the hopes of advancement and increased rewards of many employees. I am frankly pessimistic about the situation here.[73]

Canaday was correct that anti-imperialist nationalism would greatly affect BAT's future, but he was wrong about timing. In fact, the cigarette corporations regained access to markets right around the time that Canaday left. In addition, Yongtaihe offered him a new contract for a position as director of the Yongtaihe Advertising Department that gave him more autonomy from the BAT parent company than before.[74] Since Yongtaihe and BAT achieved their greatest success over the next decade, despite continued political unrest, Canaday presumably could have advanced his career had he stayed. Instead, in what must have felt like an indignity, he wrote to his brother and begged for the job he had long refused.[75] It was a disappointing end to Canaday's career in China.

Canaday's abrupt about-face makes more sense when considering his six-month language study in Beijing. Though his previous idea of China had been challenged within Yongtaihe, Beijing's foreign culture, heavily influenced by diplomats as well as financial and corporate representatives, steeped him in an elite diplomatic discourse about the follies of Chinese nationalism and the evolutionary backwardness of China. Canaday returned to drinks around the club table and also attended a festive round of balls and dinner parties. "I've been having the time of my life here in Peking this month," he wrote home, "and making up for all the dull evenings in the mud shacks of the interior. There are lots of nice people here and I'm meeting all of them."[76] Though Canaday would not move in elite diplomatic circles in the US, in Beijing the foreign culture included mid-level corporate representatives like himself. At exclusive balls, he also observed Chinese diplomats such as V. K. Wellington Koo, who had represented China at the Paris Peace Conference.[77]

From Beijing, Canaday newly dwelled in his letters on a sharp distinction between elite and nonelite Chinese people. After one ball, he wrote that some Chinese people there were "beautifully dressed in the semi-foreign style, so smart and intelligent and lovely in appearance that one did not think of them as Chinese or as of any special nationality, but simply as rare and interesting types of humanity's highest development." Having preserved the idea of human development arranged in vertical levels, Canaday conjured a new apex of civilization: "I think that at the top all races merge; it is only at the lower levels of development that the differences and the battling goes on.... The highest human qualities are essentially the same the small world over and are usually recognized by the highest type of people in every land."[78]

At the same time, Canaday demonstrated a renewed interest in rehearsing a notion of China as backward and in need of Western tutelage—that is, a narrative of modernity—all despite his firsthand experience of the proficiency of men in Yongtaihe. In an essay written at the close of his six months in Beijing, and just before the May 30th Movement, Canaday posed the unoriginal idea that Chinese people were at an evolutionary stage equivalent to medieval Europe. "Middle Ages, Still, in the Middle Kingdom" argued that the United States and Europe had made steady progress over the past several centuries, but China, "with far more of civilization's beginnings to build upon than we had in the West, has lapsed into slothful degradation and dawdled along endlessly in material filth and confusion of mind." The essay recounted Chinese superstitions, the reliance on medical "quacks and sorcerers" and the "disgusting" lack of hygiene.[79]

In its derogatory tone and faithfulness to formula, the essay is unremarkable; what deserves attention is the erasure it performs. The essay took the form of a conversation between two foreigners traveling together through "ancient" Shandong Province, accompanied by Chinese servants—not by Chinese businessmen, which is how Canaday himself had traveled through Shandong on multiple occasions. The essay's servants moved in the background, managing primitive cooking and transportation technologies and taking directions from the foreigners, who knew where to go and how to get there.[80] Reading this essay in isolation from Canaday's other documents, one might imagine that Canaday had had no opportunity to get to know Chinese people other than servants and that his racism could be explained partly by ignorance. Clearly, we need a better explanation.

In Beijing, Canaday encountered a foreign culture that melded the commitment to segregation evidenced by foreign businesses with a foreign diplomat discourse of Chinese lack of fitness for self-rule. Canaday likely encountered such ideas about Chinese governance in Shanghai as well, but in Beijing he associated with people who were empowered to make public statements and influence decisions about China's future. Canaday's newly drawn distinction between elite and other Chinese people mirrored Beijing's public diplomatic culture as well as the US Chinese exclusion policy, negotiated by US and Chinese diplomats, that made exemptions for students, businessmen, and other elites. The policy's corresponding treaty with China, as recounted in chapter 1, was a source of ongoing strain in US-China business and diplomatic relations; the exemptions were necessary for the relations to continue.[81]

In 1925, Beijing was particularly tense as diplomats held together tenuous international alliances while long-brewing conflicts reached a boiling point. In the wake of World War I, Chinese diplomats challenged imperial treaties and made claims to sovereignty, especially insisting that parts of the Shandong Peninsula be returned to Chinese jurisdiction. Areas of the Shandong Peninsula had been a concession controlled by Germany, but during the war, Japan took Germany's position, accelerating its own claims as an imperial power. China refused to sign the Treaty of Versailles because that treaty did not return Shandong to Chinese control. In 1919, Chinese students and others launched the May 4th Movement in protest. The Washington Conference of 1921–22 did nominally return jurisdiction of the peninsula to China, but left many Japanese investments across Shandong untouched, essentially reaffirming Japan's place in a multinational imperial alliance in China. Despite the US's profession of Wilsonian principles of self-determination, then, US diplomats had no intention in fully supporting Chinese sovereignty, but rather wished to mollify China while perpetuating imperial formations. Beijing's foreign diplomatic community professed support for Chinese sovereignty while rehearsing arguments for continued imperial privileges. The May 30th Movement demonstrated just how potent anti-imperial resentments had become in China and destroyed the fragile accord achieved at the Washington Conference.[82]

When the May 30th Movement disrupted Canaday's life, he had six months and counting of immersion in Beijing's foreign culture to help him make sense of the upheaval. His letters and the "Middle Ages" essay demonstrate that he had become well versed in a highly racialized view of China, an idea of China that differed both from the one that he had arrived with two-and-a-half years previously and from the idea of China he had begun to develop in Yongtaihe. By making exemptions for Chinese elites, Canaday could maintain a view of China as inherently mired in sloth and backwardness, despite his own experience. Canaday's interest in Chinese business culture and his respect for his Chinese superiors had become incompatible with modernity's exacting narrative requirements.

Canaday's story in Yongtaihe reveals that the subsidiary represented a significant restructuring of BAT but that the resulting hybrid was not a meld that abrogated imperial hierarchies, as Wilsonian development advocates would suggest. Rather, it created a parallel selling structure that reinforced divisions between Chinese and foreign, and this separation remained fundamental to BAT's hierarchy. Canaday had more opportunity than other foreigners to forge different terms for his relationships to Chinese businessmen, and he responded to these opportunities by revising his ideas about China. Nevertheless, the imperial conditions of uneven power pervaded BAT-China, Shanghai's international settlements, and the foreign community of Beijing. Ultimately, Canaday's tentative changes proved no match for the challenges of the May 30th Movement, interpreted via Beijing's foreign culture, and he adopted a view that was, if anything, more harshly sinophobic than the one he arrived with a few years earlier. Though Canaday was just one person, his experiences indicate just how vexed were the conditions for cross-racial business encounters.

Where the Races Meet

As anti-imperial nationalist movements waxed across Asia—and in China and India especially—the nature and consequences of cross-racial interactions in foreign businesses became central to public foreign policy debates, but at one level removed from the story told here. No one—clearly not even Canaday—presented to the US or European public a careful rendition of what such encounters actually entailed within BAT. Rather, highly placed interlocutors injected strategic and competing imaginaries of such encounters into public circulation, and it was these, not the actual encounters themselves, that shaped foreign policy debates. The global circulation of Marxist ideas of capitalist exploitation influenced the May 30th protests and strikes and linked unrest in China to struggles around the world. This phenomenon strengthened anti-imperial critique in the US and put companies like BAT on the defensive, having to demonstrate they were neither racist nor exploitative.

In the midst of the May 30th Movement, journalist Nathaniel Peffer and former BAT-China head James A. Thomas squared off in a debate about the impact business relationships had on foreign relations. The Quaker pacifist organization, the Fellowship of Reconciliation, hosted the contest, with Thomas and Peffer arguing diametrically opposed positions: Thomas spoke on "Economic Expansion in the Orient: A Factor for Peace and Stability," while Peffer spoke on "Economic Expansion in the Orient: A Seed-bed of War." The two men continued to counter each other in print for several years. In 1927, Peffer published *The White Man's Dilemma: Climax of the Age of Imperialism*, with a central chapter about business titled "Where the Races Meet." Thomas responded with two memoirs that countered Peffer's points without mentioning him by name. *A Pioneer Merchant in the Orient* appeared in 1928 and *Trailing Trade a Million Miles* in 1931; the latter also featured a chapter about business relationships titled "Where the Races Meet."[83]

Both Peffer and Thomas claimed to represent accurately the activities of foreign companies like BAT. Thomas based his expertise on his tenure as head of BAT-China from 1905 to 1916; his position on BAT's board of directors from 1916 to 1920, and his role as vice president of the Chinese American Bank of Commerce from 1920 to 1924. Peffer was twenty-five years Thomas's junior and a graduate of the University of Chicago. He served as the editor of the Shanghai English-language newspaper, *The China Times*, from 1915 to about 1920, when he became the Beijing correspondent for the *New York Tribune Magazine*. He also covered China-US policy at the Washington Conference in 1921 and received a Guggenheim to return to China in 1925.[84]

Peffer and Thomas agreed that business contacts were of primary importance in the development of foreign relations in part because of the cross-racial encounters that they required, though they had very different views of the nature of those encounters. Peffer argued that "it is the business men [more than the diplomat or missionary] with whom the native comes most closely into contact, and by whom the tone of the relations between native and foreigner is set."[85] Thomas concurred in part. Unlike missionaries or diplomats who had agendas to push, he argued, the business representative "who comes into intimate contact with alien peoples" was "the most practical and successful ambassador" because his "language of business, progress and prosperity is universal."[86]

Both, then, affirmed a foreign policy function to business that rested on the nature of cross-racial interactions that occurred in the pursuit of profits.

For Peffer, the problem was that white supremacy and "white man's burden" was at play in the daily level of cross-racial encounter in China, noting that imperialism is "also a system of relations between men and men." Peffer disdained segregation policies in the treaty ports, naming the exclusion of Chinese men from foreign club membership and the prohibition of Chinese people in certain hotels and restaurants. He especially lambasted the entrenched segregation of Chinese and foreign business cultures. He asserted that foreign businessmen "seldom if ever leave their semi-foreign towns, or if they do they follow the paths beaten by other foreigners. What they know they learned within four weeks of their arrival, from conversations in hotel lobbies and club lounges with other foreigners who learned what they know in the same way."[87] Peffer made segregation practices in the treaty ports and foreign businesses a matter of public scrutiny and debate.

Yet Peffer was vague about what actually happened when foreigners left the club table long enough to do their jobs, nor did he seem to consider that Chinese employees at any level might have accrued some leverage or power in these interactions. He argued simply that there were "two separate worlds, touching each other in relations of buyer and seller, employer and employee, master and servant, or in the externals of formal official intercourse; but never truly meeting, never sharing any experience, and never understanding each other." For these reasons, Peffer saw business contact zones flatly as "centers of conflict, breeding-grounds of dissension, suspicion, mutual misunderstanding and dislike." And he tried to undermine the influence of businessmen like Thomas over foreign policy by writing, "[Yet it] is through the eyes of men such as I have described that we see countries in which they live. It is they who make the official reports to our governments [and] who send the news dispatches to our newspapers."[88]

Thomas denied that BAT-China was imperialist; in fact, he cast it as a force for international peace. Thomas distanced BAT from the earlier civilizing mission of nineteenth-century imperialism, insisting that BAT did not make Chinese people conform to US ways but rather adapted to Chinese customs. "With my cigarettes I never tried to sell my particular brand of civilization. I knew that not in my lifetime could I educate a handful to my ways; I must adapt myself to theirs."[89] In a US context where Americanization programs for immigrants were still popular, Thomas appeared tolerant. Thomas was responding to a widespread anti-imperial challenge to the idea of a single hierarchy of civilization with Europe and the US at the apex. In the wake of World War I's carnage, anti-imperial nationalists in India, the Philippines, and China found the idea of European superiority absurd and built claims to sovereignty by asserting the value of their own civilizations. Even in Europe and the US, the idea of a single civilization was slowly giving way to a notion of multiple civilizations, each with their own traditions and trajectories.[90] Thomas argued that BAT's policy was congruent with this new mode. Business relationships, he asserted, provided "solid, concrete foundations ... [for] international comity."[91]

Most fascinating about Thomas's counterargument, however, is that he created an abstraction of economic exchange rather than drawing on his actual experience. The business transaction held the potential for peace and amity, Thomas argued, because of a fundamental sameness in people as economic actors. "I found the ways of people in their social, domestic, and business activities fundamentally much the same, however much they differed outwardly in procedure and formality.... The British, the Chinese, the natives of India, the tradesmen of Russia, Borneo, Egypt, Japan, the Philippine Islands, and others all give evidence of the same elemental characteristics of honesty, fair dealing, and an appreciation of reciprocal rights." Characterizing all business as essentially barter, Thomas created a scenario of mutual benefit in commercial transaction that went back to "the first prehistoric exchanges" and extended throughout the world. He then argued that "business fairly conducted has ever led to ... mutual esteem and good will." By naturalizing economic exchange as mutually beneficial, and sidestepping how one determined if business was "fairly conducted," Thomas could claim that an inherent parity between economic actors led to a "democracy of international trade."[92]

Ironically, by relying on an abstraction of the business relationship, Thomas—like Canaday—obscured his own history. Thomas could not fully deny the segregation and inequality in treaty ports, so he conceded that foreign clubs should admit Chinese members and that foreign businessmen should socialize with Chinese businessmen. "For my part," he claimed, "if I accepted the hospitality of a Chinese merchant, I always returned it by inviting him to my house."[93] In fact, Thomas, like R.H. Gregory, did have high-ranking Chinese businessmen to his home, but he gave no details about who, or why. Nor does he discuss the nature of BAT's collaboration with a plethora of Chinese sales managers, distributors, dealers, retailers, and advertisers, not to mention the armies of workers at all levels in the manufacturing and agriculture departments.

Thomas's omission of his own business relations in China is especially striking when considering that he included in his first book a photograph of Zheng Bozhao, captioned, "Mr. Cheang Park Chew [Zheng Bozhao], Man of Affairs and Chinese Largest Tobacco Merchant," but nowhere else in either of his books did he write about Zheng. Were he interested in supporting his claim that business creates relationships of mutuality and friendship, he might have mentioned the two men's long collaboration or their 1912 trip to London, New York, and North Carolina. He might have mentioned that he and Zheng had collaborated beyond cigarettes in motion pictures, automobiles, and the bank, investing in one another's ventures. Thomas might have shared that Zheng and his wife were godparents to his daughter.[94] Thomas did not divulge any of these facts, none of which would have revealed protected particulars of business strategies. He stated that Chinese businessmen were increasingly organizing their ventures via the "joint-stock company," but did not mention Yongtaihe or even that Chinese businessmen held management positions within BAT.[95] Despite working steadily to incorporate Chinese businessmen and their expertise into BAT's corporate structure, Thomas portrayed BAT-China as entirely foreign and Chinese businessmen—even Zheng—as entirely separate, on the other side of the trading table.

Anyone reading Thomas's book would expect authoritative firsthand stories about what China was like, so Thomas replaced his extensive business experience with amusing stories about naïve Chinese primitives. For example, he recounted at length the time a factory put in a freight elevator, and shortly thereafter purchased a new, heavy cigarette machine that needed to go to the third floor. After much conversation, the Chinese workers strung the massive machine onto ropes and hauled it up the stairs because they didn't want to break the elevator. Once Thomas explained to them just what the elevator could do, he had to pass a rule to keep the workers from riding up and down on it.[96] Leaving aside the fact that Thomas never worked in the factories, this story reassures readers that Chinese economic agents are endearing, primitive innocents needing Western tutelage, not entrepreneurs doing the majority of BAT's business. Thomas also opposed Chinese, Filipino, and Indian sovereignty, arguing that the independence movements of Asia were the result of people trying to "hurdle a century" of evolution in the space of a few years.[97]

In their writings, Thomas and Canaday both erased their actual experience with Chinese businessmen of Yongtaihe and replaced it with the formulaic mythos of modernity, featuring primitive Chinese people being tutored by benevolent Western businessmen. Why? If they had lived in the twenty-first century, they might have wished to gain status by fetishing difference and bragging about their authentic cross-cultural relationships. In the 1920s, however, Yongtaihe signaled both the success and the failure of modernity. BAT did make money, but Western business practices proved far less scalable than expected, that is, they could not be implemented on a larger scale without significant revision. In addition, Chinese businessmen were more fundamentally and permanently necessary than anticipated. BAT foreigners expected to need Chinese translators, traders, and workers of many kinds, but they planned to alienate and standardize their labor via rational Western hierarchies and practices. If Chinese men did everything white men did at home and then some, that is, if they were full economic subjects, then white men's role in China was more assailable: perhaps white men were there simply to extract profits. In the context of waxing anti-imperial nationalism, Yongtaihe's existence and its history exposed modernity as a sham. Thomas and Canaday, out of conscious alarm or visceral discomfort, replaced their own memories about Yongtaihe with stock fables.

Conclusion

In the interwar era, BAT-China reconstituted itself to accommodate the growing success of Zheng Bozhao. The Zheng family had gained the negotiating power to extract from the BAT hierarchy some favorable terms and innovations in form. It certainly was in BAT's interest to help generate Chinese business forms that would be compatible with BAT's continued expansion, even in difficult political times, but BAT needed Zheng too much to be able to have full control over his actions. Chinese businessmen had always been key builders of BAT, but Zheng gained a significant power to dictate terms and direct future events. Frank H. Canaday's experience as a foreign adviser within Yongtaihe reveals some of the heterogeneous ways that BAT's corporate structure was changing. By reversing the typical historical gaze and investigating Canaday's adaptability to Yongtaihe's "modernity," it is possible to break the conventions of modern history writing and perceive an interplay between business encounters on the ground, commodities in circulation, and the debates that swirled at higher altitudes.[98]

As a trading company that became a jointly owned corporate subsidiary, Yongtaihe was a predecessor of the kind of global supply chain that developed after World War II.[99] Like those global supply chains, Yongtaihe operated in a quasi-independent fashion, but its links to the global corporation could be strategically denied, especially when consumers objected to capitalist methods or foreign intrusions. And Yongtaihe businessmen proved more effective than foreign representatives at capturing Chinese resources and transforming them into corporate accumulation. Canaday's and Thomas's distancing and disavowal of the entanglement of Yongtaihe in BAT and in their own lives presage the way that global corporations deny knowledge of contractors' labor or environmental practices.

Thomas might appear to have been promoting the economic subjectivity of Chinese people because he imagined global capitalism as an economic trade or barter between two abstract merchants acting out of self-interest. "The British, the Chinese, the natives of India, the tradesmen of Russia, Borneo, Egypt, Japan, the Philippine Islands, and others," wrote Thomas, "all give evidence of the same elemental characteristics . . . when they transact business with the foreigner." Thomas tapped the well-established strategy of speaking about economics as a set of abstractions apart from history and culture. But this too obscured the hybrid nature of BAT. Indeed, as diametrically opposed as Thomas and Peffer were about the nature of cross-racial business relationships, they agreed that foreign and Chinese people in business existed in two clearly distinguished worlds. Peffer saw that separation as racist and exploitative; Thomas cast it as an inherently egalitarian and beneficial trading relationship. Peffer certainly had a point: segregation practices along lines of race, gender, and native place origin organized BAT's management and factories as well as treaty port culture and militated against true collaboration. Such segregation was designed to translate business practices across difference while still controlling the flow of power and resources.[100] But Peffer missed and Thomas obscured the fact that BAT was much messier than the modern fantasy of expansion suggested: by the 1920s, BAT's very structure had changed and Chinese people held significant power. We could see this as BAT's contamination, an unanticipated transformation and heterogeneity that deviated from the script of modernity.[101]

Thomas invited the reader to adopt a corporate imaginary that mapped corporate boundaries onto clearly defined racial and national groups, but at the same time he insisted that these differences were inconsequential, that the inherent egalitarianism of the abstract economic exchange overcame any particulars. In other words, Thomas both capitalized on and disavowed the significance of racial difference. In this way, Thomas also anticipated the pernicious ideal of being "colorblind," that is, the late twentieth-century rejection of analyses of power inequities that insists that fairness lies in treating everyone as an abstract individual, without history or culture.[102] Only by knowing what Canaday's and Thomas's business encounters were like can we understand the important elisions in the way they tell their stories. The social and cultural history of business is of critical importance, then, for it can restore a diversity of stories of capitalism that dominant corporate imaginaries erase, thereby revealing the workings of power.

CONCLUSION

Called to Account

In November 2000, expatriate Chinese artist Xu Bing installed "The Tobacco Project" at dusk on the grounds of the restored Duke family bright leaf tobacco farm near Durham, North Carolina. The first piece a viewer would come upon when walking into the exhibit was the medical charts of Xu's father, a cigarette smoker who died of lung cancer, projected against the outer log wall of a standing nineteenth-century tobacco barn. The Chinese characters danced across the rough-hewn wall in a ghostly light as a recording played of an English translation of the chart's sterile medical language. The arresting and incongruent visual effect of Chinese characters in the rural setting closed the distance between the US South and China, revealing the tobacco industry's lines of connection and complicity that tied that Southern landscape to Xu's father's life and death in China. The intrusion of Chinese characters announced a startling presence of the past. Intervening in our imaginary of global capitalism, Xu located the corporation in the rural South, rather than just on Wall Street, and personalized what so often is vast and abstract. Who was his father? How did the foreign cigarette corporation become such an intimate and consequential presence in his life? Spectators did not learn the answers but could no longer miss that the questions concerned them.

Xu's installation was powerful because it gave the corporation an intimate history. We all know that large corporations, their products and their brands, have become part of the fabric of our lives and, indeed, of our very bodies, but the public conversation about corporate capitalism tends to hew to the abstract and technical details of finance and economics, as though these operate according to natural laws rather than as human creations that emerged within time, place, particular hopes and desires, and, of course, the human capacity for exploitation and violence. Xu insisted that global capitalism took place on the personal level of the consumer. In the curing barn, blue neon lettering spelled the word "desire" in a cloud of swirling smoke, suggesting the deeply subjective nature of smoking and, perhaps, the addictive nature of cigarettes. With no human figure pictured, Xu's piece evoked a shared struggle with the products and brands that play such a huge role in collective life.

By occupying the Duke family farm with his installation, Xu's exhibit also carried an accusation. Xu explained in an interview, "Historically, American tobacco has a strong connection to China. I did research and discovered that Duke brought the cigarette to China. Before that, there were no cigarettes in China."[1] Cigarette smoking is currently a massive health crisis in China, with smoking-related illness causing approximately one million deaths each year. In part because smoking still plays normative roles in business and workplace cultures, smoking rates among Chinese men remain astronomically high: 52.9 percent in 2010. China's three hundred million smokers are nearly one-third of the world's total. The World Health Organization estimates that tobacco-related deaths in China will rise to three million annually by 2050 if smoking rates are not reduced.[2] Without getting mired in legalities of liability, Xu pointed a finger at Duke and called the US industry to account.

Xu's "Tobacco Project" was a powerful expression of global, intimate connection that drew attention to the transnational story told in this book, but in several ways it relied on an underlying and distorted narrative of modernity. Without a deeper history of the US or Chinese context, Xu focused on the entrepreneur, Duke, as architect of the global cigarette industry. The corporate imaginary he so vividly evoked reinforced the idea of a West to East commodity chain, with the US as producer and China as consumer. Xu was silent on the issue of Chinese production or Chinese agency in creating this history, and he did not seem aware of the parallel development of the industry in the US. I suspect that some North Carolinian viewers, when they viewed Xu's father's medical charts, thought of their own loved ones who had died of smoking-related diseases. Because North Carolina was once "the cigarette state," North Carolinians still smoke at higher rates than the national average, but Xu did not acknowledge the vulnerability of US workers and consumers within the exhibit. For Xu, Duke stood for the corporation, and deeper relationships between North Carolina and China remained unrepresented.

One goal of this book has been to intervene into the typical way that the corporation is imagined via the entrepreneur or the board of directors and stockholders and to suggest a more peopled approach, one truer to the way that large business corporations operated in daily life. Over a century ago, executives like Duke became metonyms for the huge corporate entities that they worked for, becoming the main character in a romance of capitalism, whether utopic or dystopic.[3] In 1937, Thurman Arnold argued that the personification of corporations as "a big man" reduced an "infinitely complicated" economic organization in a particular political context to a simple figure in a parable.[4] A singular legal entity thus gained potential social power as a singular public relations image: someone with whom we might identify, and like or dislike. The Schumpeterian theory of creative destruction heroized the entrepreneur and attributed to it a personality as a renegade and cowboy. Perhaps the theory explained how some industries changed, but its heavy-handed application to the cigarette industry has distorted history and obscured the process of corporate empowerment. Certainly, many fine histories recount the merger movement of the late nineteenth century, yet somehow the myth of Duke's entrepreneurial brilliance has remained unrevised for approximately fifty years.

In place of the entrepreneur as protagonist, this book centers the bright leaf tobacco network, an imperial circuit of hundreds of white men from North Carolina and Virginia that, through myriad encounters with thousands of employees, built the tobacco industry in the US and China. My point is different from that of Alfred Chandler, who believed that exploring the "visible hands" of managers was the best way to understand corporations' bureaucratic modernity. I do not assume that the men of the bright leaf network were uniquely skilled or brilliantly organized. Indeed, the composition of the bright leaf network was itself a result of the ways that Jim Crow segregation shaped economic expansion in the South, directing white-collar opportunities to white men only, despite the eminent and equal qualifications of many African Americans. In China, foreign representatives relied deeply and increasingly on Chinese businessmen and workers at all levels; the bright leaf network served to maintain racial hierarchies within BAT even while tapping Chinese expertise.

By following these men from North Carolina and Virginia, the relationship between Jim Crow segregation and US imperialism in the building of the transnational cigarette industry comes into view. The notion of a network is critical to this relationship. The bright leaf network served both as a corporate managerial structure and as a conduit for the flow of white-collar workers, tobacco, cigarettes, and knowledge that, in dynamic interaction, performed imperial work in China. The imperial bright leaf network's knowledge included understanding how to best produce bright leaf tobacco and develop bright leaf seeds; race management techniques related to agriculture, factories, and the home; and communal practices rooted in Southern foods and hospitality. But the network became more than the sum of its parts: it served as a mechanism for economic expansion that preserved principles of racial hierarchy forged in Jim Crow while innovating their expression through encounter with Chinese businessmen, farmers, factory workers, and servants. In other words, the bright leaf network did its work through encounter, and therefore through struggle, for Chinese workers at all levels fought for their own interests and used foreigners' considerable *lack* of knowledge against them.

But there is more. What does it mean that a corporate imperial network did much the same work at the same time in the United States and China? This fact calls us to rethink still-prevalent global imaginaries, including the idea of modernity, that cast the US and China only in diametric opposition or on a developmental time delay. Of course, the US was—and is—a very rich country compared to China. The US was part of the imperial family of nations that coerced China to sign unequal treaties, leading to a steady flow of resources and profits from China to foreign destinations, and it would be absurd to deny the vast differences between the two countries. But the assumption of modernity as an explanatory framework has occluded the ways that corporate empowerment took place on a global stage, reorganizing local economies to new rationalities in the US as well as in China and building transnational investments in race and gender hierarchies. The dichotomy between West and East—or first world and third world or global north and global south—has sublimated crucial aspects of a connected history and the mechanisms by which capitalism fostered an increasingly global race and gender system.

I began this project with a simple premise, drawn from the legacy of social and cultural history, that corporations were made by people through daily encounter. And this encounter, I discovered, was very often intimate. In both China and the United States, the bright leaf network formed and consolidated through relations of domesticity, sexuality, and servitude, realms we are accustomed to seeing as intimate. But intimate encounters also took place at the factory, where job classification and disciplinary systems operated on the body to construct gender and race hierarchies, and in business culture, where food, cigars or cigarettes, and sociality inducted young men into corporate imperial identities. Indeed, the cigarette itself became an intimate prop for multiple plotlines in corporate history, from white-collar workers negotiating business deals, to factory workers declaring their belonging to a branded baseball team, to protesters or workers destroying cigarettes on the street.

By centering a peopled history, this book argues that economics are constituted within and through culture. Entrepreneurship, business culture, labor management, branding, and sales all have a social and cultural history in which the construction of gender, sexuality, race, native place, age, and other identity categories is integral to how the economic story unfolded. This should not be particularly surprising, since the global cigarette corporations were vast organizations made through cross-cultural encounter, but somehow "diversity" as a concept seems to stick only to lower-rank workers and consumers, while imaginaries of the corporation and the idea of economy remain abstracted and depersonalized (and by default, white). My hope is that this book, through the telling of peopled stories about businessmen and entrepreneurs in conjunction with those about factory workers, servants, sex workers, and consumers, exposes this logical lacuna to create a more accurate and useful corporate imaginary. Twentieth- and twenty-first-century capitalist ideology casts an abstracted notion of the economic as the realm of freedom and asserts that politics and culture are where history hurts. We should be wary of replicating this false divide between economics and culture in our visions for social and economic justice.

The dominant construction of economics as a realm of freedom deflects attention from the role of corporations in governance. In China, large foreign corporations like BAT served as sites of imperial governance in treaty ports and large factory compounds and also organized the flow of labor resources and profits from China to the US and Britain. That BAT was "private," unlike earlier, public chartered corporations—the British East India Company, for example—should not blind us to its political and diplomatic functions and effects. We operate with a caricature of imperialism if we define it only as colonization by foreign political rule. The state established different forms of imperialism—colonization, as in the Philippines, or extraterritoriality, as in China—but in any case, it required a mechanism of ongoing resource extraction, economic expansion, and social control, and corporations have played a consistent albeit shifting role.[5] Likewise, in the United States, cigarette corporations were critical generators and beneficiaries of the Jim Crow segregation that kept African Americans economically, socially, and politically subservient to whites. Thinking of corporations as mechanisms of governance raises the question of how economic enfranchisement and democracy can be secured in the context of global corporate capitalism.

For several decades, there was considerable evidence that unions and government regulators had reined in the greatest excesses of corporate power and made the corporation more compatible with democracy in the United States. The passage of labor laws, antitrust laws, and laws governing a wide range of corporate activity, including political contributions, succeeded in making corporations more responsive to workers' and society's needs. Of course, unions only infused democracy into corporate decision-making to the extent that unions and US laws were themselves democratic. The passage of the 1964 Civil Rights Act outlawed race and gender segregation at the workplace and the union, with the immediate result in North Carolina's cigarette factories that some African Americans moved into better jobs that allowed them, like many of their white coworkers, to purchase homes and send their children to college. In recent years, however, corporate interests have succeeded in eroding the New Deal social contract and corporate regulations, while Citizens United marked a new era of big money driving politics. The New Deal and its legacies seem increasingly like the exception in US history rather than achievements in ever-unfolding story of American progress.[6]

When Xu launched "The Tobacco Project" in 2000, the tobacco industry seemed to have crashed in the United States, but it has recently made a surprising comeback and is participating fully in a resurgence of corporate plutocracy and widening inequality. In the 1990s, cigarette companies saw tumbling profits as successful, high-profile lawsuits exposed their elaborate program to obfuscate scientific proof of cigarettes' dangers. Successful cases channeled settlement money into state-level tobacco cessation programs, some of which highly publicized cigarettes' dangers. US smoking rates declined rapidly, cigarette factories in North Carolina and Virginia closed one after another, tobacco farmers struggled, and average wages in cigarette factory towns plummeted.

The decline in the US cigarette market seemed to mark a great divergence between the cigarette industries of the US and China. Though the industries had run along parallel lines in the first decades of the twentieth century, the 1949 communist revolution in China marked a sharp break. Chinese authorities nationalized BAT-China (including Yongtaihe) and set up a state-run cigarette industry. Chinese women stopped smoking almost entirely because, in a backlash against the urban consumer markets of treaty ports, women's cigarette smoking became a symbol of Western imperial decadence. Men's rates continued to increase, perhaps because Chairman Mao was rarely seen without a cigarette in his hand, making male cigarette smoking seem revolutionary. In the 1990s, foreign cigarette companies surged back into China's newly opened market, and aggressive growth in China became the tobacco industry's callous answer to declines in the US and Europe. Furthermore, tobacco cessation programs were virtually nonexistent and restrictions on cigarette advertising were few. As historian Robert Proctor put it in 2011, "The global transnationals are busy having their way with poorer parts of the world as they once did with the rich."[7] Xu Bing, then, had both contemporary and historical reasons for exposing the US cigarette industry's role in China.

What had seemed to be the death spiral of the US domestic cigarette industry, however, turned out to be a phoenix-like resurgence. As I write this, the US and China are, once again, the "world's [two] largest tobacco profit pool[s]," according to BAT's chief executive Nicandro Durante. BAT recently called the US "an exciting opportunity for long-term growth."[8] The US is a growth market for cigarettes because, though the rate of smoking has declined, the population is rising. In addition, inequality in the US means that smoking cessation has not distributed evenly across the population—some states have smoking rates as high as China's. In 2010, China's overall smoking rate was 28.1 percent; West Virginia's was 28.6 percent.[9] Furthermore, US regulations are not very costly and the tobacco lobby has kept taxes low, leaving room for companies to raise prices. Companies make the same profit from the sale of two packs in the US as they do six packs in other developed markets and thirteen packs in developing markets.[10] In 2017, BAT reinvested in the US, buying Reynolds American, headquartered in Winston-Salem, for $49 billion, making BAT once again the largest tobacco company in the world.[11]

While cigarette prices and profits are going up, wages have declined. In 2012, Reynolds American employed some twenty-five hundred workers in manufacturing jobs in Winston-Salem. Despite realizing a 7.4 percent rise in profits the previous year, the company cut 10 percent of its jobs and slashed wages, purportedly to make wages more comparable to manufacturing jobs across the southeast.[12] Despite recent efforts to organize, the industry is not unionized. Even more vulnerable are the state's tobacco agricultural workers, most of whom are African Americans, and who now mostly work on contract farms for poverty wages; many tobacco workers make just $7.25 per hour and children as young as twelve years old are allowed to work unlimited hours. The AFL-CIO Farm Labor Organizing Committee has tracked human rights abuses on BAT's contract farms in North Carolina that parallel those reported on BAT's contract farms in Bangladesh and Indonesia. In response to pressure, BAT has begun a corporate auditing process that has confirmed the assertions of labor activists, but the company refuses to agree to any system that would give farmworkers a voice in the process.[13]

The connections between the US and China in today's global economy have their origin in globalization that occurred over a century ago, though ideas of modernity have made it difficult to discern them. Corporations like BAT now loom large over obsequious nations that wish to cull their favor, and China sets the wage standards for manufacturing jobs around the world.[14] In 1956, Abram Chayes argued that our narrow definition of corporate membership, consisting of shareholders and a board of directors, portended an increasingly undemocratic economic and political system. Merely giving shareholders a truer and more effective voice in corporate decision-making would not address the root problem, he argued, "because the shareholders are not the governed of the corporation whose consent must be sought."[15] As democratic political systems struggle against right wing insurgency around the world, it is time to recognize that corporations operate as a form of governance. At the same time, corporations are made—and therefore also remade—by people through encounter.

We need corporate imaginaries that can encompass the historical and contemporary heterogeneity of capitalism, the lines of entanglement and dependence that fly in the face of modern dreams of self-containment, independence, and mastery. Developing new corporate imaginaries requires something more than reading history from the bottom up, though that remains an important project. It requires recognizing that one can decry corporate power and still overestimate corporate cohesiveness and erase the stories we need to hear. It requires recognizing that corporate structures, even large and sprawling ones, are social as well as financial organizations, formed through innumerable intimate contacts at all levels. These encounters are mixed and unpredictable, sometimes inspiring, sometimes cruel—the stuff of life itself. Telling intimate stories about the corporation can be a way to honor life in all of its complexities in the midst of the ruins of capitalism.

ACKNOWLEDGMENTS

This book began in Reidsville, North Carolina, when Kori Graves and I interviewed elderly African American and white retired cigarette factory workers. My idealistic idea was to listen to the elders and take my cue from them about which issues to explore in my project. I learned a lot from the interviews, but even more from the casual conversations that we had with the wise and warm people we came to know best: Ruby Delancy, James Neal, and Ruth and William Davis. The project took several sharp turns since then, but I knew to look into jazz music, baseball, and the Ku Klux Klan from those conversations. More than that, I learned things about segregation that books cannot teach. I hope that I have been true to the gift of those conversations.

This is a dangerous time for the humanities, so I feel especially grateful to the many institutions that aided this research, and I hope that they all survive into the future. My thanks to the archivists and financial supporters of the Rockingham County Community College Historical Collection; the University of North Carolina Southern Collection; the David M. Rubenstein Rare Book and Manuscript Library at Duke University; the J. Y. Joyner Library at East Carolina University; the Library of Congress; the Tobacco Workers International Union Papers at the University of Maryland; the Harvard Yenching Library; the NBC Collection at the Wisconsin State Historical Society; and the Shanghai Academy of Social Sciences. Likewise, I am thankful to the National Endowment for the Humanities for fellowship support. The University of Wisconsin supplied me with a critical research fund, the Vilas Associate Award, which supported my travel for research. I thank UW's Institute for Research in the Humanities and the Feminist Scholar's Award for critical time for writing.

For housing me on research trips and for their friendship, I thank Patricia Black, Tom Jackson, Kay Lovelace, Tim Tyson, Perri Morgan, Cinder Hypki, Max Grossman, Hayim Lapin, and Kathy Franz. I also thank Ken Anthony and Dorothy Phelps Jones for tips on Reidsville, and Helen B. Marrow and James Marrow for conversations about their forebears who went to China for BAT.

No one was more surprised than me that I ended up writing a book that takes place partly in China. I enjoyed and benefitted from my year of Chinese language study, but it certainly did not prepare me to conduct research in Chinese. I am grateful to the work of three Chinese translators who made it possible for me to incorporate Chinese-language sources throughout this project. Guo Jue translated portions of the four-volume compilation of sources on BAT-China and the cigarette industry compiled by Chinese historians in the 1950s. These volumes include excerpted BAT company documents, oral histories with workers, newspaper articles, and other sources. Guo Jue also worked with me in the British American Tobacco Collection at the Shanghai Academy of Social Sciences. Wang Haochen translated many of the documents gathered on that trip and many others available through interlibrary loan. Wang so often went beyond the work of translation to find other documents for me that I eventually put him on primary research tasks in English-language sources. Xu Zhanqi has helped me with additional documents as I was completing the book.

I believe East Asianists may be the friendliest of all historians—so many have helped me with this project. Brett Sheehan is a prince. He helped me figure out how to do research in Shanghai, suggested novels as well as history books for me to read, and read several chapters of the book, offering game-changing insights. Tina Chen and Zhou Yongming offered sage advice, and Joe Dennis connected me with two brilliant student translators. Elizabeth Perry offered tips on Shanghai archives. Carol Benedict and Shelly Chan generously read portions of the manuscript. Louise Young organized a group reading of the manuscript, read nearly the entire manuscript herself, and encouraged my work at crucial moments. Antonia Finnane and Andrew David Field generously answered questions over email.

It is a privilege to be at a university with a vibrant graduate program, and I hope it survives into the future. Students who performed research tasks for me include Thea Browder, Stephanie Westcott, Kori Graves, Haley Pollack, Zoe Van Orsdol, Maddy Brigell, and Brenna Greer. In the eleventh hour, Vaneesa Cook made order out of my footnotes, and Vanessa Soleil chased down elusive sources at Duke Archives. Even more than this assistance, however, the passion and intellectual engagement of so many UW graduate students fueled my own work. Thanks and great respect to all named above and also to Abby Markwyn, David Gilbert, Maia Surdham, Crystal Moten, John Hogue, Francis Gourrier, the late Doria Johnson, Faron Levesque, Dan Guadagnolo, Bree Romano, Thomas Kivi, Johanna Lanner-Cusin, Rachel Gross, Spring Greeney, Ari Eisenberg, Elena McGrath, Ayanna Drakos, Li Lin, Kate Turner, and Siobhan McGurk. It is a sign of this richness in my life that I feel quite sure I've forgotten someone who should be named on this list.

I drew inspiration and moral support from my own teachers from graduate school, David W. Noble, who passed away while this book was in press, and George Lipsitz. At the University of Wisconsin, Steve Stern, Florencia Mallon, the late Jeanne Boydston, Mary Louise Roberts, Russ Castronovo, Karma Chavez, Sara McKinnon, and Lori Lido Lopez provided crucial mentoring and/or collaborated with me in sustaining ways.

I have benefitted greatly from circulating portions of the manuscript for commentary and discussion to a diverse range of groups, including at Mars Hill College; the Affect and Economy Group at the University of Toronto; the Center for the Study of Work, Labor and Democracy at the University of Santa Barbara; the Dartmouth Geography Department; the New Materialism Conference at the University of Michigan; and the Music Race and Empire Conference at UW. Thanks to David Gilbert, Elspeth Brown, Eileen Boris, Nelson Lichtenstein, Mona Domosh, Jay Cook, Ron Radano, and Tejumola Olaniyan for these opportunities for feedback.

A million thanks to those who took time from their own work to read and comment on mine: Carol Benedict, Shelly Chan, Gregg Mitman, Pernille Ipsen, Nancy Buenger, Miranda Johnson, Steve Kantrowitz, Julia Mickenberg, Colleen Dunlavy, Elspeth Brown, and Jay Cook. Reader superheroes who read major portions or the entire manuscript include David Herzberg, Susan Cahn, Wendy Kozol, Louise Young, Brett Sheehan, Penny Von Eschen (who reviewed the manuscript for the press), Rachel Buff, and Finn Enke. An additional anonymous reviewer for the press was extremely helpful.

Writing a book requires sustenance for the heart. Snowflower Sangha has provided community and teaching. My dad, Jim Enstad, was a wonderful presence in my life as I worked on this book. He passed away just as it went to press. Patricia Enstad, Jim Mondeau, Chris Enstad, Sophie Pfeiffer, the irrepressible Noble-Olson household, the late Jim Oleson and John D'Emilio, Ruby Balotovsky, Ellie Rae Balotovsky, Joe Austin, Judy Houck, Lisa Saywell, Cinder Hypki, Susan Cahn, Claire Wendland, Mary Moore, David Herzberg, and Rachel Buff have all filled my life with love and purpose and convince me even in the darkest of times that another world is possible. Finn Enke, with their vision, their adventures, and their sturdiest of loves, makes this world possible.

NOTES

Preface

1. John D. Kelly, "Who Counts? Imperial and Corporate Structures of Governance, Decolonization and Limited Liability," in *Lessons of Empire: Imperial Histories and American Power*, ed. Craig Calhoun et al. (New York: New Press, 2006), 157–74.

2. There is a long and vast intellectual history of critique of corporate power, including the role of financialization in shifting corporate structure and purpose, the division of ownership and management, changes in legal corporate personhood, corporate social responsibility, the marginalization of workers from corporate governance, and more. Despite this, our contemporary public conversation is impoverished by an anemic corporate imaginary. For a tip of the iceberg, see Adolf Berle and Gardiner Means, *The Modern Corporation and Private Property* ([1933] New Brunswick, N.J.: Transaction Publishing, 1991); Martin J. Sklar, *The Corporate Reconstruction of American Capitalism, 1890–1916: The Market, the Law, and Politics* (London: Cambridge University Press, 1988), 49–50; Naomi R. Lamoreaux, *The Great Merger Movement in American Business, 1895–1904* (London: Cambridge University Press, 1988); Naomi R. Lamoreaux and William J. Novak, eds., *Corporations and American Democracy* (Cambridge: Harvard University Press, 2017); William G. Roy, *Socializing Capital: The Rise of the Large Industrial Corporation in America* (Princeton: Princeton University Press, 1997).

3. I have, accordingly, assembled a diverse archive that allows me to investigate the subjective, daily experiences of a variety of corporate actors. With Kori Graves, I conducted oral histories with African American and white former American Tobacco Company factory workers. I also consulted archived oral history collections with southern factory workers, white-collar workers who went to China, and Chinese factory workers. I delved into the cultural practices of corporate employees at all levels, with particular attention to food, jazz music and dance, and baseball. Also critical were personal letters, memoirs, and photographs of white-collar Americans who went to China; personal remembrances of Chinese businessmen;

and labor union records, which allowed me to investigate intimate relations between life partners, employers and servants, and businessmen and taxi dancers. With these sources, company records, public relations records, and government investigation records could be brushed against the grain.

4. Abram Chayes, "The Modern Corporation and the Rule of Law," in *The Corporation in Modern Society*, ed. Edward S. Mason (Cambridge: Harvard University Press, 1959), 19–22; Ann Laura Stoler, "Tense and Tender Ties: The Politics of Comparison in North American History and (Post) Colonial Studies," *Journal of American History* 88, no. 3 (December 2001): 829–65; Ramon Gutierrez, "What's Love Got to Do with It?" *Journal of American History* 88, no. 3 (December 2001): 866–69. Stoler's call for historians to examine the intimate contacts of empire through servitude, domesticity, and sexuality has influenced my thinking. I combine the two concepts to find intimacy considerably beyond the realms that Stoler examines.

5. Allan M. Brandt, *The Cigarette Century: The Rise, Fall and Deadly Persistence of the Product that Defined America* (New York: Basic Books, 2007); Robert N. Proctor, *Golden Holocaust: Origins of the Cigarette Catastrophe and the Case for Abolition* (Berkeley: University of California, 2011).

Introduction

1. Lee Parker, interview by Burton Beers, June 1980, transcript, Kessler Papers, Southern Historical Collections, Manuscripts Department, University of North Carolina, Chapel Hill.

2. Richard Henry Gregory Papers, James A. Thomas Papers, David M. Rubenstein Rare Book and Manuscript Library, Duke University; Ivy Riddick Papers, Southern Historical Collection, University of North Carolina, Chapel Hill; James N. Joyner Papers, J. Y. Joyner Library, East Carolina University.

3. Morton J. Horwitz, "*Santa Clara* Revisited: The Development of Corporate Theory," *West Virginia Law Review* 88, no. 2 (Winter 1985–86): 173–224; Gregory A. Mark, "The Personification of the Business Corporation in American Law," *University of Chicago Law Review* 54, no. 4 (Autumn 1987): 1441–83; David Millon, "Theories of the Corporation," *Duke Law Journal* 1990, no. 2 (April 1990): 201–62; Lawrence M. Friedman, "The Law of Corporations," in Friedman, *A History of American Law*, 3rd ed. (New York: Simon and Schuster, 2005); William G. Roy, *Socializing Capital: The Rise of the Large Industrial Corporation in America* (Princeton: Princeton University Press, 1997); Teemu Ruskola, *Legal Orientalism: China, the United States, and Modern Law* (Cambridge: Harvard University Press, 2013); Martin J. Sklar, *The Corporate Reconstruction of American Capitalism, 1890–1916: The Market, the Law, and Politics* (London: Cambridge University Press, 1988), 49–50; Naomi R. Lamoreaux, *The Great Merger Movement in American Business, 1895–1904* (London: Cambridge University Press, 1988); Naomi R. Lamoreaux and William J. Novak, eds., *Corporations and American Democracy* (Cambridge: Harvard University Press, 2017), 245–325.

4. Nannie May Tilley, *The Bright-Tobacco Industry, 1860–1929* (Chapel Hill: University of North Carolina Press, 1948); *Report of the Commissioner of Corporations on the Tobacco Industry*, vols. 1–3 (Washington, DC: Government Printing Office, 1915).

5. Sherman Cochran, *Big Business in China: Sino-Foreign Rivalry in the Cigarette Industry, 1890–1930* (Cambridge: Harvard University Press, 1980); Cochran, *Encountering Chinese Networks: Western, Japanese, and Chinese Corporations in China, 1880–1937* (Berkeley: University

of California Press, 2000); Howard Cox, *The Global Cigarette: Origins and Evolutions of British American Tobacco, 1880–1945* (Oxford: Oxford University Press, 2000); Carol Benedict, *Golden Silk Smoke: A History of Tobacco in China, 1550–2010* (Berkeley: University of California Press, 2011).

6. Paul A. Kramer, "Imperial Openings: Civilization, Exemption, and the Geopolitics of Mobility in the History of Chinese Exclusion, 1868–1910," *Journal of the Gilded Age and Progressive Era* 14 (2015), 317. My understanding of empire also draws on the notion of assemblage, whereby humans, plants, commodities, and knowledge function together to produce and mobilize power. See Aihwa Ong and Stephen J. Collier, eds., *Global Assemblages: Technology, Politics, and Ethics as Anthropological Problems* (Malden, MA: Blackwell, 2005); Bruno Latour, *Reassembling the Social: An Introduction to Actor-Network-Theory* (Oxford: Oxford University Press, 2005), 11–12; Bill Brown, "Thing Theory," *Critical Inquiry* 28, no. 1 (Autumn 2001): 1–22; Jane Bennett, *Vibrant Matter: A Political Ecology of Things* (Durham: Duke University Press, 2010), 20–38. Environmental historians typically include nonhuman actors as agents; especially influential to my project is Drew A. Swanson, *A Golden Weed: Tobacco and Environment in the Piedmont South* (New Haven: Yale University Press, 2014).

7. Joseph A. Schumpeter, *Capitalism, Socialism, and Democracy* (London: Harper & Brothers, 1942), 82–83.

8. Patrick G. Porter, "Origins of the American Tobacco Company," *Business History Review* 43, no. 1 (Spring 1969): 59.

9. The most influential historian to characterize Duke in this way was Alfred D. Chandler, Jr., first in his piece "The Beginnings of 'Big Business' in American Industry," *Business History Review* 33, no. 1 (Spring 1959): 8; and later in the same terms in his pathbreaking book, *The Visible Hand: The Managerial Revolution in American Business* (Cambridge: Belknap Press, 1977), 382–83. Chandler's 1977 account is riddled with errors. He claims that Duke, in 1885, was the first to "set up branch sales offices operated by their own salaried personnel and managers" (382), but Ginter ran a branch sales office for the Allen Company (he and Allen were partners) in 1870, fifteen years earlier. Chandler claims that the other four manufacturers were forced to merge into Duke's business structure, but Ginter's was just as formative, perhaps more so. Chandler attributes the beginning of overseas sales to Duke's salesman, Richard B. Wright, who began exploring foreign markets in 1882, at least five years after Ginter initiated selling overseas. Chandler does not mention Ginter's name in his 1959 article, and mentions it in *The Visible Hand* only once and only to note that Ginter opposed the expansion of the ATC by taking over chewing and smoking tobacco companies (387). Chandler built his narrative on Porter, "Origins of the American Tobacco Company"; Robert F. Durden, *The Dukes of Durham, 1865–1929* (Durham: Duke University Press, 1975); Richard B. Tennant, *The American Cigarette Industry: A Study in Economic Analysis and Public Policy* (New Haven: Yale University Press, 1950); and Maurice Corina, *Trust in Tobacco: The Anglo-American Struggle for Power* (New York: St. Martins Press, 1975), all of which minimized Ginter's business.

10. See, for example, Cox, *The Global Cigarette*; Richard Kluger, *Ashes to Ashes: America's Hundred-Year Cigarette War, the Public Health, and the Unabashed Triumph of Philip Morris* (New York: Vintage Books, 1996); Dolores E. Janiewski, *Sisterhood Denied: Race, Gender, and Class in a New South Community* (Philadelphia: Temple University Press, 1985), 69–70; Pamela Laird, *Advertising Progress: American Business and the Rise of Consumer Marketing*

(Baltimore: Johns Hopkins University Press, 2001), 194; Allan M. Brandt, *The Cigarette Century: The Rise, Fall, and Deadly Persistence of the Product that Defined America* (New York: Basic Books, 2007); Robert N. Proctor, *Golden Holocaust: Origins of the Cigarette Catastrophe and the Case for Abolition* (Berkeley: University of California Press, 2011).

11. David Freddoso, "James Duke Smoked 'Em Out to Dominate Tobacco Arena," *Investor's Business Daily*, January 4, 2016, http://www.investors.com/news/management/leaders-and-success/james-duke-hit-it-big-in-tobacco/; William Kremer, "James Buchanan Duke: Father of the Modern Cigarette," *BBC World Service*, November 13, 2012, http://www.bbc.com/news/magazine-20042217/; Harry McKown, "January 1890: Creation of the American Tobacco Company," *North Carolina Miscellany*, January 1, 2009, http://blogs.lib.unc.edu/ncm/index.php/2009/01/01/this_month_jan_1890/; Phil Edwards, "What Everyone Gets Wrong About the History of Cigarettes," *Vox*, April 6, 2015, https://www.vox.com/2015/3/18/8243707/cigarette-rolling-machines/.

12. Ginter has received virtually no critical scholarly attention. Zheng Bozhao has been studied in most depth by Cochran, *Encountering Chinese Networks*; see also Howard Cox, "Learning to Do Business in China: The Evolution of BAT's Cigarette Distribution Network, 1902–1941," *Business History* 39, no. 3 (1997).

13. Zvi Ben-Dor Benite, "Modernity: The Sphinx and the Historian," *American Historical Review* 116, no. 3 (June 2011): 638–52; Carol Gluck, "The End of Elsewhere: Writing Modernity Now," *American Historical Review* 116, no. 3 (June 2011): 676–87; Rebecca E. Karl, *Staging the World: Chinese Nationalism at the Turn of the Twentieth Century* (Durham: Duke University Press, 2002), 4–17; Dipesh Chakrabarty, *Provincializing Europe: Postcolonial Thought and Historical Difference* (Princeton: Princeton University Press, 2000); Johannes Fabian, *Time and the Other: How Anthropology Makes its Object* (New York: Columbia University Press, 1983); Ruskola, *Legal Orientalism*.

14. US historians speak as if with one voice on the myth of Duke's expansion, but British economic historian Leslie Hannah argues that the monopoly lagged behind Europe's industry. Hannah, "The Whig Fable of American Tobacco, 1895–1913," *Journal of Economic History* 66, no. 1 (March 2006): 42–73.

15. Relli Shechter, *Smoking, Culture, and Economy in the Middle East: The Egyptian Tobacco Market, 1850–2000* (London: I. B. Tauris, 2006); Mary C. Neuburger, *Balkan Smoke: Tobacco and the Making of Modern Bulgaria* (Ithaca: Cornell University Press, 2013).

16. Lee Parker interview June 1980 by Burton Beers, edited transcript. Eastern Carolina University. See also the unedited transcript, Lee Parker interview June 1980 by Burton Beers, Kessler Papers, Southern Historical Collections, Manuscripts Department, University of North Carolina, Chapel Hill.

17. Cochran, *Big Business in China*; Karl Gerth, *China Made: Consumer Culture and the Creation of the Nation* (Cambridge: Harvard University Press, 2003); Jane Leung Larson, "Articulating China's First Mass Movement: Kang Youwei, Liang Qichao, the Baohuanghui, and the 1905 Anti-American Boycott," *Twentieth-Century China* 33, no. 1 (November 2007): 4–26.

18. Andrew F. Jones, *Yellow Music: Media Culture and Colonial Modernity in the Chinese Jazz Age* (Durham: Duke University Press, 2001); Andrew David Field, *Shanghai's Dancing World: Cabaret Culture and Urban Politics, 1919–1954* (Hong Kong: Chinese University Press,

2010); Frederick J. Schenker, "Empire of Syncopation: Music, Race, and Labor in Colonial Asia's Jazz Age" (dissertation, University of Wisconsin, 2016); Ronald Radano and Tejumola Olaniyan, *Audible Empire: Music, Global Politics, Critique* (Durham: Duke University Press, 2016).

19. Swanson, *A Golden Weed*, 147–81; Pete R. Daniel, *Breaking the Land: The Transformation of Cotton, Tobacco, and Rice Cultures Since 1880* (Champaign-Urbana: University of Illinois Press, 1985), 23–38; Eric Foner, *Reconstruction: America's Unfinished Revolution, 1863–1877* (New York: Harper and Row, 1988), 124–76.

20. Swanson, *A Golden Weed*; Tilley, *The Bright-Tobacco Industry*. Burley tobacco was also important to the Upper South. Its history dovetailed with bright leaf in the new American cigarette in the 1920s. For a study of burley's expansion see Nicole D. Breazeale, "Kicking the Tobacco Habit: Small Farmers, Local Markets, and the Consequences of Global Production Standards in Misiones, Argentina" (dissertation, University of Wisconsin Madison, 2010).

21. Swanson, *A Golden Weed*, 148–49, 170–78, 208–13; see also Jane Dailey, *Before Jim Crow: The Politics of Race in Post-Emancipation Virginia* (Chapel Hill: University of North Carolina Press, 2000).

22. Tilley, *The Bright-Tobacco Industry*, 489–564; Robert F. Durden, *The Dukes of Durham, 1865–1929* (Durham: Duke University Press, 1987), 3–25; Nannie M. Tilley, *The R. J. Reynolds Tobacco Company* (Chapel Hill: University of North Carolina Press, 1985), 1–35; Thomas, *Pioneer Merchant in the Orient*, 3–10.

23. Swanson, *A Golden Weed*, 47–48, 112. See also Barbara Hahn, *Making Tobacco Bright: Creating an American Commodity, 1617–1937* (Baltimore: Johns Hopkins University Press, 2011), 13–14.

24. Many historians have studied the centrality of Jim Crow to the development of US capitalism and the role that Jim Crow played as the US became an empire. See especially William P. Jones, *The Tribe of Black Ulysses: African American Lumber Workers in the Jim Crow South* (Urbana: University of Illinois Press, 2005); N. D. B. Connolly, *A World More Concrete: Real Estate and the Remaking of Jim Crow South Florida* (Chicago: University of Chicago Press, 2014); Jacqueline Dowd Hall, "The Long Civil Rights Movement and the Political Uses of the Past," *Journal of American History* 91, no. 4 (March 2005): 1233–63; Robert Korstad, *Civil Rights Unionism: Tobacco Workers and the Struggle for Democracy in the Mid-Twentieth-Century South* (Chapel Hill: University of North Carolina Press, 2003); Edward Ayers, *The Promise of the New South: Life After Reconstruction* (New York: Oxford University Press, 1992); Paul A. Kramer, *The Blood of Government: Race, Empire, the United States, and the Philippines* (Chapel Hill: University of North Carolina Press, 2006); Robert Vitalis, *America's Kingdom: Mythmaking on the Saudi Oil Frontier* (Brooklyn: Verso, 2009); David Roediger and Elizabeth D. Esch, *The Production of Difference: Race and the Management of Labor in U.S. History* (New York: Oxford University Press, 2012); Elizabeth D. Esch, *The Color Line and the Assembly Line: Managing Race in the Ford Empire* (Berkeley: University of California Press, 2018).

25. Mira Wilkins, *The Emergence of Multinational Enterprise: American Business Abroad from the Colonial Era to 1914* (Cambridge: Harvard University Press, 1970), 91–92. For the relationship of US corporations to empire in Latin America see especially Jason M. Colby, *The Business of Empire: United Fruit, Race, and US Expansion in Central America* (Ithaca: Cornell University Press, 2011); John Soluri, *Banana Cultures: Agriculture, Consumption, and Environ-*

mental Change in Honduras and the United States (Austin: University of Texas Press, 2006); Greg Grandin, *Fordlandia: The Rise and Fall of Henry Ford's Forgotten Jungle City* (New York: Picador, 2009); Gilbert M. Joseph, Catherine C. LeGrand, and Ricardo D. Salvatore, *Close Encounters of Empire: Writing the Cultural History of U.S.–Latin American Relations* (Durham: Duke University Press, 1998); Julie Greene, *The Canal Builders: Making America's Empire at the Panama Canal* (New York: Penguin, 2009).

26. Kramer, *The Blood of Government*; Julian Go, *Patterns of Empire: The British and American Empires, 1688 to the Present* (New York: Cambridge University Press, 2011); Alfred W. McCoy and Francisco A. Scarano, *Colonial Crucible: Empire in the Making of the Modern American State* (Madison: University of Wisconsin Press, 2009); William Appleman Williams, *The Tragedy of American Diplomacy* (Cleveland: World Publishing Co., 1959).

27. James A. Thomas, *Trailing Trade a Million Miles* (Durham: Duke University Press, 1931), 38.

28. Robert F. Durden, "Tar Heel Tobacconist in Tokyo, 1899–1904," *North Carolina Historical Review* 53, no. 4 (October 1976): 347–63; Cox, *The Global Cigarette*, 39–43.

29. *Tobacco (London)*, 22 (262) (1902), 475, quoted in Cox, *The Global Cigarette*, 77.

30. Jane Burbank and Frederick Cooper, *Empires in World History: Power and the Politics of Difference* (Princeton: Princeton University Press, 2010); Philip J. Stern, "History and Historiography of the English East India Company: Past, Present, and Future!," *History Compass* 7, no. 4 (July 2009): 1146–80; Philip J. Stern, "The Ideology of the Imperial Corporation: 'Informal' Empire Revisited," in "Chartering Capitalism: Organizing Markets, States, and Publics," ed. Emily Erikson, special issue, *Political Power and Social Theory* 29 (2015): 15–43. The English East India Company developed a lively trade with Guangzhou, China, beginning in the eighteenth century. This trade funneled Chinese goods to Britain, the Commonwealth, and the US. When trade deficits grew, the East India Company insisted that China take opium grown in India. China's refusal led to the Opium Wars, waged by Britain to increase its imperial reach into China's economy.

31. Mrinalini Sinha, *Specters of Mother India: The Global Restructuring of an Empire* (Durham: Duke University Press, 2006), 17–19. Sinha's heuristic of the "imperial social formation" to denote the "historical role of imperialism in assembling different societies into a system" that is "globally articulated," and her way of investigating the specificity of that system, has influenced my approach. See also Catherine Lutz, "Empire Is in the Details," *American Ethnologist* 33, no. 4 (November 2006): 593–611.

32. In 1844, US diplomat Caleb Cushing went to China and extracted the Treaty of Wanghia, which matched, and even expanded, British trading and extraterritorial rights. The US and Britain, then, politically excluded China from what Teemu Ruskola calls the "Euro-American Family of Nations." Ruskola, *Legal Orientalism*, 10–112, 126–31, 138–39. See also Louise Conrad Young, "Rethinking Empire: Lessons from Imperial and Post-Imperial Japan," in *The Oxford Handbook of the Ends of Empire*, ed. Martin Thomas and Andrew Thompson (Oxford: Oxford University Press, 2017).

33. Ruskola, *Legal Orientalism*; Cochran, *Big Business in China*; Michael H. Hunt, *The Making of a Special Relationship: The United States and China to 1914* (New York: Columbia University Press, 1983).

34. Jones, *Yellow Music*, 65.

35. Jones, *Yellow Music*; Field, *Shanghai's Dancing World*; Hanchao Lu, *Beyond the Neon Lights: Everyday Shanghai in the Early Twentieth Century* (Berkeley: University of California Press, 1999); Leo Ou-Fan Lee, *Shanghai Modern: The Flowering of a New Urban Culture in China, 1930–1945* (Cambridge: Harvard University Press, 1999); Sherman Cochran, ed., *Inventing Nanjing Road: Commercial Culture in Shanghai, 1900–1945* (Ithaca: Cornell East Asia Program, 1999).

36. Lee Parker, interview by Burton Beers, June 1980, Kessler Papers, Southern Historical Collections, Manuscripts Department, University of North Carolina, Chapel Hill.

Chapter 1

1. I am combining attention to business, innovation, and commodity circulation with a range of cultural studies methods for examining "things," including brands, such as appear in Arjun Appadurai, ed., *The Social Life of Things: Commodities in Cultural Perspective* (New York: Cambridge University Press, 1986) 3–63; Bill Brown, "Thing Theory," *Critical Inquiry* 28, no. 1 (Autumn 2001): 1–22; Bill Brown, "Reification, Reanimation, and the American Uncanny," *Critical Inquiry* 32, no. 2 (Winter 2005); Modern Girl Around the World Research Group, *The Modern Girl Around the World: Consumption, Modernity, and Globalization* (Durham: Duke University Press, 2008); Sara Ahmed, "Affective Economies," *Social Text* 22, no. 2 (Summer 2004): 117–39.

2. Relli Shechter, *Smoking, Culture and Economy in the Middle East: The Egyptian Tobacco Market, 1850–2000* (London: I. B. Tauris, 2006), 54–56.

3. Mary C. Neuburger, *Balkan Smoke: Tobacco and the Making of Modern Bulgaria* (Ithaca: Cornell University Press, 2013), 55–58.

4. Kristin L. Hoganson, *Consumers' Imperium: The Global Production of American Domesticity, 1865–1920* (Chapel Hill: University of North Carolina Press, 2007); Mona Domosh, *American Commodities in an Age of Empire* (New York: Routledge 2006); Rob Wilson and Christopher Leigh Connery, eds., *The Worlding Project: Doing Cultural Studies in the Era of Globalization* (Berkeley: North Atlantic Books, 2007).

5. Nannie M. Tilley, *The Bright-Tobacco Industry, 1860–1929* (Chapel Hill: University of North Carolina Press, 1948); "The Beginning of a Trust," *Colliers* 34, no. 20 (August 10, 1907): 15. The FS Kinney Company and the Bedrossian Brothers of New York City had included bright leaf in some cigarette blends on a small scale.

6. Brian Burns, *Lewis Ginter: Richmond's Gilded Age Icon* (Charleston: History Press, 2011), 103.

7. Quoted in Tilley, *The Bright-Tobacco Industry*, 124.

8. "Lewis Ginter: Death of This Public-Spirited Citizen," *Richmond Times*, October 3, 1897, 1–3; Burns, *Lewis Ginter*, 16–38.

9. "City's Greatest Loss," *Richmond State*, October 3, 1897, 1, 3; "Lewis Ginter," *Richmond Times*, October 3, 1897, 2.

10. Burns, *Lewis Ginter*, 34–35. The wholesale store was a partnership with John F. Alvey and was called Ginter and Alvey Fancy Dry Goods. See Ann Smart Martin, *Buying into the World of Goods: Early Consumers in Backcountry Virginia* (Baltimore: Johns Hopkins University Press, 2008), for a discussion of the spread of global products through antebellum rural Virginia.

11. Burns, *Lewis Ginter*, 55–56. Ginter, a Democrat, bankrolled the white supremacist newspaper, the *Richmond Times*, whose editor was known for attacks on the interracial Readjuster movement. Duke and James A. Thomas of Reidsville were Republicans.

12. "Lewis Ginter," *Richmond Times*, October 3, 1897, 1–3.

13. Peter J. Rachleff, *Black Labor in the South: Richmond, Virginia, 1865–1890* (Philadelphia: Temple University Press, 1984), 204, 206. In the 1870 census, the Allen Tobacco Company listed only 11 employees; by 1880 Allen & Ginter employed 300 people: 200 women, 40 men, and 60 children.

14. The number of imports, branches of foreign companies, and immigrant producers continued to increase. The Eckmeyer and Co., headed by German immigrant brothers, opened in 1875; the brothers imported the famous Compagnie LaFerme brand cigarettes of St. Petersburg, Russia, and manufactured Russian- and Egyptian-style cigarettes. The Monopol European Model Tobacco Works Factory, based in St. Petersburg, Russia, opened its US branch factory in 1882, and soon added a salesroom uptown on Broadway. *Illustrated New York: The Metropolis of Today, 1888* (New York: International Publishing Company, 1888), 112, 127.

15. The industry did not grow in this decade, according to tax receipts. Tilley, *The Bright-Tobacco Industry*, 507; *United States Tobacco Journal*, May 1, 1877. The Bedrossian Brothers Company began production in 1867. See "The Beginning of a Trust," *Colliers* 34, no. 20 (August 10, 1907): 15.

16. Tilley, *The Bright-Tobacco Industry*, 508.

17. Before the 1890s, the word "cigarette" referred both to paper cigarettes and to little cigars, tobacco wrapped in tobacco leaf rather than in paper.

18. Shechter, *Smoking, Culture, and Economy*, 28.

19. Tilley, *The Bright-Tobacco Industry*, 506.

20. Joseph A. Schumpeter, *The Theory of Economic Development: An Inquiry into Profits, Capital, Credit, Interest, and the Business Cycle*, trans. Redvers Opie ([1934]; New Brunswick: Transaction Books, 1983), 78, 91–92; Joseph A. Schumpeter, *Capitalism, Socialism, and Democracy* ([1942]; London: Unwin University Books, 1965), 81–86.

21. Quoted in Tilley, *The Bright-Tobacco Industry*, 507.

22. James A. Thomas, *A Pioneer Tobacco Merchant in the Orient* (Durham: Duke University Press, 1928), 7.

23. *New York Times*, July 15, 1878, 4.

24. Robert N. Proctor, *Golden Holocaust: Origins of the Cigarette Catastrophe and the Case for Abolition* (Berkeley: University of California Press, 2012), 34–35, 231.

25. Tilley, *The Bright-Tobacco Industry*, 508.

26. John Morgan Richards, *With John Bull and Jonathan: Reminiscences of Sixty years of an American's Life in England and in the United States*, 2nd ed. (London: T. Werner Laurie, 1905), 66–67. Richards also noted that when he arrived, "there was not a single American cigarette sold in this country."

27. "Major Ginter Dead," *Richmond Dispatch*, October 3, 1897, 2.

28. Richards, *With John Bull and Jonathan*, 68.

29. Richards, *With John Bull and Jonathan*, 68; John Bain, Jr., with Carl Werner, *Cigarettes in Fact and Fancy* (Boston: H. M. Caldwell Co., 1906), 59; Mackenzie, *Sublime Tobacco*, 277. Allen & Ginter advertised the "'Ole Virginny' Cigarette and Tobacco Stores" in a Lon-

don paper, promising, "Samples given away upon application at 217, Piccadilly." *Funny Folks*, December 8, 1877, 184.

30. Quoted in Burns, *Lewis Ginter*, 103.

31. Elspeth H. Brown, "The Commodification of Aesthetic Feeling: Race, Sexuality, and the 1920s Stage Model," *Feminist Studies* 40, no. 1 (2014): 65–97; Ahmed, "Affective Economies.".

32. Amy Milne-Smith, "A Flight to Domesticity? Making a Home in the Gentlemen's Clubs of London, 1880–1914," *Journal of British Studies* 45, no. 4 (October 2006): 799–802.

33. David W. Blight, *Race and Reunion: The Civil War in American Memory* (Cambridge: Belknap Press of Harvard University Press, 2001).

34. Advertisement, *London Times*, April 16, 1878, 12; advertisement, *Funny Folks*, December 29, 1877, 212.

35. Quoted in *London Times*, April 16, 1878, 12. Bright leaf has a lower nicotine content than other tobaccos, but because it is more easily inhaled into the lungs, nicotine absorption rate is higher.

36. Milne-Smith, "A Flight to Domesticity?"; and Amy Milne-Smith, "Club Talk: Gossip, Masculinity, and the Importance of Oral Communities in late Nineteenth-Century London," *Gender and History* 21, no. 1 (April 2009): 86–106.

37. Milne-Smith, "A Flight to Domesticity?"; Judith R. Walkowitz, *Nights Out: Life in Cosmopolitan London* (New Haven: Yale University Press, 2012); Alan Sinfield, *Wilde Century: Effeminacy, Oscar Wilde, and the Queer Moment* (New York: Columbia University Press, 1994).

38. Quoted in Matthew Hilton, *Smoking in British Popular Culture, 1800–2000* (Manchester: Manchester University Press, 2000), 53.

39. Quoted in Hilton, *Smoking in British Popular Culture*, 56.

40. "American Goods in England," *New Brunswick Daily Times* (New Jersey), March 11, 1879, 3; advertisement, *New York Times*, May 28, 1880, 12.

41. Burns, *Lewis Ginter*, 15–21, 22–26.

42. "City's Greatest Loss," *Richmond State*, October 3, 1897, 3.

43. "Lewis Ginter," *Richmond Times*, October 3, 1897, 2.

44. Richards, *With John Bull and Jonathan*, 66–68; "Lewis Ginter," *Richmond Times*, October 3, 1897, 1–3; "Major Ginter Dead," *Richmond Dispatch*, October 3, 1897, 1.

45. Washington Duke to James B. Duke, October 17, 1894, Benjamin N. Duke Papers, Duke University.

46. H. G. Cocks, "Secrets, Crimes, and Diseases, 1800–1914," in *A Gay History of Britain: Love and Sex Between Men Since the Middle Ages*, ed. Matt Cook et al. (Oxford/Westport, CT: Greenwood World Publishing, 2007), 139.

47. Richards, *With John Bull and Jonathan*. 68.

48. Bain, *Cigarettes in Fact and Fancy*, 28.

49. "Lewis Ginter," *Richmond Times*, October 3, 1897, 2.

50. Advertisement, *New York Times*, June 8, 1878, 5.

51. James P. Wood, *The Industries of Richmond: Her Trade, Commerce, Manufactures and Representative Establishments* (Richmond: Metropolitan Publishing Company, 1886), 59–60, https://archive.org/stream/industriesofrich00wood#page/n1/mode/2up; Tilley, *The Bright-Tobacco Industry*, 509; "The Cigarette Manufacture," *Frank Leslie's Illustrated Magazine*, Febru-

ary 10, 1883, 419–22. See also "City News," *Richmond Times*, October 23, 1888. "Lewis Ginter," *Richmond Times*, October 3, 1897, 2, notes that Ginter sold cigarettes in Hamburg, Brussels, India, South Africa, and Australia.

52. Timothy Mitchell, *Colonising Egypt* (New York: Cambridge University Press, 1988).

53. Shechter, *Smoking, Culture, and Economy*, 58.

54. Shechter, *Smoking, Culture, and Economy*, 32.

55. Shechter, *Smoking, Culture, and Economy*, 54. Shechter argues that Egyptian producers were not only mimicking British orientalism but were, in drawing on imagery that had gained widespread recognition as referencing the Middle East, also creating a complex set of popular nationalist symbols.

56. Bain, *Cigarettes in Fact and Fancy*, 23–24.

57. John Boileau, "Voyageurs on the Nile," *Legion Magazine*, January 1, 2004, https://legionmagazine.com/en/2004/01/voyageurs-on-the-nile/.

58. Advertisement, *Illustrated London News*, October 18, 1884, 382.

59. Advertisement, *Illustrated London News*, December 1, 1884, 26. Crown Prince Frederick was the future Emperor Frederick III; the Prince of Wales was the future King Edward VII. Thanks to Wang Haochen for identifying the figures.

60. "Lewis Ginter," *Richmond Times*, October 3, 1897, 2; The same story was referenced in a more abbreviated form a decade earlier in "Richmond Virginia," *Harper's Weekly*, January 15, 1887, 42.

61. Shechter, *Smoking, Culture, and Economy*, 58.

62. Manos Haritatos and Penelope Giakoumakis, *A History of the Greek Cigarette* (Athens: ELIA The Hellenic Literary and Historical Archive, 1997), 172; Shechter, *Smoking, Culture, and Economy*, 58–59.

63. Hoganson, *Consumers' Imperium* 114, 177.

64. Hoganson, *Consumers' Imperium*, 26. Hoganson notes that orientalist enthusiasm raged for things "Moorish, Turkish, Chinese, Japanese, or a combination thereof."

65. Richard B. Tennant, *The American Cigarette Industry: A Study in Economic Analysis and Public Policy* (New Haven: Yale University Press, 1950), 44–45.

66. Haritatos and Giakoumakis, *A History of the Greek Cigarette*, 147.

67. Bain, *Cigarettes in Fact and Fancy*, 63–64.

68. Leslie Hannah, "The Whig Fable of American Tobacco, 1895–1913," *Journal of Economic History* 66, no. 1 (March 2006): 50.

69. Bain, *Cigarettes in Fact and Fancy*, 64. See also Haritatos and Giakoumakis, *A History of the Greek Cigarette*, 174.

70. The ATC in 1902 and BAT in 1908 tried to gain control of the tobacco supply in Egypt. Shechter, *Smoking, Culture, and Economy* 85. See also Neuburger, *Balkan Smoke*; *New York Times*, January 25, 1903, 4.

71. *Report of the Commissioner of Corporations on the Tobacco Industry*, Part III (Washington, DC: Government Printing Office, 1915), 325–28.

72. *New York Times*, December 1, 1904, 2.

73. Bain, *Cigarettes in Fact and Fancy*, 24.

74. Andrew Steinmetz, *The Smoker's Guide, Philosopher and Friend: What to Smoke, What

to Smoke With, and the Whole 'What's What' of Tobacco (London: Hardwicke and Bogue, 1876), 48, 62.

75. Mrinalini Sinha, *Colonial Masculinity: The "Manly Englishman" and the "Effeminate Bengali" in the Late Nineteenth Century* (New York: Manchester University Press, 1995); Heather Streets-Salter, *Martial Races: The Military, Race, and Masculinity in British Imperial Culture, 1857–1914* (New York: Manchester University Press, 2004); James Eli Adams, *Dandies and Desert Saints: Styles of Victorian Masculinity* (Ithaca: Cornell University Press, 1995); Jeffrey Dyer, "Desert Saints or Lions Without Teeth? British Portrayals of Bedouin Masculinity in the Nineteenth-Century Arabian Peninsula," *Arab Studies Journal* 17, no. 1 (Spring 2009): 85–97.

76. Quoted in Howard Cox, *The Global Cigarette: Origins and Evolution of British American Tobacco, 1880–1945* (Oxford: Oxford University Press, 2000), 47.

77. "Major Ginter Dead," *Richmond Dispatch*, October 3, 1897, 1.

78. "Lewis Ginter," *Richmond Times*, October 3, 1897, 2.

79. "Mr. John Pope Dead," *Richmond Dispatch*, April 9, 1896, 1.

80. Quoted in Burns, *Lewis Ginter*, 128.

81. "Lewis Ginter," *Richmond Times*, October 3, 1897, 3.

82. London's West End provided men increasingly predictable and semipublic places to find male sexual partners. See Walkowitz, *Nights Out*; Cocks, "Secrets, Crimes and Diseases, 1800–1914," 117, 170.

83. Sinfield, *The Wilde Century*, 67–75. See also Hilton, *Smoking in British Popular Culture*, 53–56.

84. Matthew Hilton labels the image of Frederick Gustavus Burnaby smoking a cigarette "feminine," based on twentieth-century dichotomies, and then cannot understand how Burnaby's military prowess was simultaneously celebrated. I am avoiding words like "effete" and "feminine" as descriptors because they shift historically; to use them as transparent descriptors obstructs such analysis. See Hilton, *Smoking in British Popular Culture*, 54–55.

85. *New York Times*, January 28, 1892, 1.

86. W. B. Taylor to J. B. Duke, August 23, 1922. JB Duke Papers, Duke University.

87. Carol Benedict, *Golden-Silk Smoke: A History of Tobacco in China, 1550–2010* (Berkeley: University of California Press, 2011), 134; Fang Xiantang, *Shanghai jindai minzu juanyan gongye* [Shanghai's modern domestic cigarette industry] (Shanghai: Shanghai shehui kexueyuan chubanshe, 1989), 6.

88. Fang, *Shanghai jindai minzu juanyan gongye*, 6. Fang notes that Little Beauties were introduced in 1885, though he misidentified this Allen & Ginter brand as a Duke brand.

89. Robert Franklin Durden, *The Dukes of Durham, 1865–1929* (Durham: Duke University Press, 1975), 19, 29.

90. Advertisement for Peacock Cigarettes, *Shenbao*, May 25, 1896, 8a Advertisement for Melachrino is drawn from *Tobacco* (London) from 1895 and is reprinted in Cox, *The Global Cigarette*, plate 5.

91. Fang, *Shanghai jindai minzu juanyan gongye*, 6.

92. The American Trading Company opened in 1892, three years before the practice was legal; the Mercantile Tobacco Company opened in 1893; the American Cigarette Company in 1897; the Taipei cigarette factory in 1898; and the Murai cigarette factory in 1897. See Fang,

Shanghai jindai minzu juanyan gongye, 10; "Ghost Head advertisement," *Shenbao*, November 8, 1900,; "Orchid advertisement," *Shenbao*, November 29, 1900.

93. Benedict, *Golden-Silk Smoke*, 135.

94. Fang, *Shanghai jindai minzu juanyan gongye*, 10.

95. The American Tobacco Company, Wills, and British American Tobacco all engaged in this practice. See, for example, *North China Herald*, February 21, 1900, 338; May 23, 1900, 944; April 24, 1901, 818; September 27, 1907, 762–65.

96. Benedict, *Golden-Silk Smoke*, 134.

97. Catherine Yeh, *Shanghai Love: Courtesans, Intellectuals, and Entertainment Culture, 1850–1910* (Seattle: University of Washington Press, 2006), 32–37, 91, 94.

98. Antonia Finnane, *Changing Clothes in China: Fashion, History, Nation* (New York: Columbia University Press, 2008), 101.

99. Timothy Mitchell makes this point about Cairo in *Colonising Egypt*, 161–64.

100. Yeh, *Shanghai Love*, 32, 91, 94.

101. Gail Hershatter, *Dangerous Pleasures: Prostitution and Modernity in Twentieth-Century Shanghai* (Berkeley: University of California Press, 1997).

102. James Lafayette Hutchinson, *China Hand* (Boston: Lothrop, Lee and Shepard Co., 1936), 98.

103. Lee Parker and Ruth Dorval Jones, *China and the Golden Weed* (Ahoskie, NC: Herald Publishing Co., 1976), 122–23.

104. Benedict, *Golden-Silk Smoke*, 134.

105. Yeh, *Shanghai Love*, 26, 31, 32. There were several classes of female sex workers in Qing China; these courtesans were of the highest class. When they did form sexual relationships with elite clients, such clients had to follow extensive rituals and rules and pay large sums of money in order for the proprietor to accept them. Proprietors virtually never accepted foreigners as clients. Foreigners had no problem finding sex workers in China, but they were of another class.

106. Yeh, *Shanghai Love*, 50.

107. *Shenbao*, June 12, 1912.

108. Yeh, *Shanghai Love*, 68.

109. Rickshaw drivers also drew commentary as early smokers of cigarettes in public. Unlike high-class courtesans, rickshaw drivers were very low on the ladder of income and status in China. *Shenbao*, May 12, 1911, 12.

110. Michael H. Hunt, *The Making of a Special Relationship: The United States and China to 1914* (New York: Columbia University Press, 1983), 237–42; Sherman Cochran, *Big Business in China: Sino-Foreign Rivalry in the Cigarette Industry, 1890–1930* (Cambridge: Harvard University Press, 1980), 45–52; Erika Lee, *At America's Gates: Chinese Immigration During the Exclusion Era, 1882–1943* (Chapel Hill: University of North Carolina Press, 2003); Karl Gerth, *China Made: Consumer Culture and the Creation of the Nation* (Cambridge: Harvard University Asia Center, Harvard University Press, 2003), 127–28; Jane Leung Larson, "Articulating China's First Mass Movement: Kang Youwei, Liang Qichao, the Baohuanghui, and the 1905 Anti-American Boycott," *Twentieth-Century China* 33, no. 1 (November 2007): 4–26.

111. Quoted in Cochran, *Big Business in China*, 47–48.

112. Cheng Renjie, "Ying Mei Yan Gongsi Maiban Zheng Bozhao" [British American

Tobacco Corporation's comprador Zheng Bozhao], in *Zhonghua Wenshi Ziliao Wenku* [*Chinese Literary and Historical Material Collection, Volume 14*] (Beijing: Zhongguo Wenshi Chubanshe, 1996), 741–51.

Chapter 2

1. "Virginia Affairs," *Baltimore Sun*, December 21, 1889; George Arents testimony, transcript, Chancery of New Jersey: John P. Stockton and American Tobacco Company, 18–19, James Buchanan Duke Papers, David M. Rubenstein Rare Book & Manuscript Library, Duke University (hereafter Duke Papers). This first-chartered ATC would own stocks in the existing five cigarette companies, originally outside the capabilities of corporations. Lawrence Friedman notes that legislatures did grant special charters that allowed this. See Lawrence M. Friedman, *A History of American Law*, 3rd ed. (New York: Touchstone Publishers, 2001), 395.

2. George Arents testimony, transcript, Chancery of New Jersey: John P. Stockton and American Tobacco Company, 19, Duke Papers.

3. Philip J. Stern, "The Ideology of the Imperial Corporation: 'Informal' Empire Revisited," in "Chartering Capitalism: Organizing Markets, States and Publics," ed. Emily Erikson, special issue, *Political Power and Social Theory* 29 (2015): 15–43; Ann Laura Stoler, "On Degrees of Imperial Sovereignty," *Public Culture* 18, no. 1 (Winter 2006): 125–46.

4. Morton J. Horwitz, "Santa Clara Revisited: The Development of Corporate Theory," *West Virginia Law Review* 88, no. 2 (1985–86): 173–224; Gregory A. Mark, "The Personification of the Business Corporation in American Law," *University of Chicago Law Review* 54, no. 4 (Autumn 1987): 1441–83; David Millon, "Theories of the Corporation," *Duke Law Journal* 1990, no. 2 (April 1990): 201–62; Lawrence M. Friedman, "The Law of Corporations," in Friedman, *A History of American Law*, 3rd ed. (New York: Touchstone, 2001); William G. Roy, *Socializing Capital: The Rise of the Large Industrial Corporation in America* (Princeton: Princeton University Press, 1997); Teemu Ruskola, *Legal Orientalism: China, the United States, and Modern Law* (Cambridge: Harvard University Press, 2013); Ruth H. Bloch and Naomi Lamoreaux, "Corporations and the Fourteenth Amendment," in Naomi R. Lamoreaux and William J. Novak, eds., *Corporations and American Democracy* (Cambridge: Harvard University Press, 2017), 286–325.

5. Martin J. Sklar, *The Corporate Reconstruction of American Capitalism, 1890–1916: The Market, the Law, and Politics* (New York: Cambridge University Press, 1988), 49–50.

6. Max Weber, *The Protestant Ethic and the Spirit of Capitalism*, trans. Talcott Parsons ([1904–5] Boston: Unwin Hyman, 1930); Max Weber, *General Economic History*, trans. Frank H. Knight ([1927]; repr., Mineola, NY: Dover Publications, 2003); Andrew Zimmerman, *Alabama in Africa: Booker T. Washington, the German Empire, and the Globalization of the New South* (Princeton: Princeton University Press, 2010), 73, 213; Ruskola, *Legal Orientalism*, 43–47, 63, 83; Arjun Appadurai, "The Spirit of Calculation," *Cambridge Journal of Anthropology* 30: 1 (Spring 2012): 8–9; Bill Maurer, "Repressed Futures: Financial Derivatives' Theological Unconscious," *Economy and Society* 31, no. 1 (February 2002): 15–36.

7. Alfred D. Chandler, Jr., *The Visible Hand: The Managerial Revolution in American Business* (Cambridge: Belknap Press, 1977); Robert H. Wiebe, *The Search for Order, 1877–1920* (New York: Hill and Wang, 1967).

8. Edward W. Said, *Orientalism* (New York: Pantheon Books, 1978). My perspective here draws on Dipesh Chakrabarty, *Provincializing Europe: Postcolonial Thought and Historical Difference* (Princeton: Princeton University Press, 2000); Johannes Fabian, *Time and the Other: How Anthropology Makes Its Object* (New York: Columbia University Press, 1983); Gayatri Chakravorty Spivak, "Can the Subaltern Speak?," in Marxism and the Interpretation of Culture, ed. C. Nelson and L. Grossberg (Basingstoke: Macmillan Education, 1988): 271–313; Mrinalini Sinha, *Specters of Mother India: The Global Restructuring of an Empire* (Durham: Duke University Press, 2006).

9. Important exceptions are Bruno Latour, *We Have Never Been Modern*, trans. Catherine Porter (Cambridge: Harvard University Press, 1993); Appadurai, "The Spirit of Calculation," 3–17; Mary Poovey, *Genres of the Credit Economy: Mediating Value in Eighteenth- and Nineteenth-Century Britain* (Chicago: University of Chicago Press, 2008).

10. Brian Burns, *Lewis Ginter: Richmond's Gilded Age Icon* (Charleston, SC: History Press, 2011), 103.

11. Burns, *Lewis Ginter*, 90.

12. For example, Bayard Rustin adopted his partner, Walter Naegle, who was thirty-seven years his junior, in order "to strengthen Naegle's claims as heir and closest relative." See John D'Emilio, *Lost Prophet: The Life and Times of Bayard Rustin* (New York: The Free Press, 2003), 507. For other examples, see Kao Beck, "How Marriage Inequality Prompts Gay Partners to Adopt One Another," *Atlantic*, November 27, 2013. http://www.theatlantic.com/national/archive/2013/11/how-marriage-inequality-prompts-gay-partners-to-adopt-one-another/281546/.

13. "The Cigarette Deal," *Richmond Dispatch*, January 5, 1890, 8; "Allen and Ginter Sell," *Richmond Dispatch*, January 4, 1890, 1.

14. "Lewis Ginter," *Richmond Times*, October 3, 1897, 2.

15. Joseph A. Schumpeter, *Capitalism, Socialism, and Democracy* (New York: Harper and Row, 1942). The story of Duke had not always been told this way, nor has there been such tight consensus. For example, "The Beginnings of a Trust," *Colliers*, August 10, 1907, 15–16, discusses Duke's "quick grasp" of the advantages of the cigarette machine, but Earl Mayo, "The Tobacco War," *Leslie's Magazine*, March 1903, 517–26, tells the story of Duke's career without mention of the cigarette machine.

16. Joseph A. Schumpeter, "Economic Theory and Entrepreneurial History," in *Change and the Entrepreneur: Postulates and the Patterns of Entrepreneurial History*, Research Center in Entrepreneurial History (Cambridge: Harvard University Press, 1949), 75–83; Joseph A. Schumpeter, "The Creative Response in Economic History," *Journal of Economic History* 7, no. 2 (November 1947): 149.

17. Alvin J. Silk and Louis William Stern, "The Changing Nature of Innovation in Marketing: A Study of Selected Business Leaders, 1852–1958," *Business History Review* 37, no. 3 (Autumn 1963): 182–99.

18. Patrick G. Porter, "Origins of the American Tobacco Company," *Business History Review* 43, no. 1 (Spring 1969): 59–76.

19. Chandler, *The Visible Hand*, 382–91. Chandler's overemphasis on Duke predated Porter's article. Alfred D. Chandler, Jr., "The Beginnings of 'Big Business' in American Industry," *Business History Review* 33, no. 1 (Spring 1959): 8–9.

20. Milton Alexander and Edward M. Mazze, eds., *Sales Management: Theory and Practice* (New York: Pitman Publishing, 1965), 32–34; Robert Sobel, *The Entrepreneurs: An American Adventure* (New York: Houghton Mifflin Harcourt, 1986); Maury Klein, *The Change Makers: From Carnegie to Gates: How Great Entrepreneurs Transformed Ideas into Industries* (New York: Times Books, 2003); E. Wayne Nafziger, *Economic Development* (Cambridge: Cambridge University Press, 2012), 383. Many historians repeat the myth without mentioning Schumpeter, including Allan M. Brandt, *The Cigarette Century: The Rise, Fall, and Deadly Persistence of the Product that Defined America* (New York: Basic Books, 2007); Robert N. Proctor, *The Golden Holocaust: Origins of the Cigarette Catastrophe and the Case for Abolition* (Berkeley: University of California Press, 2011).

21. Nannie Tilley, *The Bright-Tobacco Industry, 1860–1929* (Chapel Hill: University of North Carolina Press, 1948), 569–70.

22. Equivalence determined using measuringworth.com.

23. Richard B. Tennant, *The American Cigarette Industry: A Study in Economic Analysis and Public Policy* (New Haven: Yale University Press, 1950), 18; *Report of the Commissioner of Corporations on the Tobacco Industry Part I* (Washington DC, 1909–15), 67; B. W. E. Alford, *W. D. and H. O. Wills and the Development of the U.K. Tobacco Industry, 1786–1965* (London: Methuen and Co., Ltd., 1973), 140.

24. Tilley, *The Bright-Tobacco Industry*, 570–73; quote is from 573.

25. *Report of the Commissioner of Corporations on the Tobacco Industry Part I*, 66; the Everitt and the Cowman were both patented in 1882, before Duke used any machine. Tennant, *The American Cigarette Industry*, 147; Alford, *W. D. and H. O. Wills*, 141; The Cowman Manufacturing Company, US Patent 262177A, filed/issued August 1, 1882, www.google.com/patents/US262177/. Leslie Hannah is the one scholar who has rigorously challenged Duke's outsized reputation, though he focuses mostly on a later period and his relationship to European industry. Hannah calls cigarette machines an example of "multiple, simultaneous invention," in "The Whig Fable of American Tobacco, 1895–1913," *Journal of Economic History* 66, no. 1 (March 2006): 64.

26. Tilley, *The Bright-Tobacco Industry*, 573.

27. Robert F. Durden, *The Dukes of Durham, 1865–1929* ([1975]; Durham: Duke University Press, 1987), 32; James B. Duke testimony, *US v. American Tobacco Company* (1908), copy in Duke Papers; "Girls on a Strike," *New York Times*, June 6, 1888, 6; Tilley, *The Bright-Tobacco Industry*, 574.

28. For descriptions of deskilled handrolling, see *New York Times*, August 16, 1875, 2; B. W. C. Roberts and Richard F. Knapp, "Paving the Way for the Tobacco Trust: From Hand Rolling to Mechanized Production by W. Duke, Sons and Company," *North Carolina Historical Review* 69, no. 3 (July 1992): 260.

29. James P. Wood, *The Industries of Richmond: Her Trade, Commerce, Manufactures and Representative Establishments* (Richmond: Metropolitan Publishing Company, 1886), 59–60.

30. Patricia A. Cooper, *Once a Cigar Maker: Men, Women, and Work Culture in American Cigar Factories, 1900–1919* (Urbana: University of Illinois Press, 1987), 12–15, 18, 25–29; *Richmond Times*, October 23, 1888, 1.

31. "Cigarette-Rollers on Strike," *New York Times*, May 3, 1883, 8; "City and Suburban News," *New York Times*, November 26, 1885, 8.

32. Tilley, *The Bright-Tobacco Industry*, 557. Duke made only nine million cigarettes that year, not enough to register on the radar of his large competitors.

33. Durham's population: http://www.historync.org/NCCityPopulations1800s.htm; Richmond's population: (2004), "Historical Census Browser," accessed December 29, 2015, University of Virginia, Geospatial and Statistical Data Center,: http://mapserver.lib.virginia .edu/; and New York City's population: http://www.demographia.com/db-nyc2000.htm. Gavin Wright, *Old South, New South: Revolutions in the Southern Economy Since the Civil War* (New York: Basic Books, 1986), 7–10.

34. "Working Girls Badly Frightened," *New York Times*, December 27, 1884, 2. Brother artisanal cigarette rollers David and J. M. Siegel ran the Durham cigarette departments at Blackwell Tobacco Company (1880) and W. Duke and Sons (1881) respectively. Both had previously worked for New York companies and recruited former coworkers to Durham. By 1885, the brothers started their own Durham cigarette company. Hiram V. Paul, *History of the Town of Durham, NC* (Raleigh: Edwards, Broughton and Company, 1884), 117–88.

35. "C.M.P.U. No. 27, Durham, N.C., Chartered!" *Progress* 3, no. 2 (September 26, 1884): 1; "Correspondence," *Progress* 3, no. 6 (January 24, 1885): 3; *Progress* 3, no. 10 (May 29, 1885): 3; Tilley, *The Bright-Tobacco Industry*, 519.

36. Quoted in Tilley, *The Bright-Tobacco Industry*, 519.

37. "Working in a Tobacco Factory: Interview with Laura Cox in the Durham *Morning Herald*, January 17, 1926," *Learn NC*, www.learnnc.org/lp/pages/4701, accessed December 14, 2010.

38. *New York Times*, October 28, 1886.

39. One source of historians' exaggeration of Duke's advantage is the *Report of the Commissioner of Corporations on the Tobacco Industry Part II* (Washington DC, 1909–15), 95–97. Studying the 1889 domestic sales records and company assets of the five companies that merged to become the ATC, the commission noted that W. Duke and Sons sold 834 million cigarettes to Allen & Ginter's 517 million in 1889. It also noted that Ginter put up only 8.7 percent of his capital in tangible assets, with the rest being goodwill, while Duke put up 14.1 percent of his tangible assets, and correspondingly less goodwill. This baffled the commission. Why was Ginter's goodwill worth more when he had lower sales? Why did he and Duke have an equal share in the corporation? The commission concluded that the businessmen did not know what they were doing, and that the industry was too undeveloped to have generated significant brand value. However, the commission failed to consider, or even acknowledge, overseas markets, where Ginter had a much more extensive business. The misleading statistics and erroneous interpretation are repeated in Tennant, *The American Cigarette Industry*, 24.

40. Durden, *Dukes of Durham*, 32, 36.

41. *Richmond Dispatch*, March 16, 1890, 3.

42. William Henry Harrison Cowles, "James Albert Bonsack," 49 H. R. 2950, House and Senate Reports, Reports on Public Bill (March 5, 1886); downloadable from ProQuest Congressional.

43. H. D. and H. O. Wills installed Bonsack machines in January 1884, over a year before Duke did so. See Alford, *W. D. and H. O. Wills*, 148, 154–55.

44. *Tobacco* (London) 4, no. 38 (1884), 18, cited in Howard Cox, *The Global Cigarette:*

Origins and Evolution of the British Tobacco Company (London: Oxford University Press, 2000), 21.

45. James B. Duke to D. B. Strouse, March 27, 1889, James B. Duke Papers, Duke University.

46. Durden, *Dukes of Durham*, 41.

47. Durden, *Dukes of Durham*, 42.

48. Durden, *Dukes of Durham*, 28–36, 50.

49. James B. Duke to D. B. Strouse, March 20, 1889, Duke Papers.

50. James B. Duke to D. B. Strouse, March 27, 1889, Duke Papers.

51. James B. Duke to D. B. Strouse, March 20, 1889, Duke Papers.

52. D. B. Strouse to James B. Duke, July 8, 1887, Duke Papers.

53. Durden, *Dukes of Durham*, 41; *Report of the Commissioner of Corporations on the Tobacco Industry Part I*, 4, found that a motive for forming the ATC was to gain control over the Bonsack and other machines.

54. D. B. Strouse to James B. Duke, March 23, 1889; James B. Duke to D. B. Strouse, March 27, 1889, Duke Papers.

55. "Virginia Affairs," *Baltimore Sun*, December 20, 1889, supplement 2.

56. *Baltimore Sun*, December 21, 1889; *Raleigh News and Observer*, December 25, 1889, 16.

57. "Virginia Affairs: Excitement over the Incorporation of the American Tobacco Company," *Baltimore Sun*, December 21, 1889, 9; *Raleigh News and Observer*, December 25, 1889, 16; Tilley, *The Bright-Tobacco Industry*, 269–70.

58. "Allen and Ginter Sell," *Richmond Dispatch*, January 4, 1890, 1; *Baltimore Sun*, December 21, 1889, 9; *New York Times*, February 1, 1890, 2.

59. "The Transfer Made," *Richmond Dispatch*, March 18, 1890, 1.

60. Arents testimony, 12, 19–24, Duke Papers.

61. Arents testimony, 37, 43–49, Duke Papers.

62. Horwitz, "*Santa Clara* Revisited," 194; Christopher Grandy, *New Jersey and the Fiscal Origins of Modern Corporation Law* (New York: Garland Publishing, 1993), 13–15, 40–53.

63. Horwitz, "*Santa Clara* Revisited," 187; see also Thomas R. Navin and Marian V. Sears, "The Rise of a Market for Industrial Securities, 1887–1902," *Business History Review* 26, no. 2 (June 1955): 115–16.

64. "Will Another Trust Be Formed?," *Reidsville Review*, March 25, 1891. The motivation to form mergers was enhanced by high tariffs. See Colleen A. Dunlavy, "Why Did American Business Get So Big?," *Audacity*, Spring 1994, 42–49.

65. W. B. Taylor to James B. Duke, August 23, 1922, Duke Papers.

66. "Blow to a Great Trust," *New York Times*, November 19, 1895, 1; "Indicted Under the Trust Law," *New York Times*, May 6, 1890, 1; "Trusts Attacked in Virginia," *New York Times*, February 1, 1892, 1; "Fighting the Tobacco Trust," *New York Times*, March 17, 1893, 10; "American Tobacco Company," *New York Times*, May 16, 1897, 1; "Tobacco Trust Attacked," *New York Times*, November 17, 1893, 2. Georgia passed an antimonopoly law in 1897 that made ATC contracts in the state invalid. "Georgia after Trusts," *Wall Street Journal*, January 8, 1897, 1.

67. John Wilbur Jenkins, *James B. Duke, Master Builder* (New York: George H. Doran Co., 1927), 96–97.

68. "Fighting the Tobacco Trust," *New York Times*, March 17, 1893, 10.

69. "Iowa Wholesale Grocers Refuse to Handle ATC Products," *Wall Street Journal*, April 10, 1897, 1.

70. Richard White, *Railroaded: The Transcontinentals and the Making of Modern America* (New York: W. W. Norton & Co., 2011), 111.

71. White, *Railroaded*, 513.

72. Jeffrey Sklansky, "William Leggett and the Melodrama of the Market," in *Capitalism Takes Command: The Social Transformation of Nineteenth-Century America*, ed. Michael Zakim and Gary J. Kornblith (Chicago: University of Chicago Press, 2012), 199–222.

73. *Report of the Commissioner of Corporations on the Tobacco Industry* (Washington, DC, 1909–15), III, 49; *Report of the Commissioner of Corporations on the Tobacco Industry* (Washington, DC, 1909–15), I, xxxi.

74. *Report of the Commissioner of Corporations on the Tobacco Industry I*, 326. See also "To Fight the Tobacco Trust," *New York Times*, September 23, 1898, 1.

75. Ginter opposed the form of a trust during the 1889 ATC negotiations. Brief for defendants and points in support of demurer / [*by Richard V. Lindabury, W. W. Fuller, Joseph H. Choate*], Newark, N.J., 1896, downloadable from *The Making of Modern Law* (Gale, Cengage Learning); George Arents, testimony, 18, Duke Papers.

76. "Mr. Butler's Charges Denied," *New York Times*, November 27, 1893, 3.

77. "Secretary Arents' Denial," *Richmond Dispatch*, February 20, 1895, 1.

78. James B. Duke, testimony, circuit court Mss, 1908, Duke Papers; Durden, *Dukes of Durham*, 63.

79. Leslie Hannah makes this argument in detail. See "The Whig Fable of American Tobacco," 47–54.

80. Jenkins, *James B. Duke*, 95–96.

81. Ginter and his allies tried unsuccessful to shift the composition of the board of directors in 1896; they also appeared to be in independent negotiations with St. Louis plug manufacturers Liggett and Myers and Drummond, the main companies that held out against the ATC's takeover attempts. Duke's biographer claims, without documentation, that Ginter convinced Oliver Payne to buy up ATC stock but then Payne voted with Duke rather than Ginter at the meeting. Though possible, it appears that Payne did not become a major stockholder until after Ginter's death. It's likely that Ginter was working with other financiers, possibly Moore and Schley or DeCordova and Henry Allen and Co., who did buy up stock in March 1896, when Ginter was making his most strenuous move against Duke. Pope's death in April 1896 may have spoiled these plans. *Wall Street Journal*, January 18, 1896, 1; "Tobacco," *Wall Street Journal*, March 11, 1896, 1; *Wall Street Journal*, March 17, 1896; *Wall Street Journal*, March 20, 1896; Jenkins, *James B. Duke*, 100–102.

82. Quote from *Milwaukee Journal*, February 21, 1896; "John Pope," *Virginia Magazine of History and Biography* 4, no. 3 (January 1897), 320–21; Burns, *Lewis Ginter*, 180–83.

83. *Wall Street Journal*, May 1, 1897, 2. See also *New York Times*, May 2, 1897. The other original incorporators were also out of the way. Kinney was already off the board, and Kimball had died in 1895.

84. "Lewis Ginter," *Richmond Times*, October 3, 1897, 1.

85. *Progress*, 3, no. 10 (May 29, 1885): 3.

86. A. E. Watkin, "Baron, Bernhard (1850–1929)," rev. Christine Clark, in *Oxford Dictionary of National Biography*, ed. H. C. G. Matthew and Brian Harrison (Oxford: Oxford University Press, 2004), http//www.oforddnb.come/view/article/30614; Christine Clark, "Baron, Bernhard," in *Dictionary of Business Biography: A Biographical Dictionary of Business Leaders Active in Britain in the Period 1860–1980*, ed. David J. Jeremy, ed. (London: Butterworths, 1984), 177–80; Report of the Commission of Corporations on Tobacco, part 2, 117, notes that the NCTC's brands had no goodwill, indicating that the brands had not established a presence in the market.

87. "Can Use the Cigarette Machines," *New York Times*, May 2, 1894, 8.

88. See http://www.measuringworth.com; "To Get Salary of $50,000 a Year," *New York Times*, February 12, 1895, 6.

89. "Mr. Butler's Charges Denied," *New York Times*, November 27, 1893, 3.

90. "Mr. Butler's Charges Denied," *New York Times*, November 27, 1893, 3.

91. *Wall Street Journal*, June 18, 1895, 1.

92. Brief for defendants and points in support of demurer / [by Richard V. Lindabury, W.W. Fuller, Joseph H. Choate] Newark, N.J., 1896, downloadable from *The Making of Modern Law* (Gale, Cengage Learning). See also *New York Times*, December 28, 1892, 10; "Tobacco Trust Rights," *New York Times*, November 18, 1896, 1.

93. "Tobacco Trust Attacked," *New York Times*, November 17, 1893, 2.

94. *New York Times*, December 13, 1896, 7; March 13, 1897, 14; November 18, 1896, 1.

95. "Blow to a Great Trust," *New York Times*, November 19, 1895, 3.

96. A demurrer is a formal legal objection that asserts that the facts presented by the opponent, while true, do not entitle the plaintiff to prevail in the suit. http://legal-dictionary.thefreedictionary.com/demurrer.

97. *Brief for defendants and points in support of demurer*, 58, 57, 100; quotation is on p. 57.

98. "Tobacco Trust Victory," *New York Times*, March 13, 1897, 14.

99. "Tobacco Men to Be Tried," *New York Times*, January 23, 1897, 4; *New York Times* November 18, 1896, 13.

100. "Tobacco Jurors Disagree," *New York Times*, June 30, 1897, 12.

101. "Narrow Escape of the Combine," *Weekly News and Courier* (Charleston, SC), July 7, 1897, 13.

102. Mark, "The Personification of the Business Corporation in American Law," 1443, 1472.

103. Teemu Ruskola, "Ted Turner's Clitoridectomy, Etc.: Gendering and Queering the Corporate Person," Berkshire Conference on Women's History, Toronto, May 23, 2014.

104. Ruth H. Bloch and Naomi Lamoreaux rightly push against misconceptions about corporate personhood in "Corporations and the Fourteenth Amendment," in Naomi R. Lamoreaux and William J. Novak, eds., *Corporations and American Democracy*, 286–325. The Supreme Court did not award Fourteenth Amendment "liberty rights" to corporations until the second third of the twentieth century. The authors recognize that corporations tried to claim Fourteenth Amendment rights ever since its passage, succeeding at times though not at the Supreme Court. The ATC's cases are particular to the New Jersey incorporation law and its validity; the line they are engaging is whether corporations are public or private. Other mecha-

nisms did arise to regulate corporations, including antitrust laws, and the Supreme Court disbanded the ATC in 1911. However, this was after twenty-one years of operation.

105. Michael A. Ross, "Justice Miller's Reconstruction: The Slaughter-House Cases, Health Codes, and Civil Rights in New Orleans, 1861–1873," *Journal of Southern History* 64, no. 4 (November 1998): 649–76; Horwitz, "Santa Clara Revisited," 173–224.

106. Mark, "The Personification of the Business Corporation in American Law," 1482–83.

107. Amy Dru Stanley, *From Bondage to Contract: Wage Labor, Marriage, and the Market in the Age of Slave Emancipation* (Cambridge: Cambridge University Press, 1998), x, x–xii, 55–56.

108. Barbara Young Welke, *Law and the Borders of Belonging in the Long Nineteenth Century United States* (New York: Cambridge University Press, 2010), 3–4, 22, 141–43, 146.

109. Welke, *Law and the Borders of Belonging*, 119–40; David S. Cecelski and Timothy B. Tyson, eds., *Democracy Betrayed: The Wilmington Race Riot of 1898 and Its Legacy* (Chapel Hill: University of North Carolina Press, 1998); Glenda Elizabeth Gilmore, *Gender and Jim Crow: Women and the Politics of White Supremacy in North Carolina, 1896–1920* (Chapel Hill: University of North Carolina Press, 1996); Nancy MacLean, *Freedom Is Not Enough: The Opening of the American Workplace* (Cambridge: Harvard University Press, 2006).

110. Welke, *Law and the Borders of Belonging*, 103; Nayan Shah, *Contagious Divides: Epidemics and Race in San Francisco's Chinatown* (Berkeley: University of California Press, 2001), 65–66.

111. Ruskola, *Legal Orientalism*, 141–48; Shah, *Contagious Divides*, 166; Moon-Ho Jung, *Coolies and Cane: Race, Labor, and Sugar in the Age of Emancipation* (Baltimore: Johns Hopkins University Press, 2006), 4–5, 12, 85.

112. Navin and Sears, "The Rise of a Market for Industrial Securities," 106–10; Sklar, *The Corporate Reconstruction of American Capitalism*, 85.

113. Quotes from Sklar, *The Corporate Reconstruction of American Capitalism*, 49, 50.

114. Susan Strasser, *Satisfaction Guaranteed: The Making of the American Mass Market* (New York: Pantheon Books, 1989), 43–44, 47; Tilley, *The Bright-Tobacco Industry*, 522–23.

115. *Milwaukee Journal*, February 24, 1897, 7.

116. *Report of the Commissioner of Corporations on the Tobacco Industry Part II*, 6–7, 87–97, 216, 282; Tennant, *The American Cigarette Industry*, 36–37.

117. Strasser, *Satisfaction Guaranteed*, 47.

118. Financiers gained control of the ATC board of directors in the 1890s, consolidated power and profits in their hands, and raised cash for foreign expansion. Thomas F. Ryan, P. A. B. Widener, A. N. Brady, and tobacconist William Butler formed the Union Tobacco Company, acquired several of Duke's competitors, and offered to sell to Duke with a very high price tag, including places for financiers on the board of directors. In 1901, Duke and the financiers created the Consolidated Tobacco Company and offered stockholders of the ATC and Continental to exchange their stocks for bonds that guaranteed a return on investment. Continental had never earned dividends, so stockholders turned in their stock. The financiers had inside knowledge, however, that a sizable tobacco tax was about to be revoked, which would generate a spike in revenue. This move allowed just six men to control a massive amount of stock and to reserve the gains after the tax revocation for themselves. It also made $40 million available in cash for foreign expansion. *Report of the Commissioner of Corporations on the Tobacco Industry*, part I, 73–74, 114–25; Tennant, *The American Cigarette Industry*, 35; Howard

Cox, *The Global Cigarette: Origins and Evolution of British American Tobacco 1880–1945* (New York: Oxford University Press, 2000), 65–66; "American Tobacco Litigation," *New York Times*, April 22, 1896, 12.

119. Ruskola, *Legal Orientalism*, 125–39.

120. Paul A. Kramer, *The Blood of Government: Race, Empire, the United States, and the Philippines* (Chapel Hill: University of North Carolina Press, 2006).

121. Mira Wilkins, *The Emergence of Multinational Enterprise: American Business Abroad from the Colonial Era to 1914* (Cambridge: Harvard University Press, 1970), 91–92.

122. Robert F. Durden, "Tar Heel Tobacconist in Tokyo, 1899–1904," *North Carolina Historical Review* 53, no. 4 (October 1976): 347–49; Cox, *The Global Cigarette*, 40, 73; Letter of Appointment, April 24, 1900, Edward J. Parrish Papers, Duke University.

123. Quoted in Durden, "Tar Heel Tobacconist in Tokyo, 1899–1904," 358; Sherman Cochran, *Big Business in China: Sino-Foreign Rivalry in the Cigarette Industry, 1890–1930* (Cambridge: Harvard University Press, 1980), 40–41.

124. Quoted in Cox, *The Global Cigarette*, 77.

125. Ruskola, *Legal Orientalism*, 5–6.

126. Trevor K. Plante, "US Marines in the Boxer Rebellion," *Prologue Magazine* 31, no. 4 (Winter 1999), National Archives and Records Administration Website, https://www.archives.gov/publications/prologue/1999/winter/boxer-rebellion-1.html/.

127. Among the many scholars who critique the term "informal empire" are Ann Laura Stoler, "On Degrees of Imperial Sovereignty," *Public Culture* 18, no. 1 (Winter 2006): 125–46; Ruskola, *Legal Orientalism*; Philip J. Stern, "The Ideology of the Imperial Corporation: 'Informal' Empire Revisited," in "Chartering Capitalism: Organizing Markets, States and Publics," ed. Emily Erikson, special issue, *Political Power and Social Theory* 29 (2015): 15–43.

Chapter 3

1. Drew A. Swanson, *A Golden Weed: Tobacco and Environment in the Piedmont South* (New Haven: Yale University Press, 2014); Barbara Hahn, *Making Tobacco Bright: Creating an American Commodity, 1617–1937* (Baltimore: Johns Hopkins University Press, 2011); Howard Cox, *The Global Cigarette: Origins and Evolutions of British American Tobacco, 1880–1945* (Oxford: Oxford University Press, 2000), 202; Mary C. Neuburger, *Balkan Smoke: Tobacco and the Making of Modern Bulgaria* (Ithaca: Cornell University Press, 2013), 222; Steven C. Rupert, *A Most Promising Weed: A History of Tobacco Farming and Labor in Colonial Zimbabwe, 1890–1945* (Athens: Ohio University Press, 1998), 3–25; Nannie May Tilley, *The Bright-Tobacco Industry 1860–1929* (Chapel Hill: University of North Carolina Press, 1948), 385–86. Bright leaf would spread to Australia, India, Brazil, Zimbabwe, Canada, and Bulgaria and beyond. Efforts began early in Zimbabwe, where European settlers initiated the farming of bright leaf, and in Australia, England, and Germany.

2. Sources informing my understanding of the bright leaf network's assemblage of human and nonhuman elements include Paul A. Kramer, "Imperial Openings: Civilization, Exemption, and the Geopolitics of Mobility in the History of Chinese Exclusion, 1868–1910," *Journal of the Gilded Age and Progressive Era* 14 (2015): 317; Aihwa Ong and Stephen J. Collier, eds., *Global Assemblages: Technology, Politics, and Ethics as Anthropological Problems* (Malden, MA: Wiley-Blackwell, 2005); Bruno Latour, *Reassembling the Social: An Introduction to Actor-*

Network-Theory (Oxford: Oxford University Press, 2005), 11–12; Bill Brown, "Thing Theory," *Critical Inquiry* 28, no. 1 (Autumn 2001): 1–22; Swanson, *A Golden Weed*. For a critique of the idea of exporting Jim Crow, see Paul A. Kramer, *The Blood of Government: Race, Empire, the United States, and the Philippines* (Chapel Hill: University of North Carolina Press, 2006), 21.

3. Elizabeth D. Esch, *The Color Line and the Assembly Line: Managing Race in the Ford Empire* (Berkeley: University of California Press, 2018); David R. Roediger and Elizabeth D. Esch, *The Production of Difference: Race and the Management of Labor in US History* (New York: Oxford University Press, 2012).

4. Grace Elizabeth Hale, *Making Whiteness: The Culture of Segregation in the South, 1890–1940* (New York: Pantheon, 1998), 85–119; Micki McElya, *Clinging to Mammy: The Faithful Slave in Twentieth-Century America* (Cambridge: Harvard University Press, 2007), 38–73; Jane Hunter, *The Gospel of Gentility: American Women Missionaries in Turn-of-the-Century China* (New Haven: Yale University Press, 1984), 128–73; Amy Kaplan, "Manifest Domesticity," *American Literature* 70, no. 3 (September, 1998): 581–606; Ann Laura Stoler, *Carnal Knowledge and Imperial Power: Race and the Intimate in Colonial Rule* (Berkeley: University of California Press, 2002).

5. John Gary Anderson, *Autobiography* (Rock Hill, SC; Lakeland, FL: privately printed, 1936), 106.

6. James A. Thomas, *A Pioneer Tobacco Merchant in the Orient* (Durham: Duke University Press, 1928), 5–6; Anderson, *Autobiography*, 106.

7. "An Outline of the Business of Messrs. A. Motley & Co.," *Reidsville Review*, February 27, 1889, 6; *Reidsville Review*, June 29, 1884, 4; *Reidsville Review*, March 4, 1891, 3; Thomas, *A Pioneer Tobacco Merchant*, 7–10.

8. *Report of the Commissioner of Corporations on the Tobacco Industry I*, 326. See also "To Fight the Tobacco Trust," *New York Times*, September 23, 1898, 1.

9. Thomas, *A Pioneer Tobacco Merchant*, 23–24.

10. North Carolina Council: Jane Gregory Marrow Oral History Interview, June 1, 1980, East Carolina Manuscripts Collection, J. Y. Joyner Library, East Carolina University (hereafter cited as East Carolina Manuscripts Collection); Jane Gregory Marrow Oral History Interview, January 26, 1976, East Carolina Manuscripts Collection.

11. North Carolina Council: Jane Gregory Marrow Oral History Interview, June 1, 1980, East Carolina Manuscripts Collection; Jane Gregory Marrow Oral History Interview, January 26, 1976, East Carolina Manuscripts Collection; Thomas, *A Pioneer Tobacco Merchant*, 308–9.

12. R. H. Gregory, diary, Richard Henry Gregory Papers, Duke University Library (hereafter cited as Richard Henry Gregory Papers).

13. R. H. Gregory to Kate Arrington, January 12, 1906, Richard Henry Gregory Papers.

14. Sherman Cochran, *Big Business in China: Sino-Foreign Rivalry in the Cigarette Industry, 1890–1930* (Cambridge: Harvard University Press, 1980), 46.

15. Thomas, *A Pioneer Tobacco Merchant*, 46; R. H. Gregory, diary, June 1906, Richard Henry Gregory Papers.

16. Thomas, *A Pioneer Tobacco Merchant*, 43; R. H. Gregory, diary, June 1906.

17. Swanson, *A Golden Weed*, 50–51.

18. R. H. Gregory to Kate Arrington, January 12, 1906, Richard Henry Gregory Papers.

19. R. H. Gregory, diary, June 1906, Richard Henry Gregory Papers.

20. R. H. Gregory, diary, June 16, 1906, Richard Henry Gregory Papers.

21. Southern farmers only picked bright leaf tobacco leaves when they ripened, rather than harvesting the entire plant at once. R. H. Gregory diary, August 7, 1906, Richard Henry Gregory Papers.

22. Thomas, *A Pioneer Tobacco Merchant*, 43. Sherman Cochran argues that Chinese tobacco farmers did not have to learn new techniques to grow bright leaf tobacco. Sherman Cochran, *Big Business in China: Sino-Foreign Rivalry in the Cigarette Industry, 1890–1930* (Cambridge: Harvard University Press, 1980), 26–27.

23. BAT continued to experiment with Chinese seeds, according to Chen Han-seng's 1937 study of Chinese bright leaf tobacco cultivation in Shandong, Anhui and Henan Provinces. Chen Han-seng [Chen Hansheng], *Industrial Capital and Chinese Peasants: A Study of the Livelihood of Chinese Tobacco Cultivators* (Shanghai: Kelly and Walsh, Ltd, 1939), 6.

24. Swanson, *A Golden Weed*, 52.

25. Thomas, *A Pioneer Tobacco Merchant*, 45.

26. Thomas, *A Pioneer Tobacco Merchant*, 44.

27. Sherman Cochran, *Encountering Chinese Networks: Western, Japanese and Chinese Corporations in China, 1880–1937* (Berkeley: University of California Press, 2000), 53.

28. R. H. Gregory to A. G. Jeffress, November 12, 1915, James Augustus Thomas Papers, David M. Rubenstein Rare Book & Manuscript Library, Duke University (hereafter cited as James A. Thomas Papers).

29. Chen, *Industrial Capital and Chinese Peasants*. See also Hsu Yung-sui [Xu Yongsui], "Tobacco Marketing in Eastern Shantung," *Institute of Pacific Relations* 74, U-4-5 (1935–1937): 171–75; Chang Ka-tu [Zhang Jiatuo], "Impressions from Three Months Observation in the Eastern Shantung Tobacco Region," *Dong Fang Za Zhi* 32, no. 6 (March 1936): 109–13. It is possible that Chen's given name was Cifang and that Wang's given name was Yanci.

30. Chen, *Industrial Capital and Chinese Peasants*, 7–9; Hsu, "Tobacco Marketing in Eastern Shantung," 172.

31. Hsu, "Tobacco Marketing in Eastern Shantung," 171, 172, 174.

32. Chen, *Industrial Capital and Chinese Peasants*, 7–9; Hsu, "Tobacco Marketing in Eastern Shantung," 172.

33. R. H. Gregory to A. G. Jeffress, November 12, 1915, James A. Thomas Papers.

34. Chen, *Industrial Capital and Chinese Peasants*, 11; Hsu, "Tobacco Marketing in Eastern Shantung," 172.

35. Chen, *Industrial Capital and Chinese Peasants*, 5–11.

36. Before sharecropping, owners pushed for the delayed-wage system, another system based in debt. Swanson, *A Golden Weed*, 162–63, 170–79, 181. See also Pete R. Daniel, *Breaking the Land: The Transformation of Cotton, Tobacco, and Rice Cultures Since 1880* (Champaign-Urbana: University of Illinois Press, 1985), 23–38; Eric Foner, *Reconstruction: America's Unfinished Revolution, 1863–1877* (New York: Harper and Row, 1988), 124–76.

37. Hsu, "Tobacco Marketing in Eastern Shantung," 175.

38. R. H. Gregory to A. G. Jeffress, November 12, 1915, James A. Thomas Papers, Duke University.

39. The Hong'an Real Estate Corporation was formed by nine Chinese businessmen in order to purchase for BAT. Though their connection to BAT was well known, they negotiated

with the necessary officials to be registered as a Chinese company. See Cochran, *Big Business in China*, 141; Cheng Renjie, "Ying Mei Yan Gongsi Maiban Zheng Bozhao" ["British American Tobacco Corporation's comprador Zheng Bozhao"], in *Zhonghua Wenshi Ziliao Wenku* [Chinese Literary and Historical Material Collection, Volume 14] (Beijing: Zhongguo Wenshi Chubanshe, 1996), 747–48.

40. Chang, "Impressions from Three Months Observation in the Eastern Shantung Tobacco Region," 109.

41. Chen, *Industrial Capital and Chinese Peasants*, 24–27. For a discussion of a Chinese capitalist's efforts to keep Shandong warlords at bay see Brett Sheehan, *Industrial Eden: A Chinese Capitalist Vision* (Cambridge: Harvard University Press, 2015), 39–44.

42. Cochran, *Big Business in China*, 144.

43. C.W. Pettitt to James A. Thomas, March 14, 1921, James A. Thomas Papers.

44. R. H. Gregory to A. G. Jeffress, cc. James A. Thomas and Thomas Cobbs, December 7, 1915, James A. Thomas Papers.

45. Cochran, *Big Business in China*, 145, 199.

46. Thomas, *A Pioneer Tobacco Merchant*, 85–86.

47. Quoted in Cochran, *Big Business in China*, 75.

48. R. H. Gregory to A. G. Jeffress, November 11, 1915, James A. Thomas Papers, .

49. Chang, "Impressions from Three Months Observation," 111.

50. R. H. Gregory to A. G. Jeffress, cc. James A. Thomas and Thomas Cobbs, December 7, 1915, James A. Thomas Papers.

51. James Lafayette Hutchinson, *China Hand* (Boston: Lothrop, Lee and Shepard Co., 1936), 4–5. Hutchinson received $25 per week plus expenses.

52. Lee Parker, interview by Burton Beers, June 1980, Kessler Papers, Southern Historical Collection, Manuscripts Department, Wilson Library, University of North Carolina at Chapel Hill.

53. Lee Parker Interview, Kessler Papers.

54. Irwin S. Smith Oral History Interview, July 28, 1982, East Carolina Manuscript Collection.

55. Thomas, *A Pioneer Tobacco Merchant*, 86.

56. Hutchinson, *China Hand*, 4. See Tina Chen, *Double Agency: Acts of Impersonation in Asian American Literature and Culture* (Stanford: Stanford University Press, 2005), 35–59.

57. Hutchinson, *China Hand*, 14.

58. Contract, James N. Joyner Papers, East Carolina Manuscript Collection. See also oral history interviews with Gordon Cheatham, Robert Bostick, and James C. Richardson, East Carolina Manuscript Collection. After four years of service, marriage was allowed. For a discussion of the marriage restrictions in Delhi imposed by tobacco companies in the nineteenth century, see Stoler, *Carnal Knowledge and Imperial Power*, 29.

59. R. H. Gregory to Kate Arrington, January 12, 12, 1906, Richard Henry Gregory Papers.

60. Lee Parker and Ruth Dorval Jones, *China and the Golden Weed* (Ahoskie, NC: Herald Publishing Company, 1976), 13–15, 29.

61. Hutchinson, *China Hand*, 12.

62. C. Stuart Carr Oral History Interview, September 15, 1980, East Carolina Manuscript Collection; Hattie Gregory to Kate Arrington, September 7, 1913, Hattie A. Gregory Papers,

East Carolina Manuscript Collection, J. Y. Joyner Library, East Carolina University (hereafter Hattie A. Gregory Papers).

63. Fae Ceridwen Dussart, "'That Unit of Civilisation' and 'the Talent Peculiar to Women': British Employers and Their Servants in the Nineteenth-Century Indian Empire," *Identities: Global Studies in Culture and Power* 22, no. 6 (2015): 706–21.

64. North Carolina Council: Jane Gregory Marrow Oral History Interview, June 1, 1980, East Carolina Manuscript Collection.

65. Stoler, *Carnal Knowledge and Imperial Power*, 23; Hunter, *The Gospel of Gentility*, 166–67. Hunter discussed the "domestic empire" that missionaries maintained in China, noting that one directly compared their work to the disciplining of workers on Southern plantations.

66. Hattie Gregory to Kate Arrington, April 14, 1913, Hattie A. Gregory Papers.

67. Hattie Gregory to Kate Arrington, December 23, 1912, Hattie A. Gregory Papers.

68. Elizabeth Clark-Lewis, *Living In, Living Out: African American Domestics in Washington DC, 1910–1940* (Washington, DC: Smithsonian Institution Press, 1994); Tera W. Hunter, *To 'Joy My Freedom: Southern Black Women's Lives and Labors after the Civil War* (Cambridge: Harvard University Press, 1997).

69. Hattie Gregory to Mamma, September 7, 1913, Hattie A. Gregory Papers.

70. Hattie Gregory to Mamma, September 7, 1913, Hattie A. Gregory Papers.

71. Hattie Gregory to Mamma, September 4, 1918, Hattie A. Gregory Papers.

72. Hattie Gregory to Kate Arrington, December 17, 1914, Hattie A. Gregory Papers.

73. Clyde Gore Oral History Interview, April 28, 1977, East Carolina Manuscript Collection.

74. Hattie Gregory to Mamma, September 4, 1918, Hattie A. Gregory Papers.

75. Amy Kaplan, *The Anarchy of Empire in the Making of US Culture* (Cambridge: Harvard University Press, 2002), 25–26. See also Laura Wexler, *Tender Violence: Domestic Visions in an Age of U.S. Imperialism* (Chapel Hill: University of North Carolina Press, 2000); Ann Laura Stoler, ed., *Haunted by Empire: Geographies of Intimacy in North American History* (Durham: Duke University Press, 2006).

76. North Carolina Council: Jane Gregory Marrow Oral History Interview, June 1, 1980, East Carolina Manuscript Collection.

77. Hale, *Making Whiteness*, 244–50; Jennifer Lynn Ritterhouse, *Growing Up Jim Crow: How Black and White Southern Children Learned Race* (Chapel Hill: University of North Carolina Press, 2006).

78. Hattie Gregory to Kate Arrington, January 24, 1918, Hattie A. Gregory Papers.

79. Hattie Gregory to Kate Arrington, February 13, 1913, Hattie A. Gregory Papers.

80. North Carolina Council: Jane Gregory Marrow Oral History Interview, June 1, 1980, East Carolina Manuscript Collection.

81. Hattie Gregory to Kate Arrington, December 17, 1914, Hattie A. Gregory Papers.

82. John Duffy, *The Sanitarians: A History of American Public Health* (Urbana: University of Illinois Press, 1990); Ruth Rogaski, *Hygenic Modernity: Meanings of Health and Disease in Treaty-Port China* (Berkeley: University of California Press, 2004).

83. Mark M. Smith, *How Race Is Made: Slavery, Segregation, and the Senses* (Chapel Hill: University of North Carolina Press, 2006).

84. Quote from Edward L. Ayers, *Southern Crossing: A History of the American South, 1877–1906* (New York: Oxford University Press, 1995), 100; Hale, *Making Whiteness*, 87; see also Jacquelyn Dowd Hall, "The Long Civil Rights Movement and the Political Uses of the Past," *Journal of American History* 91, no. 4 (March 2005): 1233–63.

85. Hattie Gregory to Kate Arrington, December 17, 1914, Hattie A. Gregory Papers.

86. Donald E. MacInnis, "Francis Lister Hawks Pott," Biographical Dictionary of Chinese Christianity, http://www.bdcconline.net/en/stories/p/pott-francis-lister-hawks.php/.

87. Hattie Gregory to Kate Arrington, December 17, 1914, Hattie A. Gregory Papers.

88. Irwin Smith Oral History Interview, July 28, 1982, East Carolina Manuscript Collection.

89. Gregory J. Seigworth and Melissa Gregg, "An Inventory of Shimmers," and Ben Highmore, "Bitter after Taste: Affect, Food, and Social Aesthetics," in *The Affect Theory Reader*, ed. Gregory J. Seigworth and Melissa Gregg (Durham: Duke University Press, 2010), 1–25, 118–37.

90. Thomas, *A Pioneer Tobacco Merchant*, 78–79; James A. Thomas to E. S. Bowling, October 26, 1915, James A. Thomas Papers; Hattie Gregory to Kate Arrington, February 13, 1913; Hattie Gregory to Kate Arrington, December 23, 1912, Hattie A. Gregory Papers.

91. Hattie Gregory to Kate Arrington, February 13, 1913, Hattie A. Gregory Papers.

92. Hattie Gregory to Kate Arrington, April 14, 1913; Hattie Gregory to Kate Arrington, May 19, 1913, Hattie A. Gregory Papers.

93. North Carolina Council: Jane Gregory Marrow Oral History Interview, June 1, 1980, East Carolina Manuscript Collection.

94. Mayfair Mei-hui Yang, *Gifts, Favors, and Banquets: The Art of Social Relationships in China* (Ithaca: Cornell University Press, 1994), examines the post-1949 period, not the Republican era. Nevertheless, it provides valuable insight into the possible meaning of gift practices in relationship to banquets.

95. Frank H. Canaday, unpublished memoir, 40, Frank H. Canaday Papers, Harvard Yenching Library, Harvard University.

96. Parker and Jones, *China and the Golden Weed*, 122–23.

97. Hutchinson, *China Hand*, 70.

98. Hattie Gregory to Kate Arrington, December 23, 1912, Hattie A. Gregory Papers.

99. Hattie Gregory to Kate Arrington, December 23, 1912, Hattie A. Gregory Papers.

100. Hattie Gregory to Mamma, September 7, 1913, Hattie A. Gregory Papers.

101. Paul A. Kramer, "Power and Connection: Imperial Histories of the United States in the World," *American Historical Review* 116, no. 5 (December 2011): 1350.

Chapter 4

1. Ruby Delancy, interview with author and Kori Graves, June 19, 2001; Qian Meifeng, interview transcript excerpt, July 18, 1963, Economic Research Institute in Shanghai Academy of Social Sciences, eds., *Yingmeiyan gongsi zaihua qiye ziliao huibian* [Materials on the British American Tobacco Enterprises in China] (Beijing: Zhonghua shuju, 1983), 3, no. 1051; Qian Meifeng interview transcript, July 18, 1963, British American Tobacco Collection, Shanghai Academy of Social Sciences (hereafter cited as BAT Collection, SASS); Elizabeth J. Perry,

Shanghai on Strike: The Politics of Chinese Labor (Stanford: Stanford University Press, 1993), 149–50.

2. See Stephen J. Collier and Aihwa Ong, "Global Assemblages, Anthropological Problems," in *Global Assemblages: Technology, Politics, and Ethnics as Anthropological Problems*, ed. Stephen J. Collier and Aihwa Ong (Malden, MA: Blackwell, 2005), 11–14; Mona Domosh, "Labor Geographies in a Time of Early Globalization: Strikes Against Singer in Scotland and Russia in the Early 20th Century," *Geoforum* 39 (2008): 1676–86.

3. Ivy G. Riddick to William Morris, March 19, 1934, Ivy G. Riddick Papers, Southern Collection, University of North Carolina, Chapel Hill (hereafter cited as Riddick Papers).

4. Dipesh Chakrabarty, *Provincializing Europe: Postcolonial Thought and Historical Difference* (Princeton: Princeton University Press, 2000), 51–62; Michael Denning, "Representing Global Labor," *Social Text* 25: 3 (Fall 2007): 125–45; Lisa Lowe, "Work, Immigration, Gender: New Subjects of Cultural Politics," in Janice A. Radway, Barry Shank, Penny Von Eschen, and Kevin Gaines, eds., *American Studies: An Anthology* (Malden, MA: Wiley-Blackwell, 2009), 177–84; David R. Roediger and Elizabeth D. Esch, introduction to *The Production of Difference: Race and the Management of Labor in US History* (New York: Oxford University Press, 2012).

5. The foundational study of job segregation in the tobacco industry is Charles S. Johnson, *Patterns of Negro Segregation* (New York: Harper and Brothers, 1943). See also Dolores Janiewski, "Southern Honor, Southern Dishonor: Managerial Ideology and the Construction of Gender, Race, and Class Relations in Southern Industry," in *Work Engendered: Toward a New History of American Labor*, ed. Ava Baron (Ithaca: Cornell University Press, 1991), 70–91; Dolores E. Janiewski, *Sisterhood Denied: Race, Gender, and Class in a New South Community* (Philadelphia: Temple University Press, 1985); Robert Rodgers Korstad, *Civil Rights Unionism: Tobacco Workers and the Struggle for Democracy in the Mid-Twentieth-Century South* (Chapel Hill: University of North Carolina Press, 2003).

6. Sherman Cochran, *Big Business in China: Sino-Foreign Rivalry in the Cigarette Industry, 1890–1930* (Cambridge: Harvard University Press, 1980), 16, 129, 164, 137.

7. Daniel E. Bender and Jana K. Lipman, *Making the Empire Work: Labor and United States Imperialism* (New York: New York University Press, 2015); Elizabeth D. Esch, *The Color Line and the Assembly Line: Managing Race in the Ford Empire* (Berkeley: University of California Press, 2018); Robert Vitalis, *America's Kingdom: Mythmaking on the Saudi Oil Frontier* (Stanford: Stanford University Press, 2007); Julie Greene, *The Canal Builders: Making America's Empire at the Panama Canal* (London: Penguin Press, 2009); Jason M. Colby, *The Business of Empire: United Fruit, Race, and US Expansion in Central America* (Ithaca: Cornell University Press, 2011), 79–148.

8. Cochran, *Big Business in China*, 42–43, 45.

9. John Barrett, "America in China: Our Position and Opportunity," *North American Review* 175, no. 552 (November 1902): 660. See also Thomas J. McCormick, *China Market: America's Quest for Informal Empire, 1893–1901* (Chicago: Quadrangle Books, 1967).

10. Emily Honig, *Sisters and Strangers: Women in the Shanghai Cotton Mills, 1919–1949* (Stanford: Stanford University Press, 1986), 16; John Eperjesi, "The American Asiatic Association and the Imperialist Imaginary of the American Pacific," *boundary 2* 28, no. 1 (2001): 195–219.

11. Edward J. Parrish to James B. Duke, May 16, 1904, Edward J. Parrish Papers, Duke University, Durham. Parrish sent a carbon copy to London. Parrish had opposed a factory because the Japanese government reportedly held off from nationalizing earlier because of taxes it gained from Murai's East Asian exports.

12. Quoted in Howard Cox, "Learning to do Business in China: The Evolution of BAT's Cigarette Distribution Network, 1902–41," *Business History* 39, no. 3 (1997): 61.

13. Moon-Ho Jung, *Coolies and Cane: Race, Labor, and Sugar in the Age of Emancipation* (Baltimore: Johns Hopkins University Press, 2006), 5–13; Elliot Young, "Chinese Coolies, Universal Rights, and the Limits of Liberalism in an Age of Empire," *Past and Present* 227, no. 1 (May 2015): 121–49.

14. Nayan Shah, *Contagious Divides: Epidemics and Race in San Francisco's Chinatown* (Berkeley: University of California Press, 2001); Alexander Saxton, *The Indispensable Enemy: Labor and the Anti-Chinese Movement in California* ([1971]; Berkeley: University of California Press, 1995); Lawrence Glickman, "Inventing the 'American Standard of Living': Gender, Race, and Working-Class Identity, 1880–1925," *Labor History* 34 (1993): 221–35.

15. *Some Reasons for Chinese Exclusion: Meat vs. Rice, American Manhood Against Asiatic Coolieism, Which Shall Survive?* 57th Cong., 1st sess., S. Doc. No. 137 (1902) (statement of Samuel Gompers). Washington, DC: Government Printing Office.

16. Quoted in Michael Omi, review of *Thinking Orientals: Migration, Contact, and Exoticism in Modern America*, by Henry Yu, *Journal of Asian American Studies* 5, no. 2 (June 2002): 181.

17. *Report of the Mission to China of the Blackburn Chamber of Commerce, 1896–7: FSA Bourne's Section* (Blackburn: North-East Lancashire Press Company, 1898), 231.

18. James A. Thomas to Willard Straight, July 20, 1915, James A. Thomas Papers, Duke University, Durham (hereafter cited as Thomas Papers). As a member of the American Asiatic Association, Thomas opposed the Chinese Exclusion Act because it complicated business relationships in China. Eperjesi, "The American Asiatic Association and the Imperialist Imaginary of the American Pacific," 195–219; Paul A. Kramer, "Imperial Openings: Civilization, Exemption, and the Geopolitics of Mobility in the History of Chinese Exclusion," *Journal of the Gilded Age and Progressive Era* 14 (2015): 317–47.

19. James A. Thomas to R. L. Watt, July 31, 1915, Thomas Papers.

20. Julean Arnold and William H. Gale, "Labor and Industrial Conditions in China," *Trade Information Bulletin*, no. 75, October 30, 1922, 1.

21. New York produced 60 percent of the company's cigarettes in 1905; New Orleans produced 10 percent. Meyer Jacobstein, "The Tobacco Industry in the United States," *Columbia University Studies in History, Economics, and Public Law* 26, no. 3 (1907): 97.

22. This is the argument of Leslie Hannah, "The Whig Fable of American Tobacco, 1895–1913," *Journal of Economic History* 66, no. 1 (March 2006): 42–73, see especially page 45.

23. Tobacco historians have couched this as a "return" of the factory to the Reynolds family, for the ATC had taken over a Reynolds factory at the same site in 1899. However, in the subsequent twelve years the ATC had used Reynolds to consolidate dozens of chewing and smoking tobacco companies in the vicinity. When the Supreme Court split the ATC into four companies, RJ Reynolds received far more than the original family business. Nannie M. Tilley, *RJ Reynolds Tobacco Company* (Chapel Hill: University of North Carolina Press, 1985).

24. "Charles Penn Dies in New York," *Reidsville Review*, October 23, 1931, 1; "Reidsville Pays Honor to Its Beloved Son," *Reidsville Review*, October 26, 1931, 1. Penn was a philanthropist in Reidsville, giving the founding endowment for the Annie Penn hospital and contributing to many other causes.

25. The cigarette industry followed the much larger textile industry. See Jennifer Guglielmo, *Living the Revolution: Italian Women's Resistance and Radicalism in New York City, 1880–1945* (Chapel Hill: University of North Carolina Press, 2010), 44–78.

26. Quoted in Nannie M. Tilley, *The Bright-Tobacco Industry, 1860–1929* (Chapel Hill: University of North Carolina Press, 1948), 45.

27. Tilley, *RJ Reynolds*, 271–72.

28. Martha Gena Harris, interview by Dolores Janiewski, April 29, 1977, Southern Historical Collection, University of North Carolina, Chapel Hill.

29. *United States Tobacco Journal* 89, no. 2 (January 12, 1918): 5.

30. "Manufacture of Cigarettes to Commence Immediately," *Reidsville Review*, October 20, 1916, 1; "The Cigarette Factory an Accomplished Fact," *Reidsville Review*, October 24, 1916, 1. By 1924, the factory hired men to operate the machines and women to catch the cigarettes. "Reidsville Plan Turns Out 15 Million Cigarettes Daily," *Southern Tobacco Journal* 44, no. 29 (July 17, 1923): 2.

31. Korstad, *Civil Rights Unionism*, 43, 44, 70; "Remember When," *Reidsville Review*, August 19, 1932, 2; population figure from http://library.guilford.edu/c.php?g=142981&p=1220753.

32. "Our Oldest Tobacco Factory: The Forerunner of a Most Important Industry," *Reidsville Review*, February 27, 1889, 4.

33. Ruby Delancy, interview with author and Kori Graves, June 19, 2001.

34. Ruby Delancy, interview with author and Kori Graves, June 19, 2001.

35. Perry, *Shanghai on Strike*, 17.

36. Perry, *Shanghai on Strike*, 19, 26–27; Gail Hershatter, *Dangerous Pleasures: Prostitution and Modernity in Twentieth-Century Shanghai* (Berkeley: University of California Press, 1997).

37. US Bureau of Foreign and Domestic Commerce, Supplement to Commerce Reports Trade Information Bulletin No. 61 "Industrial Conditions in China," September 18, 1922, published by the Bureau of Foreign and Domestic Commerce, US Department of Commerce; Perry, *Shanghai on Strike*, 58.

38. Perry, *Shanghai on Strike*, 140. At Nanyang, stemmers received $1.00 per day, while cigarette workers received $0.85.

39. The word "foreman" is used in both the US and China for these jobs; I substitute the gender-neutral word "supervisor." In China, many supervisors were women.

40. I. G. Riddick, "Factory Report," April 1932, BAT Collection, SASS; Li Xinbao, interview transcript, BAT Collection, SASS; see also Perry, *Shanghai on Strike*, 26–29, 33, 39, 60, 140.

41. Dolores Janiewski, "Southern Honor, Southern Dishonor: Managerial Ideology and the Construction of Gender, Race and Class Relations in Southern Industry," in *Work Engendered*, 70–91; Perry, *Shanghai on Strike*, 137–42.

42. Perry, *Shanghai on Strike*, 138.

43. Wage Record Book, BAT Collection, SASS. The wage record book had no names, but

listed workers by number. The record was checked and a worker's fingerprint added when the wage was paid. See also *Eastern Times*, June 5, 1929, which refers to workers injured by the company police as "Nos 627 and 943."

44. Riddick, "Factory Report," April 1932, BAT Collection, SASS. BAT managers claimed they could not eliminate child labor because of this practice.

45. Dolores E. Janiewski, *Sisterhood Denied: Race, Gender, and Class in a New South Community* (Philadelphia: Temple University Press, 1985), 104.

46. Ruby Jones, quoted in Korstad, *Civil Rights Unionism*, 110.

47. Elizabeth "Lib" Chaney, interview with the author and Kori Graves; William Davis, interview with author and Kori Graves, July 5, 2001.

48. Perry, *Shanghai on Strike*, 138.

49. Evelyn Farthing, interviewed by Dawn K. Parrish, April 20, 1996, Historical Collection, Rockingham Community College.

50. Jonathan Daniels, *Tar Heels: A Portrait of North Carolina* (New York: Dodd, Mead and Company, 1941), 134.

51. Marion Troxler, interview with author and Kori Graves, July 13, 2001.

52. Patsy Cheatham, interview by Beverly Jones, July 9, 1979, Southern Historical Collection, University of North Carolina, Chapel Hill.

53. Ruby Delancy, interview with author and Kori Graves, June 19, 2001.

54. Perry, *Shanghai on Strike*, 142; *North China Herald*, November 15, 1925.

55. Perry, *Shanghai on Strike*, 141.

56. "No Strike," *New York Times*, February 9, 1880, 5.

57. L. I. Strickland interview, April 24, 1976, conducted by Treva S. Nunnally, Rockingham Community College, Historical Collection; untitled, *Reidsville Review*, November 6, 1903, 3.

58. Rice allowances came to be regulated by union contracts and were a regular and expected aspect of payment. See "Conversation at Mr. T. V. Soong's House Preliminary to the Signing of the Agreement Between the Shanghai BAT Factories and Their Workers on January 16th, 1928," BAT Collection, SASS.

59. Ruby Delancy, interview with author and Kori Graves, June 19, 2001.

60. Riddick, "Factory Report," April 1932, BAT Collection, SASS. Riddick's report contains the cost of living in Pudong, and noted that the problem of pilfering "may be largely attributed to those employees who draw less than 80 cents per day." Male workers started at 55 cents per day.

61. Qiao Jinding, interview transcript, July 16, 1963, BAT Collection, SASS; "Fracas Between the Workmen and Police of the BAT New Factory," *Eastern Times*, June 5, 1929, reports that a male worker was found with four cigarettes in the washroom. The attempt to bring him to the office resulted in a fight with numerous workers and other police, resulting in injuries. *North China Herald*, June 8, 1929.

62. *North China Herald*, February 2, 1929.

63. Zhang Yongsheng, interview transcript, August 7, 1958, BAT Collection, SASS; [family name illegible] Zikun, interview transcript, September 2, 1958, BAT Collection, SASS; Chen [given name illegible], interview transcript, July 12, 1963, BAT Collection, SASS; Hao

Lixiang, interview transcript, August 1958; *Yingmeiyan gongsi zaihua qiye ziliao huibian* (Beijing: Zhonghua shuju, 1983), 1115; see also "Excerpt about overseers in the factories narrated by former workers," 1116–18.

64. "Factory Workers' Grievance," *North China Herald*, June 26, 1920, BAT Collection, SASS.

65. Perry, *Shanghai on Strike*, 50.

66. Qian Meifeng, interview transcript, July 18, 1963, BAT Collection, SASS.

67. Perry, *Shanghai on Strike*, 143, 145.

68. Riddick, "Factory Report," April 1932, 6, BAT Collection, SASS.

69. Chen [given name illegible], interview transcript, July 12, 1963, BAT Collection, SASS.

70. Li Xinbao, interview transcript, August 11, 1958; Ma Wenyuan, interview transcript, no date, BAT Collection, SASS.

71. Zheng Yongsheng, interview transcript, August 7, 1958, BAT Collection; Hao Lixiang, interview transcript, no date, BAT Collection, SASS.

72. Mo Jianchun, interview transcript, no date, BAT Collection, SASS. Mo notes that some of the British foreign managers were transferred from Liverpool. The one manager that he names, however, was Ivy Riddick, who was from North Carolina and would later be in charge of the Shanghai factories.

73. Huang Zhihao, interview transcript, July 23, 1963, BAT Collection, SASS.

74. Zhang Yongsheng, interview transcript, August 7, 1958, BAT Collection, SASS.

75. Mu Guilan, interview transcript, no date, BAT Collection, SASS.

76. "Terms for Mutual Assistance Agreed Upon Between the BAT Factory Workers and the BAT Factory Authorities at Shanghai," January 16, 1928, BAT Collection, SASS.

77. These interviews with workers were conducted by Mao-era historians who were writing state-sponsored histories of capitalist tyranny. The historians clearly asked interviewees for condemnation of foreigners so I have used these sources with care. I have only included information that is specific and confirmed by more than two narrators. The commentary on foreigners is much more vague ("very cruel") than the stories about conflicts with Chinese supervisors. Workers did report in large numbers that foreigners "scolded," "hit," and pulled them by their hair, which is certainly believable.

78. Zhu Quanfa, interview transcript, June 12, 1958, BAT Collection, SASS; Zhao Qizhang, interview transcript, BAT Collection; Zhang Yongsheng, interview transcript, July 7, 1958, BAT Collection, SASS.

79. Cochran, *Big Business in China*, 2; the US Department of Commerce celebrated in 1922 that "in China there is a total lack of State regulations applying to labor." US Department of Commerce, "Industrial Conditions in China," Supplement to Commerce Reports Trade Information, Bulletin No. 61, September 18, 1922, Bureau of Foreign and Domestic Commerce.

80. Perry, *Shanghai on Strike*, 136.

81. "Terms for Mutual Assistance Agreed Upon Between the BAT Factory Workers and the BAT Factory Authorities at Shanghai," January 16, 1928, BAT Collection, SASS.

82. Perry, *Shanghai on Strike*, 79.

83. Perry, *Shanghai on Strike*, 136, 145–56; Cochran, *Big Business in China*, 180–85.

84. Elizabeth J. Perry, *Patrolling the Revolution: Worker Militias, Citizenship, and the Modern Chinese State* (New York: Rowman and Littlefield, 2006), 59–104.

85. James N. Joyner to James Y. Joyner, June 14, 1929, East Carolina Manuscripts Collection, J. Y. Joyner Library, East Carolina University.

86. Janet Irons, *Testing the New Deal: The General Textile Strike of 1934 in the American South* (Urbana: University of Illinois Press, 2000), 110–19.

87. Dora Scott Miller, interview conducted by Beverly W. Jones, June 6, 1979, Southern Historical Collection, 13; Martha Gina Harris, interview conducted by Dolores Janiewski, April 29, 1977, Southern Historical Collection, UNC; see also Tilley, *RJ Reynolds*, 278–79.

88. Dora Scott Miller, interview conducted by Beverly W. Jones, June 6, 1979, Southern Historical Collection; Tilley, *RJ Reynolds*, 149.

89. Bertie Pratt, interview, Southern Collection; Ruby Jones quoted in Korstad, *Civil Rights Unionism*, 111; Tilley, *RJ Reynolds*, 272; E. V. Boswell to E. Lewis Evans, September 1, 1935, Tobacco Workers International Union Records, Hornbake Library, University of Maryland (hereafter cited as TWIU Records); Ruby Delancy, interview with author and Kori Graves, June 19, 2001.

90. Dora Scott Miller, interview; Anne Mack Barbee, interview; Black quoted in Korstad, *Civil Rights Unionism*, 110; Danielle McGuire, *At the Dark End of the Street: Black Women, Rape, and Resistance—A New History of the Civil Rights Movement from Rosa Parks to the Rise of Black Power* (New York: Knopf, 2010).

91. William and Ruth Davis, interview with author and Kori Graves, Reidsville, NC, July 10, 2001; James Neal, interview with author and Kori Graves, July 6 2001; Radford Powell to E. Lewis Evans, June 20, 1937, TWIU Records.

92. Will Huff to E. Lewis Evans, February 16, 1935; "History of Local No. 191"; Pete Jeffers to E. Lewis Evans, March 9, 1935, TWIU Records.

93. Alice Williamson to E. Lewis Evans, March 9, 1935; S. M. Johnson to E. Lewis Evans, March 9, 1935; Earnest Gwynn to E. Lewis Evans, March 9, 1935; C. C. Caldwell to E. Lewis Evans, March 9, 1935; Henry Jones to E. Lewis Evans, March 9, 1935; Esther Jones to E. Lewis Evans, March 9, 1935; Russell Dill to E. Lewis Evans, March 9, 1935; E. Lewis Evans to Pete Jeffers, March 12, 1935, TWIU Records.

94. Radford G. Powell, "My Life as I Remember it with Local Union 192," 4, vertical file "American Tobacco Company," Rockingham Community College Library.

95. E. V. Boswell to E. Lewis Evans, May 4, 1935, TWIU Records.

96. "History of Local No. 191," n.d., TWIU Records.

97. Drew A. Swanson, *A Golden Weed: Tobacco and Environment in the Piedmont South* (New Haven: Yale University Press, 2014), 173.

98. Rev. William R. Jones, "Reminiscences of Growing Up on Rockhouse Creek, Rockingham County, North Carolina," *Journal of Rockingham County History and Genealogy* 20, no. 1 (June 1995): 33.

99. James Neal, interview with author and Kori Graves, July 6, 2001. For a history of the Klan in the 1920s see Linda Gordon, *The Second Coming of the KKK: The Ku Klux Klan of the 1920s and the American Political Tradition* (New York: W. W. Norton and Co., 2017), 107–10. Gordon argues that the KKK could take up the side of unions or owners, though it usually

supported owners. In Reidsville, the Klan seemed to include both union members and the owner class.

100. *Reidsville Review*, September 23, 1931, 1; *Reidsville Review*, August 8, 1935, 1; *Reidsville Review*, September 1, 1921, 2; Harry Harrison Kroll, *Riders in the Night* (Philadelphia: University of Pennsylvania Press, 1965); Gordon, *The Second Coming of the KKK*, 107.

101. "Local Citizens in Mass Meeting This Morning," *Reidsville Review*, July 2, 1935, 1. See also "Union Official Make Denial," *Reidsville Review*, July 3, 1935, 1; Powell, "My Life as I Remember It," 5; E. Lewis Evans to William Herrod, July 12, 1935, TWIU Records; Stuart Bruce Kaufman, *Challenge and Change: The History of the Tobacco Workers International Union* (Kensington, MD: Bakers, Confectionery, and Tobacco Workers International Union, 1986), 83–85.

102. Powell, "My Life as I Remember It," 5.

103. E. V. Boswell to E. Lewis Evans, September 1, 1935, TWIU Records; *Reidsville Review*, October 7, 1935, 8; National Labor Relations Board, Case No. R-32, Transcript and Decision, September 1, 1936, TWIU Records.

104. Powell, "My Life as I Remember It," 11.

105. E. Lewis Evans to Radford Powell, June 2, 1937, TWIU records; Kaufman, *Challenge and Change*, 45–46.

Chapter 5

1. I have modified the phrase "structuring elements of our everyday lives" from Melissa Aronczyk and Devon Powers, eds., introduction to *Blowing Up the Brand: Critical Perspectives on Promotional Culture* (New York: Peter Lang, 2010), 3. In changing "elements" to "agents," I am thinking of Bill Brown, introduction to "Thing Theory," *Critical Inquiry* 28, no. 1 (Autumn 2001): 1–22.

2. Sherman Cochran, *Big Business in China: Sino-Foreign Rivalry in the Cigarette Industry, 1890–1930* (Cambridge: Harvard University Press, 1980), 132.

3. Susan Strasser, *Satisfaction Guaranteed: The Making of the American Mass Market* (New York: Pantheon, 1989); Charles F. McGovern, *Sold American: Consumption and Citizenship, 1890–1945* (Chapel Hill: University of North Carolina Press, 2006); Michael Schudson, *Advertising, the Uneasy Persuasion: Its Dubious Impact on American Society* (New York: Basic Books, 1984); Roland Marchand, *Advertising the American Dream: Making Way for Modernity, 1920–1940* (Berkeley: University of California Press, 1985); Sherman Cochran, ed., *Inventing Nanjing Road: Commercial Culture in Shanghai, 1900–1945* ([1999]; Ithaca: Cornell University Press, 2010).

4. Carol Benedict, *Golden-Silk Smoke: A History of Tobacco in China, 1550–2010* (Berkeley: University of California Press, 2011), 161; David Embrey Fraser, "Smoking Out the Enemy: The National Goods Movement and the Advertising of Nationalism in China, 1880–1937," Ph.D. diss., University of California, Berkeley, 1999, 146–48.

5. Nannie M. Tilley, *The RJ Reynolds Tobacco Company* (Chapel Hill: University of North Carolina Press, 1985), 213.

6. My conception of the brand is influenced by Naomi Klein, *No Logo: Taking Aim at the Brand Bullies* (Toronto: Knopf Canada, 2000); Celia Lury, *Brands: The Logos of the Global Economy* (London: Routledge, 2004); Melissa Aronczyk and Devon Powers, eds., *Blowing Up*

the Brand: Critical Perspectives on Promotional Culture (New York: Peter Lang, 2010); Sarah Banet-Weiser, Authentic TM: Politics and Ambivalence in a Brand Culture (New York: New York University Press, 2012). These superb studies all date key transformations in the brand five to seven decades after they occurred, which is a good argument for US historians to weigh in again on this topic in the history of capitalism.

7. For a discussion of the social circulation of the commodity, see Nan Enstad, *Ladies of Labor, Girls of Adventure: Working Women, Popular Culture, and Labor Politics at the Turn of the Twentieth Century* (New York: Columbia University Press, 1999); Aronczyk and Powers, introduction to *Blowing Up the Brand*, 11.

8. Cheng Renjie, "Ying Mei Yan Gongsi Maiban Zheng Bozhao" ["British American Tobacco Corporation's comprador Zheng Bozhao,"] in *Zhonghua Wenshi Ziliao Wenku, [Chinese Literary and Historical Material Collection, Volume 14]* (Beijing: Zhongguo Wenshi Chubanshe, 1996), 741–51. See also Sherman Cochran, *Encountering Chinese Networks: Western, Japanese, and Chinese Corporations in China, 1880–1937* (Berkeley: University of California Press, 2000), 44–69; Howard Cox, "Learning to Do Business in China: The Evolution of BAT's Cigarette Distribution Network, 1902–41," *Business History* 39, no. 3 (July 1997): 30–64.

9. Cheng, "Ying Mei Yan Gongsi Maiban Zheng Bozhao," 742; Thomas Cobbs to Cheang Park Chew [Zheng Bozhao], May 16, 1919, James A. Thomas Papers, David M. Rubenstein Rare Book and Manuscript Library, Duke University; Cochran, *Big Business in China*, 15–17; Anna Lowenhaupt Tsing, *The Mushroom at the End of the World: On the Possibility of Life in Capitalist Ruins* (Princeton: Princeton University Press, 2015), 38–40.

10. Miao Lihua, "Wu Tingsheng Yu Yingmeiyan Gongsi" (Wu Tingsheng and British American Tobacco Corporation) Shanghai Wenshi Ziliao Xuanji, 56 Ji: Jiushanghai De Waishang Yu Maiban (*Shanghai Literary and Historical Material Collection, Volume 56: Foreign Businessmen and Compradors of the Old Shanghai*) (Shanghai: Shanghai Renmin Chubanshe, 1987), 145–48; Cochran, *Encountering Chinese Networks*, 54.

11. Cheng, "Ying Mei Yan Gongsi Maiban Zheng Bozhao," 741.

12. Arthur Wilhelm Madsen, *The State as Manufacturer and Trader; an Example Based on the Commercial, Industrial and Fiscal Results Obtained from Government Tobacco Monopolies* (London: Unwin, 1916), 16–23.

13. Miao, "Wu Tingsheng," 145–55; Cochran, *Encountering Chinese Networks*, 54–56.

14. Pipe tobacco had been sold by such merchant networks for generations. Benedict, *Golden-Silk Smoke*; see also Gary G. Hamilton and Chi-kong Lai, "Consumerism Without Capitalism: Consumption and Brand Names in Late Imperial China," in *The Social Economy of Consumption*, ed. Henry J. Rutz and Benjamin S. Orlove (Boston: University Press of America, 1989).

15. Cheng Renjie, "Ying Mei Yan Gongsi Maiban Zheng Bozhao," 743.

16. Frank H. Canaday to Arthur Bassett, October 3, 1923, Frank H. Canaday Papers, Harvard Yenching Library, Harvard University (hereafter Canaday Papers).

17. Benedict, *Golden-Silk Smoke*, 158–63.

18. Sherman Cochran, *Chinese Medicine Men: Consumer Culture in China and Southeast Asia* (Cambridge: Harvard University Press, 2006).

19. Carl Crow, *Foreign Devils in the Flowery Kingdom* (New York: Harper and Brothers, 1940), 58; Sherman Cochran, "Transnational Origins of Advertising in Early Twentieth-Century China," in Cochran ed., *Inventing Nanjing Road*, 40.

20. James Lafayette Hutchinson, *China Hand* (Boston: Lothrop, Lee and Shepard Co., 1936), 266; Zhu Hanying (former worker in the Shanghai Tobacco Printing Factory), interview, November 1963, Yingmeiyan gongsi zaihua qiye ziliao huibian [Collection of Business Documents of the British American Tobacco Company Ltd.] (China), (Beijing: Zhonghua shuju, 1983), 1, no. 234–35.

21. Hutchinson, *China Hand*, 53, 102.

22. Pearl Buck, quoted in Cochran, *Big Business in China*, 134.

23. Hutchinson, *China Hand*, 53.

24. Hutchinson, *China Hand*, 267.

25. Frank H. Canaday to Ward Canaday, August 30, 1924, Canaday Papers.

26. Yongtaihe Tobacco Corporations Inventory of Advertising Stock Poster, Hanger and Handbills etc for 31st October 1926, Canaday Papers.

27. Frank H. Canaday to Arthur Bassett, October 3, 1923, Canaday Papers.

28. Frank H. Canaday to Ward Canaday, August 30, 1924, Canaday Papers.

29. Ellen Johnston Liang, *Selling Happiness: Calendar Posters and Visual Culture in Twentieth Century Shanghai* (Honolulu: University of Hawai'i Press, 2004).

30. Hutchinson, *China Hand*, 267.

31. Cochran, *Encountering Chinese Networks*, 40.

32. Quoted in Sherman Cochran, "Transnational Origins of Advertising in Early Twentieth-Century China," in *Inventing Nanjing Road*, 57.

33. Unknown to A. Rose, June 17, 1924, BAT Collection, SASS.

34. Benedict, *Golden-Silk Smoke*, 161; Fraser, "Smoking Out the Enemy," 146–48.

35. Cochran, "Transnational Origins of Advertising in Early Twentieth-Century China," in *Inventing Nanjing Road*; Laing, *Selling Happiness*, 61–73.

36. My main primary source for this section is the *Report of the Commissioner of Corporations on the Tobacco Industry*, Part III (Washington, DC: Government Printing Office, 1915), augmented by RJ Reynolds Tobacco Company records accessed through the Truth Tobacco Industry Documents collection coordinated by the University of California, San Francisco library, https://www.industrydocumentslibrary.ucsf.edu/tobacco/.

37. *Report of the Commissioner of Corporations on the Tobacco Industry*, Part III, 261; "Dissolution No Gain to Tobacco Users," *New York Times*, May 10, 1915, 23.

38. RJ Reynolds Collection, Truth Tobacco Industry Documents, Bates # 502591708A/1710.

39. *Report of the Commissioner of Corporations*, Part III, 323, 251.

40. The ATC monopoly produced five smoking tobacco brands of different types at the RJR branch, but only Prince Albert made a good showing, representing 3.4 percent of the ATC's smoking tobacco production by 1911. *Report of the Commissioner of Corporations*, Part III, 279; Tilley, *RJ Reynolds*, 161.

41. *Report of the Commissioner of Corporations*, Part III, 257.

42. *Report of the Commissioner of Corporations*, Part III, 257–58.

43. *Report of the Commissioner of Corporations*, Part III, 259. Richard Tennant argues that "concentrated advertising was probably the result of accident and of practical necessity [rather] than of considered principle." Richard B. Tennant, *The American Cigarette Industry: A Study in Economic Analysis and Public Policy* (New Haven: Yale University Press, 1950), 83.

44. Tilley, *RJ Reynolds*, 210.

45. *Report of the Commissioner of Corporations*, Part III, 328, 332.

46. Tilley, *RJ Reynolds*, 219–20.

47. *Report of the Commissioner of Corporations*, Part III, 325–28.

48. *Report of the Commissioner of Corporations*, Part III, 328.

49. Tilley, *RJ Reynolds*, 213–14.

50. Tilley, *RJ Reynolds*, 220.

51. RJ Reynolds Collection, Truth Tobacco Industry Documents, Bates # 502591708A/1710; Bates # 502591718; *US Tobacco Journal* (December 5, 1914): 28.

52. Tennant, *The American Cigarette Industry*, 76.

53. Tennant, *The American Cigarette Industry*, 76.

54. RJ Reynolds Collection, Truth Tobacco Industry Documents, Bates # 50246668; Bates # 500567013/7020; Bates # 500227677/7684. See also Allan Brandt, *Cigarette Century: The Rise, Fall, and Deadly Persistence of the Product that Defined America* (New York: Basic Books, 2007), 69–101; Catherine Gudis, *Buyways: Billboards, Automobiles, and the American Landscape* (New York: Routledge, 2004); T. J. Jackson Lears, *Fables of Abundance: A Cultural History of Advertising in America* (New York: Basic Books, 1995), 196–234.

55. Jeffrey Wasserstrom, *Student Protests in Twentieth Century China: The View From Shanghai* (Stanford: Stanford University Press, 1991), 101.

56. Wasserstrom, *Student Protests*; Fraser, "Smoking Out the Enemy"; Karl Gerth, *China Made: Consumer Culture and the Creation of the Nation* (Cambridge: Harvard University Press, 2003).

57. Elizabeth J. Perry, *Shanghai on Strike: The Politics of Chinese Labor* (Stanford: Stanford University Press, 1993), 81–84, 148–51.

58. Fraser, "Smoking Out the Enemy," 120.

59. Economic Research Institution in Shanghai Social Academy, ed., *Yingmeiyan gongsi zaihua qiye ziliao huibian* [Collection of Business Documents of British-American Tobacco Co. Ltd. (china)], Series of Documents of Modern Chinese Economic History (Beijing: Zhonghua shuju, 1983), 4, no. 1323–24, table 43.

60. Frank H. Canaday to Arthur Bassett, June 26, 1925, Canaday Papers.

61. *Yingmeiyan gongsi zaihua qiye ziliao huibian* 4, no. 1323–24, table 43. This occurred in many places including Tongzhou and Nantuong, Jiangsu; Wenzhou and Xiaoshan, Zhejiang; Hezhao, Anhui; and Nanchong, Jiangxi.

62. Unnamed newspaper in Wenzhou, August 16, 1925, *Yingmeiyan gongsi zaihua qiye ziliao huibian*, 4, no. 1457.

63. Four such cases are noted in *Yingmeiyan gongsi zaihua qiye ziliao huibian* 4, no. 1323–24, table 43.

64. Frank H. Canaday to Arthur Bassett, June 26, 1925, Canaday Papers.

65. *Daily Republic*, September 2, 1925, *Yingmeiyan gongsi zaihua qiye ziliao huibian* 4, no. 1457–58.

66. Frank H. Canaday to Arthur Bassett, June 26, 1925, Canaday Papers.

67. Leaflet, *Yingmeiyan gongsi zaihua qiye ziliao huibian*, 4, no. 1325.

68. *Yingmeiyan gongsi zaihua qiye ziliao huibian* 4, no. 1323–24, table 43; Shanghai Shi Tongzhi Qikan 1934, *Yingmeiyan gongsi zaihua qiye ziliao huibian* 4, no. 1439–40.

69. Frank H. Canaday to Arthur Bassett, June 26, 1925, Canaday Papers.

70. "American Consular Service to Wing Tai Vo Tobacco Corporation," BAT Collection, SASS.

71. W. Scott to unknown, July 1, 1925, BAT Collection; *Yingmeiyan gongsi zaihua qiye ziliao huibian*, 4, no. 1390–91.

72. *Shanghai Journal of Commerce* (July 19, 1925), BAT Collection, SASS.

73. Frank H. Canaday to Bassett, June 26, 1925, Canaday Papers.

74. Leaflet, Hangzhou, *Yingmeiyan gongsi zaihua qiye ziliao huibian*, 4, no. 1459.

75. *Ch'an Daily*, August 27, 1925, *Yingmeiyan gongsi zaihua qiye ziliao huibian*, 4, no. 1467.

76. Leaflet, Suzhou, *Yingmeiyan gongsi zaihua qiye ziliao huibian*, 4, no. 1459–60.

77. *Ch'an Daily*, August 26, 1925, *Yingmeiyan gongsi zaihua qiye ziliao huibian*, 4, no. 1466–67.

78. Leaflet, Suzhou, *Yingmeiyan gongsi zaihua qiye ziliao huibian*, 4, no. 1459–60.

79. Poster, BAT Collection, SASS.

80. Special volume on the May 30th Movement, Yee Tsoong Archive, *Yingmeiyan gongsi zaihua qiye ziliao huibian*, 4, no. 1447–48.

81. Unnamed newspaper, Wenzhou, August 16, 1925, *Yingmeiyan gongsi zaihua qiye ziliao huibian*, 4, no. 1457.

82. Fraser, "Smoking Out the Enemy," 102.

83. Nanyang Brothers Tobacco Company, oral history, *Yingmeiyan gongsi zaihua qiye ziliao huibian*, 4, no. 1350; *Daily Republic*, January 5, 1929, *Yingmeiyan gongsi zaihua qiye ziliao huibian* 4, no. 1351. See also Karl Gerth, *China Made: Consumer Culture and the Creation of the Nation* (Cambridge: Harvard University Press, 2004). There is another story of Chinese companies' marketing efforts through cigarette cards and ties to other culture industries. BAT competed with Nanyang and Huacheng, but there were also smaller companies that worked on a different scale than BAT. See Benedict, *Golden Silk Smoke*, 131–48.

84. "Over 25 Billion Made Annually in Reidsville," *Reidsville Review*, January 6, 1930, 1; "Local News," *Reidsville Review*, January 14, 1925, 5; untitled, *Reidsville Review*, February 11, 1931, 5; untitled, *Reidsville Review*, February 20, 1933, 5; Lindley S. Butler, *Rockingham County: A Brief History* (Raleigh: North Carolina Division of Archives and History, 1982), 78.

85. Adrian Burgos Jr., *Playing America's Game: Baseball, Latinos, and the Color Line* (Berkeley: University of California Press, 2007), 1–8, 71–73; Paul A. Kramer, *The Blood of Government: Race, Empire, the United States, and the Philippines* (Chapel Hill: University of North Carolina Press, 2006), 286, 371; Linda Espana-Maram, *Creating Masculinity in Los Angeles's Little Manila: Working-Class Filipinos and Popular Culture, 1920s–1950s* (New York: Columbia University Press, 2006).

86. "Briggs Brothers, Hoggard Were Instrumental in Early Baseball," *Reidsville Review*, April 21, 1988, Rockingham Community College Library, "Sports-History" vertical file.

87. Jacquelyn Dowd Hall et al., *Like a Family: The Making of a Southern Mill World*

(Chapel Hill: University of North Carolina Press, 1987), 135; Patrick Richards, "Textile Mill Baseball in the 1930s: Workers' Negotiation of Identities" (unpublished paper in possession of author); J. Chris Holaday, *Professional Baseball in North Carolina: An Illustrated City-by-City History, 1901–1996* (Jefferson, NC: McFarland, 1998), 200; Jim L. Sumner, *Separating the Men from the Boys: The First Half-Century of the Carolina League* (Winston-Salem: John F. Blair, 1994), 6–12.

88. Katy Thomas, "The Reidsville Luckies: A Team of Our Own," unpublished, edited interview transcript, April 1993, "Sports-History" vertical file, Rockingham Community College Library.

89. "The Briggs Brothers," *Reidsville Review*, August 7, 1994, 1B. See also Jeannine Manning Hutson, "Glory Days: Briggs Back on the Diamond," *The Ledger* [Eden, NC] August 25, 1993, 1, "Sports-History" vertical file, Rockingham Community College Library.

90. "The Briggs Brothers," *Reidsville Review*, August 7, 1994, 1B.

91. Thomas, "The Reidsville Luckies: A Team of Our Own."

92. "The Briggs Brothers," *Reidsville Review*, August 7, 1994, 1B.

93. William Davis, interview by author and Kori Graves, July 10, 2001.

94. McGovern, *Sold American*, 43.

95. Holaday, *Professional Baseball in North Carolina*, 221.

96. Micki McElya, *Clinging to Mammy: The Faithful Slave in Twentieth-Century America* (Cambridge: Harvard University Press, 2007).

97. Paul K. Edwards, *The Southern Urban Negro As a Consumer* (New York: Prentice-Hall, Inc., 1932), 232–40.

98. Jason Chambers, *Madison Avenue and the Color Line: African Americans in the Advertising Industry* (Philadelphia: University of Pennsylvania Press, 2008), 86; see also Robert E. Weems, *Desegregating the Dollar: African American Consumerism in the Twentieth Century* (New York: New York University Press, 1998), 7–30; Brenna Greer, "Selling Liberia: Moss H. Kendrix, the Liberian Centennial Commission, and the Post–World War II Trade in Black Progress," *Enterprise and Society* 14, no. 2 (June 2013): 303–26; Lizabeth Cohen, *Consumers' Republic: The Politics of Mass Consumption in Postwar America* (New York: Knopf, 2003). African Americans' use of the Lucky Strike brand anticipates some of the dynamics that Cohen discusses.

99. Edward Bernays to O. V. Richards, January 21, 1931; Claude A. Barnett to George W. Hill, January 30, 1931, Edward Bernays Papers, Library of Congress, Washington, DC.

100. The most famous of Bernays's schemes were the Reach for a Lucky campaign and the Torches of Freedom campaign. See Stuart Ewen, *PR! A Social History of Spin* (New York: Basic Books, 1996); Larry Tye, *The Father of Spin: Edward L. Bernays and the Birth of Public Relations* (New York: Crown, 1998).

101. Roy C. Flannagan, *The Story of Lucky Strike* (New York: American Tobacco Company, 1938), facing p. 46. Flannagan published a series of related articles in the white *Richmond News Leader*.

102. William Davis, interview by author and Kori Graves, July 10, 2001; Dale Hagwood, "William Davis Remembers the Negro Leagues," *Reidsville Review*, date unknown; Jim Eastridge, "Ball Days of Yore: Former Player Recalls Years on Negro Team," *Reidsville Review*, date unknown [clippings provided by Davis].

103. William Davis, interview by author and Kori Graves, July 10, 2001. See Bruce Adelson, *Brushing Back Jim Crow: The Integration of Minor-League Baseball in the American South* (Charlottesville: University Press of Virginia, 1999), 39–41; Dick Clark and Larry Lester, eds., *The Negro Leagues Book: A Monumental Work from the Negro Leagues Committee of the Society for American Baseball Research* (Cleveland: Society for American Baseball Research, 1994).

104. Alain Locke, quoted in Marlon B. Ross, *Manning the Race: Reforming Black Men in the Jim Crow Era* (New York: New York University Press, 2004), 80. See also Ross's discussion of Locke, 77–89.

105. Stanford Research into the Impact of Tobacco Advertising (SRITA)'s collection of World War I–related tobacco advertisements reveals that the four successor companies advertised a range of brands explicitly to and for soldiers, including the Turkish and Turkish-blend brands, not just or primarily Camels, Lucky Strikes, and Chesterfields. In addition, the government took over the ATC's entire supply of Bull Durham pipe tobacco for military use in 1918, indicating that pipe smoking and hand-rolled cigarettes remained important. http://tobacco.stanford.edu/tobacco_main/images.php?token2=fm_st187.php&token1=fm_img10965.php&theme_file=fm_mt023.php&theme_name=War%20&%20Aviation&subtheme_name=World%20War%20I/; "Government Take Output Bull Durham," *Reidsville Review*, April 9, 1918, 1.

Chapter 6

1. David W. Gilbert, *The Product of Our Souls: Ragtime, Race, and the Birth of the Manhattan Musical Marketplace* (Chapel Hill: University of North Carolina Press, 2015); Lewis A. Erenberg, *Steppin' Out: New York Nightlife and the Transformation of American Culture, 1890–1930* ([1981]; Chicago: University of Chicago Press, 1984), 146–75, 233–64; Shane Vogel, *The Scene of Harlem Cabaret: Race, Sexuality, Performance* (Chicago: University of Chicago Press, 2009); Andrew David Field, *Shanghai's Dancing World: Cabaret Culture and Urban Politics, 1919–1954* (Hong Kong: Chinese University Press, 2010); Andrew F. Jones, *Yellow Music: Media Culture and Colonial Modernity in the Chinese Jazz Age* (Durham: Duke University Press, 2001).

2. My treatment of cigarettes and jazz as entwined is indebted to the notion of assemblage as it is often used in affect theory to relate things to bodies. See Jane Bennett, *Vibrant Matter: A Political Ecology of Things* (Durham: Duke University Press, 2010), 20–38; Sara Ahmed, "Affective Economies," *Social Text* 2, no. 2 (Summer 2004): 117–39; Melissa Gregg and Gregory J. Seigworth, eds., *The Affect Theory Reader* (Durham: Duke University Press, 2010), 1–25; Kathleen Stewart, *Ordinary Affects* (Durham: Duke University Press, 2007), 1–2.

3. Philip K. Eberly, *Music in the Air: America's Changing Tastes in Popular Music, 1920–1980* (New York: Hastings House Publishers, 1982), 32–33, 114–17; Jim Cox, *Music Radio: The Great Performers and Programs of the 1920s through Early 1960s* (Jefferson, NC: McFarland and Company, 2005).

4. Quote is from *Straits Times* (Singapore), November 30, 1928, 10. See also Frederick J. Schenker, "Empire of Syncopation: Music, Race, and Labor in Colonial Asia's Jazz Age" (dissertation, University of Wisconsin, 2016); Bruce Johnson, "The Jazz Diaspora," in *The Cambridge Companion to Jazz*, ed. Marvyn Cooke and David Horn (Cambridge: Cambridge University Press, 2003), 33–54; Jairo Moreno, "Imperial Aurality: Jazz, the Archive, and U.S.

Empire," in *Audible Empire: Music, Global Politics, Critique*, ed. Ronald Rodano and Tejumola Olaniyan (Durham: Duke University Press, 2016), 135–60.

5. Brenda Dixon Gottschild, *The Black Dancing Body: A Geography From Coon to Cool* (New York: Palgrave MacMillan, 2003); Susan C. Cook, "Passionless Dancing and Passionate Reform: Respectability, Modernism, and the Social Dancing of Irene and Vernon Castle," in *The Passion of Music and Dance: Body, Gender, and Sexuality*, ed. William Washabaugh (Oxford, UK: Berg Publishers, 1998), 133–50; Field, *Shanghai's Dancing World*, 53–82, 119–52.

6. Alys Eve Weinbaum et al., eds., *The Modern Girl Around the World: Consumption, Modernity, and Globalization* (Durham: Duke University Press, 2008); Ellen Johnston Liang, *Selling Happiness: Calendar Posters and Visual Culture in Twentieth Century Shanghai* (Honolulu: University of Hawai'i Press, 2004); Francesca Dal Lago, "Crossed Legs in 1930s Shanghai: How 'Modern' the Modern Woman?" *East Asian History* 19 (2000), 103–44; Antonia Finnane, *Changing Clothes in China: Fashion, History, Nation* (New York: Columbia University Press, 2008), 139–200; Tani Barlow, "'What Is a Poem?' History and the Modern Girl," in David Palumbo-Liu, Bruce Robbins, and Nirvana Tanoukhi, eds., *Immanuel Wallerstein and the Problem of the World: System, Scale, Culture* (Durham: Duke University Press, 2011), 155–72; Linda Mizejewski, *Ziegfeld Girl: Image and Icon in Culture and Cinema* (Durham: Duke University Press, 1999).

7. Field, *Shanghai's Dancing World*, 21–25.

8. Jones, *Yellow Music*, 65.

9. *All About Shanghai: A Standard Guidebook* (Shanghai: University Press, 1935; Hong Kong: Oxford University Press, 1983), 73.

10. Irwin S. Smith Oral History, July 28, 1982, East Carolina Manuscript Collection.

11. Schenker, "Empire of Syncopation: Music, Race, and Labor in Colonial Asia's Jazz Age," 46–99; Bradley Shope, "'They Treat Us White Folks Fine': African American Musicians and the Popular Music Terrain in Late Colonial India," *South Asian Popular Culture* 5, no. 2 (November 2007): 105; "Performer Tells of His Travels and Thrills of Oriental Climes," *Chicago Defender*, August 23, 1930, 5.

12. E. Taylor Atkins, *Blue Nippon: Authenticating Jazz in Japan* (Durham: Duke University Press, 2001), 19–44.

13. Jones, *Yellow Music*; Schenker, "Empire of Syncopation"; Whitey Smith with C. L. McDermott, *I Didn't Make a Million* (Manila: Philippine Education Company, 1956).

14. See Vogel, *The Scene of Harlem Cabaret*, 62–73; Field, *Shanghai's Dancing World*, 100–101; 107–9; Jones, *Yellow Music*.

15. James Lafayette Hutchinson, *China Hand* (Boston: Lothrop, Lee and Shepard Co., 1936), 314.

16. Gottschild, *The Black Dancing Body*, 12–40.

17. *All About Shanghai: A Standard Guidebook*, 76.

18. Atkins, *Blue Nippon*, 19–44; David Suisman, *Selling Sounds: The Commercial Revolution in American Music* (Cambridge: Harvard University Press, 2009), 207–9.

19. *Straits Times* (Singapore), November 30, 1928, 10; Buck Clayton assisted by Nancy Miller Elliott, *Buck Clayton's Jazz World* (New York: Oxford University Press, 1987), 50.

20. Mu Shiying, "Craven 'A'" (1932), trans. Andrew David Field in Andrew David Field, *Mu Shiying: China's Lost Modernist* (Hong Kong: Hong Kong University Press, 2014), 78. . See

also Shumei Shi, *The Lure of the Modern: Writing Modernism in Semicolonial China, 1917–1937* (Berkeley: University of California Press, 2001), 317–22.

21. See Ronald Radano, *Lying Up a Nation: Race and Black Music* (Chicago: University of Chicago Press, 2003), 234–37; Cook, "Passionless Dancing and Passionate Reform," 133–37.

22. Matthew Hilton, *Smoking in British Popular Culture, 1880–2000: Perfect Pleasures* (Manchester: Manchester University Press, 2000), 138–50; Cassandra Tate, *Cigarette Wars: The Triumph of 'the Little White Slaver'* (New York: Oxford University Press, 1999), 93–118; Benedict, *Golden-Silk Smoke*, 12–13, 200, 210.

23. Over thirty years ago, Michael Schudson argued that US women had started smoking cigarettes in significant numbers before companies began marketing directly to them. Cigarette advertising to women, he insisted, followed a shift in trend rather than created that trend, but no one since then has explained what did cause the trend. The synergy between cigarettes and jazz dance are a likely contributing cause. Michael Schudson, *Advertising, the Uneasy Persuasion: Its Dubious Impact on American Society* (New York: Basic Books, 1984), 179–83.

24. Men and women also attended theater, motion pictures, and amusement parks together. Kathy Peiss, *Cheap Amusements: Working Women and Leisure in Turn of the Century New York* (Philadelphia: Temple University Press, 1985), 11–33.

25. Alys Eve Weinbaum et al., eds., *The Modern Girl Around the World: Consumption, Modernity, and Globalization* (Durham: Duke University Press, 2008).

26. David W. Gilbert, *The Product of Our Souls: Ragtime, Race, and the Birth of the Manhattan Musical Marketplace* (Chapel Hill: University of North Carolina Press, 2015); Cook, "Passionless Dance and Passionate Reform," 133–51; Lewis A. Erenberg, *Steppin' Out: New York Nightlife and the Transformation of American Culture, 1890–1930* ([1981]; Chicago: University of Chicago Press, 1984), 146–75, 233–64.

27. Radano, *Lying Up a Nation*, 234–36; Neil Leonard, "The Reactions to Ragtime," in *Ragtime: Its History, Composers and Music*, ed. John Edward Hasse (New York: Schirmer Books, 1985), 102–16; Cook, "Passionless Dance and Passionate Reform," 133–37.

28. David R. Roediger, *The Wages of Whiteness: Race and the Making of the American Working Class* (New York: Verso, 1991), 115–32; Eric Lott, *Love and Theft: Blackface Minstrelsy and the American Working Class* (New York: Oxford University Press, 1995).

29. Radano, *Lying Up a Nation*, 234–36.

30. *Variety*, quoted in Vogel, *The Scene of Harlem Cabaret*, 2.

31. *The Messenger*, quoted in Vogel, *The Scene of Harlem Cabaret*, 6.

32. Michele Hilmes, *Radio Voices: American Broadcasting, 1922–1952* (Minneapolis: University of Minnesota Press, 1997), 119.

33. Hutchinson, *China Hand*, 377–78. Shanghai radio stations also broadcast jazz via live feed from the clubs, though I have not discovered whether BAT was directly involved in this practice. Schenker, "Empire of Syncopation," 475.

34. Philip K. Eberly, *Music in the Air: America's Changing Tastes in Popular Music, 1920–1980* (New York: Hastings House Publishers, 1982), 32–33, 114–17; Jim Cox, *Music Radio: The Great Performers and Programs of the 1920s through Early 1960s* (Jefferson, North Carolina: McFarland and Company, 2005), 25–26, 35, 39, 48, 50.

35. Hilmes, *Radio Voices*, 45; Suisman, *Selling Sounds*, 174–75.

36. Quoted in Eberly, *Music in the Air*, 117.

37. Suisman, *Selling Sounds*, 267.

38. Suisman, *Selling Sounds*, 259.

39. NBC Script WEAF, March 23, 1929, American Tobacco Company files, NBC Papers, Wisconsin State Historical Society, University of Wisconsin, Madison.

40. Eberly, *Music in the Air*, 117.

41. John Gunther, *Taken at the Flood: The Story of Albert D. Lasker* (New York: Harper and Brothers, 1960), 196–97.

42. "Music Formula for the Lucky Strike Radio Hour," March 6, 1931, Edward Bernays Papers, Library of Congress.

43. "Music Formula for the Lucky Strike Radio Hour," March 6, 1931, Edward Bernays Papers, Library of Congress.

44. Robert D. Heinl, "Off the Antenna," *Washington Post*, December 29, 1929, A5; Heinl, "Off the Antenna," *Washington Post*, September 23, 1928, S10.

45. "Your Hit Parade (Formula)," n.d., American Tobacco Company Files, NBC Papers.

46. "Your Hit Parade (Formula)," n.d., American Tobacco Company Files, NBC Papers.

47. Radano, *Lying Up a Nation*, 237.

48. "Musical Formula for the Lucky Strike Radio Hour," memo, March 6, 1931, American Tobacco Company Files, Edward Bernays Papers, Library of Congress.

49. Pierre Key to Merlin H. Aylesworth (president of NBC), December 20, 1933, American Tobacco Company files, NBC Papers. The letter included an article to be published in *Musical Digest* in January 1934.

50. Quoted in Cox, *Music Radio*, 319.

51. "Music Formula for the Lucky Strike Radio Hour," March 6, 1931, Edward Bernays Papers, Library of Congress.

52. Edward Bernays to O. V. Richards, March 25, 1929, American Tobacco Files, Bernays Papers, Library of Congress.

53. Suisman, *Selling Sounds*, 40–14, 210, 235–37; Karl Hagstrom Miller, *Inventing Folk and Pop Music in the Age of Jim Crow* (Durham: Duke University Press, 2010), 187–213.

54. Suisman, *Selling Sounds*, 174–75; Hilmes, *Radio Voices*, 45–49; Clifford J. Doerksen, *American Babel: Rogue Radio Broadcasters of the Jazz Age* (Philadelphia: University of Pennsylvania Press, 2005), 33–48.

55. Doerksen, *American Babel*, 23–25, 31–33.

56. Suisman, *Selling Sounds*, 264; Hilmes, *Radio Voices*, 183; Alexander Russo, *Points on the Dial: Golden Age Radio Beyond the Networks* (Durham: Duke University Press, 2010), 28.

57. African Americans did influence the sounds of the cigarette companies' shows, especially the *Camel Caravan*, featuring Benny Goodman's band. Mary Lou Williams composed the show's theme song, "Camel Hop"; Fletcher Henderson did most of the arrangements; and Goodman regularly hired African American musicians. Linda Dahl, *Morning Glory: A Biography of Mary Lou Williams* (New York: Random House, 1999), 110, 433; "New Song Hit," *Atlanta Daily World*, July 3, 1938, 2. Lucky Strike Hit Parade Orchestra director Mark Warnow hired Benny Carter as an arranger in 1942, an event the *Pittsburgh Courier* called "perhaps one of the most important events in the progress of Negro music." "Benny Carter Is Signed to Write for Lucky Strike's Commercial," *Pittsburgh Courier*, February 7, 1942, 21.

58. "Cab Calloway Goes on Lucky Strike Hour Dec. 29," *Atlanta World*, December 16, 1931, 6.

59. "Schedule of Orchestras on Lucky Broadcasts," American Tobacco Company files, NBC Papers, Wisconsin State Historical Society.

60. "Negro Orchestra Was on Ether from Plane," *Atlanta Daily World*, May 27, 1932, 6. One reader of the *Pittsburgh Courier* wrote in to ask "why the American Negro continues to buy products like Lucky Strike cigarettes . . . when [the companies] will not give Negro artists a chance on their radio programs." Ivan Browning, quoted in Earl J. Morris, "Ivan Browning 'Debunks' Hollywood," *Pittsburgh Courier*, March 16, 1940, 21.

61. Irwin S. Smith Oral History, July 28, 1982, East Carolina Manuscript Collection.

62. Paul Goalby Cressey, *The Taxi-Dance Hall: A Sociological Study in Commercialized Recreation and City Life* (Chicago: University of Chicago Press, 1932); Linda Espana-Maram, *Creating Masculinity in Los Angeles's Little Manila: Working-Class Filipinos and Popular Culture, 1920s–1950s* (New York: Columbia University Press, 2006); Field, *Shanghai's Dancing World*, 119–52.

63. Hutchinson, *China Hand*, 351.

64. Frank Canaday, diary, June 26, 1923, July 18, 1923, Frank Canaday Papers, Yenching Library, Harvard University (hereafter cited as Canaday Papers).

65. Frank Canaday, unpublished memoir, Canaday Papers.

66. Frank Canaday, diary, August 18, 1923, Canaday Papers.

67. James A. Thomas, for example, traveled and lived abroad for the ATC or BAT through his adult life and did not marry until he was in his fifties. However, his records and his memoirs show no hint of courtship or sexual activity. Likewise, memoirs of men who went to China typically discuss the prevalence of sexual opportunity but admit to none of their own.

68. James N. Joyner to F. A. Perry, April 22, 1926, James N. Joyner Papers, East Carolina Manuscript Collection, J. Y. Joyner Library, East Carolina University (hereafter James N. Joyner Papers).

69. Liu Lilin to James N. Joyner, February 23, 1936, James N. Joyner Papers.

70. Field, *Shanghai's Dancing World*, 131.

71. Liu Lilin to James N. Joyner, May 22, 1934; Liu Lilin to James N. Joyner, May 14, 1934; subscription payment to *Cosmopolitan* magazine, 1933, James N. Joyner Papers. Again, such gifts were typical of ongoing relationships between taxi-dancers and clients. Field, *Shanghai's Dancing World*, p. 136.

72. James N. Joyner to Liu Lilin, June 7, 1934, James N. Joyner Papers.

73. James N. Joyner to Liu Lilin, June 7, 1934, James N. Joyner Papers.

74. Liu Lilin to James N. Joyner, February 23, 1936, James N. Joyner Papers.

75. James N. Joyner to Tang Foshu, undated, James N. Joyner Papers.

76. Liu Lilin to James N. Joyner, September 11, 1935; Liu Lilin to James N. Joyner, December 8, 1935; Liu Lilin to James N. Joyner, February 23, 1936; BAT foreign staff person (name illegible) to James N. Joyner, September 17, 1935, James N. Joyner Papers.

77. Field, *Shanghai's Dancing World*, 8–12.

78. Field discusses the Dong-Feng Company in *Shanghai's Dancing World*, 92–96, 295. The company was run by Dong Yunlong and Feng Yixiang.

79. Francoise Lionnet and Shu-mei Shih call this kind of formation "minor transnationalism," that is, transnational exchange between colonized or formerly colonized people that circumvented the imperial powers. Francoise Lionnet and Shu-mei Shih, eds., *Minor Transnationalism* (Durham: Duke University Press, 2005), 1–21.

80. S. James Staley, "Is It True What They Say About China?," *Metronome*, December 1936, 47; Smith with McDermott, *I Didn't Make a Million*, 24.

81. Schenker, "Empire of Syncopation," 333–45; 472–88; *Chicago Defender*, July 31, 1926, 6. Teddy Weatherford seemed to be booking bands and dancers into all of Dong and Feng's cabaret properties. Thelma Porter attributed her six-month gig at the Canidrome to Weatherford, and Weatherford himself was known for playing in multiple venues each night he appeared, drawing a special round of applause in each place. See Thelma Porter, "Stage Stars in China Tell of Fine Treatment Abroad," *Chicago Defender* (September 29, 1934), 9; Shope, "'They Treat Us White Folks Fine,'" 105; Langston Hughes, *I Wonder as I Wander: An Autobiographical Journey* (New York: Rinehart, 1956), 251.

82. *Straits Times*, April 22, 1937, 13.

83. Buck Clayton assisted by Nancy Miller Elliott, *Buck Clayton's Jazz World* (New York: Oxford University Press, 1987), 67.

84. "China Resembles US as Bands Pack the Theaters," *Chicago Defender*, July 14, 1934, 8; *Chicago Defender*, February 4, 1922, 15; *Chicago Defender*, April 28, 1934, 8. See Jones, *Yellow Music*, 103, for Chinese views of jazz.

85. *Chicago Defender*, July 14, 1934, 8; *Pittsburgh Courier*, August 17, 1935, A7; Clayton, *Buck Clayton's Jazz World*, 70; *New York Amsterdam News*, December 11, 1943, 1A; Clayton, *Buck Clayton's Jazz World*, 78; *New York Amsterdam News*, July 14, 1945; "India Is Colorful; Sometimes Shocking," *Baltimore Afro-American*, October 21, 1944, 12.

86. Clayton, *Buck Clayton's Jazz World*, 75–76.

87. Hughes, *I Wonder as I Wander*, 252–53.

88. Hughes, *I Wonder as I Wander*, 257–59.

89. Clayton's band members also each hired a servant in China, as did virtually all foreigners and elite Chinese families, which gave them an investment in imperial hierarchies. Thelma Porter indicated that there was one servant per household. Thelma Porter, "Stage Stars in China Tell of Fine Treatment Abroad," *Chicago Defender*, September 29, 1934, 9.

90. Clayton, *Buck Clayton's Jazz World*, 70.

91. Burton W. Peretti, *The Creation of Jazz: Music, Race, and Culture in Urban America* (Urbana: University of Illinois Press, 1992), is structured around exactly this narrative.

92. James B. Neal, interview with author and Kori Graves, July 6, 2001.

93. Durham's African American newspaper, *The Carolina Times*, carried weekly advertisements for the bands that would play at the Durham armory.

94. Jacqui Malone, "Jazz Music in Motion: Dancers and Big Bands," in *The Jazz Cadence of American Culture*, ed. Robert G. O'Meally (New York: Columbia University Press, 1998), 278–97. Quote from Durham is on pages 292–93.

95. Richard O. Boyer, "The Hot Bach," in *The Duke Ellington Reader*, ed. Mark Tucker (New York: Oxford, 1992), 238. Originally published in *The New Yorker* in 1944.

96. "Ellington on Swing and Its Critics," *The Duke Ellington Reader*, 140. Originally published in *Downbeat* in 1939.

97. Quoted in Boyer, "The Hot Bach," 233.

98. William and Ruth Davis, interview with author and Kori Graves, Reidsville, North Carolina, July 10, 2001.

99. James B. Neal, interview with author and Kori Graves, July 6, 2001. See also *New York Times*, March 28, 1933, 23, for a discussion of this practice.

100. Marion Snow, interview with author and Kori Graves, Reidsville, North Carolina, July 7, 2001.

101. Quoted in Kevin Kelly Gaines, "Duke Ellington, 'Black, Brown and Beige,' and the Cultural Politics of Race," in *Music and the Racial Imagination*, ed. Ron Radano and Philip V. Bohlman (Chicago: University of Chicago Press, 2000), 599.

102. Wynton Marsalis and Robert G. O'Meally, "Duke Ellington: 'Music Like a Big Hot Pot of Good Gumbo,'" in *The Jazz Cadence of American Culture*, 144.

103. James B. Neal, interview with author and Kori Graves, July 6, 2001.

104. Ruth and William Davis, interview with author and Kori Graves, July 10, 2001.

105. Ralph Ellison, "Homage to Duke Ellington on His Birthday," in *Living With Music: Ralph Ellison's Jazz Writings*, ed. Robert G. O'Meally (New York: Modern Library, 2001), 81.

Chapter 7

1. Frank H. Canaday, unpublished memoir, 43, Frank H. Canaday Papers, Harvard Yenching Library, Harvard University (hereafter cited as Canaday Papers).

2. Cheng Renjie, "Ying Mei Yan Gongsi Maiban Zheng Bozhao" ["British American Tobacco Corporation's comprador Zheng Bozhao"], in *Zhonghua Wenshi Ziliao Wenku [Chinese Literary and Historical Material Collection, Volume 14]* (Beijing: Zhongguo Wenshi Chubanshe, 1996), 741–42.

3. Ann Laura Stoler, "On Degrees of Imperial Sovereignty," *Public Culture* 18, no. 1 (Winter 2006): 142. Stoler calls upon historians to discover agents of empire that are emergent.

4. Paul A. Kramer, "Power and Connection: Imperial Histories of the United States in the World," *American Historical Review* 116, no. 5 (December 2011): 1375.

5. Until recently, the US historiography on the corporation in China has asked why Chinese businesses were slow to adopt the corporation form after it became available (in the 1880s), a question invested in the ideology of modernity. In the interwar era, however, some Chinese businessmen did incorporate as a way to raise capital. Many of those who did not incorporate, like BAT's competitor, the Nanyang Brothers, stated that they were worried about losing managerial control, a reasonable fear, as BAT tried to buy Nanyang on two occasions. See William C. Kirby, "China Unincorporated: Company Law and Business Enterprise in Twentieth-Century China," *Journal of Asian Studies* 54, no. 1 (February 1995): 43–63; Brett Sheehan, *Industrial Eden: A Chinese Capitalist Vision* (Cambridge: Harvard University Press, 2015), 7–8; Sherman Cochran, *Big Business in China: Sino-Foreign Rivalry in the Cigarette Industry, 1890–1930* (Cambridge: Harvard University Press, 1980), 88; Teemu Ruskola, *Legal Orientalism: China, the United States, and Modern Law* (Cambridge: Harvard University Press, 2013), 60–107.

6. Sherman Cochran, *Encountering Chinese Networks: Western, Japanese, and Chinese Corporations in China, 1880–1937* (Berkeley: University of California Press, 2000), 44–69.

7. Akira Iriye, *After Imperialism: The Search for a New Order in the Far East, 1921–1931*

(Cambridge: Harvard University Press, 1965); Susan Pedersen, *The Guardians: The League of Nations and the Crisis of Empire* (Oxford: Oxford University Press, 2015), 1–17; Erez Manela, *The Wilsonian Moment: Self Determination and the International Origins of Anticolonial Nationalism* (Oxford: Oxford University Press, 2009), 3–14; Emily S. Rosenberg, *Financial Missionaries to the World: The Politics and Culture of Dollar Diplomacy, 1900–1930* (Durham: Duke University Press, 2003), 70–96, 122–23.

8. Anna Lowenhaupt Tsing, *The Mushroom at the End of the World: On the Possibility of Life in Capitalist Ruins* (Princeton: Princeton University Press, 2015), 38.

9. Frank H. Canaday, unpublished memoir, 35, 38, 43, Canaday Papers.

10. Frank H. Canaday, unpublished memoir, 38–40 (quote is on 40), Canaday Papers.

11. Thomas F. Cobbs to Mr. Cheang Park Chew (Zheng Bozhao), May 16, 1919, James Augustus Thomas Papers, David M. Rubenstein Rare Book & Manuscript Library, Duke University (hereafter James A. Thomas Papers); Cheng Renjie, "Ying Mei Yan Gongsi Maiban Zheng Bozhao," 741; Howard Cox, "Learning to Do Business in China: The Evolution of BAT's Cigarette Distribution Network, 1902–41," *Business History* 39, no. 3 (1997): 49–50; Cochran, *Encountering Chinese Networks*, 57–58.

12. Cheng Renjie, "Ying Mei Yan Gongsi Maiban Zheng Bozhao," 741–43; Howard Cox, "Learning to Do Business in China: The Evolution of BAT's Cigarette Distribution Network, 1902–41," *Business History* 39, no. 3 (1997): 49–50; Cochran, *Encountering Chinese Networks*, 57–62.

13. Frank H. Canaday to Baldwin, November 30, 1923, Canaday Papers.

14. Frank H. Canaday, unpublished memoir, 118, Canaday Papers; Cheng Renjie, "Ying Mei Yan Gongsi Maiban Zheng Bozhao," 743; Cochran, *Encountering Chinese Networks*, 58.

15. Cheng Renjie, "Ying Mei Yan Gongsi Maiban Zheng Bozhao," 747–48.

16. Thomas F. Cobbs to Mr. Cheang Park Chew (Zheng Bozhao), May 16, 1919, James A. Thomas Papers.

17. Thomas Cobb to Arthur Bassett, May 17, 1919, BAT Collection.

18. Frank H. Canaday to Baldwin, November 30, 1923, Canaday Papers.

19. Cox, "Learning to Do Business in China," 37, 49.

20. The stated reason was that Thomas was beginning his vice presidency of the China-American Bank of Commerce, but Thomas resisted, arguing that the bank would not disrupt but assist his duties as director. Assistant secretary to James A. Thomas, December 21, 1920, James A. Thomas Papers; Noel H. Pugach, *Same Bed, Different Dreams: A History of the Chinese American Bank of Commerce, 1919–1937* (Hong Kong: Centre of Asian Studies, 1997), 66.

21. Frank H. Canaday, unpublished memoir, 35–38; James Lafayette Hutchinson, *China Hand* (Boston: Lothrop, Lee and Shepard Co., 1936), 272–95; Cochran, *Big Business in China*, 163–64. Duke also resigned as chairman of BAT in 1923.

22. James A. Thomas, "Selling and Civilization: Some Principles of an Open Sesame to Big Business Success in the East," *Asia* 23, no. 12 (December 1923): 949, says that "four of five kept their jobs," but in *A Pioneer Tobacco Merchant*, 106, Thomas wrote that the number was "two out of five" after one year on the job.

23. This quote is compiled from two letters recounting the same story. Frank H. Canaday to Ward Canaday, n.d.; Frank H. Canaday to Baldwin, March 20, 1923, Canaday Papers.

24. David Yau-fai Ho, "On the Concept of Face," *American Journal of Sociology* 81 (Janu-

ary 1976): 867–84; Mayfair Mei-hui Yang, *Gifts, Favors, and Banquets: the Art of Social Relationships in China* (Ithaca: Cornell University Press, 1994).

25. Robert A. Bickers and Jeffrey N. Wasserstrom, "Shanghai's 'Dogs and Chinese Not Admitted' Sign: Legend, History, and Contemporary Symbol," *China Quarterly* 142 (June 1995): 444–66.

26. This is Sherman Cochran's insight. See Cochran, *Encountering Chinese Networks*, 50; Cheng Renjie, "Ying Mei Yan Gongsi Maiban Zheng Bozhao," 744–45.

27. Peffer, *The White Man's Dilemma*, 147; Antonia Finnane, *Changing Clothes in China: Fashion, History, Nation* (New York: Columbia University Press, 2008), 177–88.

28. Cheng Renjie, "Ying Mei Yan Gongsi Maiban Zheng Bozhao," 744–45; Cochran, *Encountering Chinese Networks*, 50.

29. Frank H. Canaday, unpublished memoir, 35, 119, Canaday Papers.

30. For description of BAT's typical training of foreigners in the sales department, see Hutchinson, *China Hand*; Lee Parker and Ruth Dorval Jones, *China and the Golden Weed* (Ahoskie, NC: Herald Publishing Co., 1976).

31. Frank H. Canaday, unpublished memoir, 47–50, Canaday Papers.

32. Thomas, "Selling and Civilization," 948.

33. Frank H. Canaday to Ward Canaday, May 21, 1923, Canaday Papers.

34. Frank H. Canaday, unpublished memoir, 99, Canaday Papers.

35. Frank H. Canaday, unpublished memoir, 82, Canaday Papers.

36. Frank H. Canaday to Mother, February 3, 1923; Frank H. Canaday, unpublished memoir, 34, 47, Canaday Papers.

37. For example, see *Some Reasons for Chinese Exclusion: Meat vs. Rice, American Manhood Against Asiatic Coolieism, Which Shall Survive?* 57th Cong., 1st sess., S. Doc. No. 137 (1902) (statement of Samuel Gompers). Washington, DC: Government Printing Office. See also John W. Dower, *War Without Mercy: Race and Power in the Pacific* (New York: Pantheon, 1986).

38. Frank H. Canaday to Mariam Canaday, June 2, 1923, Canaday Papers. The Powhattan Club name reflected a larger appropriation of the indigenous origins of tobacco in the Americas. The ATC's telegraph address was "Powhattan," and its letterhead featured a drawing of an Indian in Plains headdress.

39. The American Club rules read, "No ladies are allowed to visit the Club or to be introduced into the Club under any pretext whatever except on special occasions by consent of the Committee." American Club Membership Documents, December 14, 1923, Canaday Papers.

40. Frank H. Canaday to Mariam Canaday, June 2, 1923, Canaday Papers.

41. Frank H. Canaday to Baldwin, September 24, 1922, Canaday Papers. Facts about Canaday are drawn from Baldwin Letter of Recommendation to British American Tobacco Co., November 27, 1922; Frank H. Canaday to Ward Canaday, October 23, 1922, Canaday Papers.

42. Frank H. Canaday to unknown, March 16, 1923, Canaday Papers.

43. Hutchinson, *China Hand*, 12–14; Thomas's memoir was entitled *A Pioneer Tobacco Merchant in the Orient*.

44. Ralph Waldo Emerson said this about joint-stock companies in 1841. Ralph Waldo Emerson, *The Essay on Self Reliance* (East Aurora, NY: Roycrofters, 1908), 14. See also Angel Kwollek-Folland, *Engendering Business: Men and Women in the Corporate Office, 1870–1930*

(Baltimore: Johns Hopkins University Press, 1994), 181–82; Michael Zakim, *Accounting for Capitalism: The World the Clerk Made* (Chicago: University of Chicago Press, 2018).

45. Ruskola, *Legal Orientalism*, 98.

46. Cheng Renjie, "Ying Mei Yan Gongsi Maiban Zheng Bozhao," 743.

47. Sheehan, *Industrial Eden*, 1–15; Ruskola, *Legal Orientalism*, 60–107.

48. Frank H. Canaday to Ward Canaday, May 21, 1923, Canaday Papers.

49. Hutchinson, *China Hand*, 221.

50. Hutchinson, *China Hand*, 37.

51. Hutchinson, *China Hand*, 45, 52–53.

52. Frank H. Canaday, diary, October 18, 1923, Canaday Papers.

53. BAT's and Yongtaihe's procedures on sales trips is described in Frank H. Canaday to C. L. Conrady, October 12, 1923, Canaday Papers; Hutchinson, *China Hand*, 33–53.

54. Dipesh Chakrabarty, *Provincializing Europe: Postcolonial Thought and Historical Difference* (Princeton: Princeton University Press, 2000), 8–10, 65.

55. Frank H. Canaday to Mariam Canaday, October 14, 1923, Canaday Papers.

56. Frank H. Canaday to C. L. Conrady, October 12, 1923, Canaday Papers.

57. Frank H. Canaday to unknown, October 4, 1923, Canaday Papers.

58. Frank H. Canaday to unknown, October 4, 1923, Canaday Papers.

59. Frank H. Canaday to Mariam Canaday, October 14, 1923, Canaday Papers.

60. Frank H. Canaday to Arthur Bassett, October 14, 1923; Frank H. Canaday to C. Y. Yik [Yi], October 14, 1923; Frank H. Canaday to C. L. Conrady, October 12, 1923; unknown (BAT headquarters) to Frank H. Canaday, October 23, 1923, Canaday Papers.

61. Frank H. Canaday to Mariam Canaday, October 14, 1923, Canaday Papers.

62. Frank H. Canaday, small diary, n.d., Canaday Papers.

63. Frank H. Canaday to Mr. Baldwin, November 30, 1923, Canaday Papers.

64. Frank H. Canaday to Mr. Baldwin, August 31, 1924; Frank H. Canaday to Ward Canaday, December 21, 1924; C. L. Conrady to Frank H. Canaday, January 4, 1926; Canaday Papers.

65. Frank H. Canaday to Ward Canaday, February 17, 1925, Canaday Papers.

66. Frank H. Canaday to Mr. Baldwin, August 31, 1924, Canaday Papers.

67. Frank H. Canaday to Mr. Baldwin, February 17, 1925, Canaday Papers.

68. Frank H. Canaday to Ward Canaday, February 17, 1925, Canaday Papers.

69. Emerson, *The Essay on Self Reliance*, 15, 16, 51, 13.

70. Ho, "On the Concept of Face," 867–84; Yang, *Gifts, Favors, and Banquets*, 1994.

71. The law had grandfathered in foreign businesses formed by 1916, but allowed no new ones. In order do business, the subsidiary had to remove the Yongtaihe name from cigarette packages, disband its own superb distribution system and operate through a new Chinese dealer with no foreign ties. Canaday to W. B. Christian, BAT, June 10, 1925; Canaday to Bassett, June 11, 1925, Canaday Papers.

72. Canaday to Bassett, June 16, 1925, Canaday Papers.

73. Excerpt from letter in scrapbook, dated 1925, Canaday Papers.

74. C. L. Conrady to Frank H. Canaday, January 4, 1926, Canaday Papers.

75. Frank H. Canaday to Ward Canaday, n.d. (1926); Ward Canaday to Frank H. Canaday, January 6, 1926; Canaday Papers.

76. Frank H. Canaday to unknown, November 11, 1924, Canaday Papers.

77. Frank H. Canaday to Mariam Canaday, November 23, 1924, Canaday Papers.

78. Frank H. Canaday to Mariam Canaday, November 23, 1924, Canaday Papers.

79. Frank H. Canaday, "Middle Ages, Still, in 'Middle Kingdom,'" unpublished essay, May 1925, Canaday papers.

80. Canaday, "Middle Ages, Still, in 'Middle Kingdom.'"

81. Paul A. Kramer, "Imperial Openings: Civilization, Exemption, and the Geopolitics of Mobility in the History of Chinese Exclusion, 1868–1910," *Journal of the Gilded Age and Progressive Era* 14 (2015): 317–47.

82. Akira Iriye, *After Imperialism: The Search for a New Order in the Far East 1921–1931* ([1965]; New York: Atheneum, 1969), 18–21, 57; Prasenjit Duara, *Sovereignty and Authenticity: Manchukuo and the East Asian Modern* (Lanham: Rowman and Littlefield Publishers Inc., 2003), 10; Jane Burbank and Frederick Cooper, *Empires in World History: Power and the Politics of Difference* (Princeton: Princeton University Press, 2010), 386–89.

83. Fellowship of Reconciliation Conference Program, 1925, James A. Thomas Papers; Nathaniel Peffer, *The White Man's Dilemma: Climax of the Age of Imperialism* (New York: John Day Co. Inc., 1927); James A. Thomas, *A Pioneer Tobacco Merchant in the Orient* (Durham: Duke University Press, 1928); James A. Thomas, *Trailing Trade a Million Miles* (Durham: Duke University Press, 1931).

84. "Nathaniel Peffer of Columbia, Expert on the Far East, Dies," *New York Times*, April 14, 1964, 37.

85. Peffer, *The White Man's Dilemma*, 174.

86. Thomas, *Trailing Trade*, 5.

87. Peffer, *The White Man's Dilemma*, 157, 180.

88. Peffer, *The White Man's Dilemma*, 149–50, 178, 180.

89. Thomas, "Selling and Civilization," 896–99, 948–49.

90. Louise Conrad Young, "Rethinking Empire: Lessons from Imperial and Post-Imperial Japan," in *The Oxford Handbook of the Ends of Empire*, eds. Martin Thomas and Andrew Thompson (Oxford: Oxford University Press, 2017); Paul A. Kramer, *The Blood of Government: Race, Empire, the United States, and the Philippines* (Chapel Hill: University of North Carolina Press, 2006), 382; Duara, *Sovereignty and Authenticity*, 91–92.

91. Thomas, "Selling and Civilization," 949.

92. Thomas, *Trailing Trade*, 4, 5, 54.

93. Thomas, *A Pioneer Tobacco Merchant*, 112–13.

94. Zheng Bozhao to James A. Thomas, November 18, 1926; James A. Thomas to Anderson Motor Company, January 14, 1922, James A. Thomas Papers; Cheng Renjie, "Ying Mei Yan Gongsi Maiban Zheng Bozhao," 749.

95. Thomas, *A Pioneer Tobacco Merchant*, 111.

96. Thomas, *A Pioneer Tobacco Merchant*, 140.

97. James A. Thomas, "Topics for Discussion: James A. Thomas Observes As Follows," James A. Thomas Papers; Thomas, *A Pioneer Tobacco Merchant*, 239; Thomas, *Trailing Trade*, 41.

98. Tani E. Barlow, "History and the Border," *Journal of Women's History* 18, no. 2 (2009): 8–32.

99. Tsing, *The Mushroom at the End of the World*, 113.

100. Tsing, *The Mushroom at the End of the World*, 62.

101. I borrow the concept of contamination from Tsing. She argues the process of "making both humans and nonhumans into resources for investment" has caused us to "imbue both people and things with alienation, that is, the ability to stand alone, as if the entanglements of life did not matter." Tsing argues that "collaboration means working across difference, which leads to contamination." She calls on us to move beyond "fantasies of self-containment" that inhere in the concept of homo economicus. Tsing, *The Mushroom at the End of the World*, 4–5, 28.

102. George Lipsitz, *The Possessive Investment in Whiteness: How White People Profit from Identity Politics* (Philadelphia: Temple University Press, 1998); Tim White, *Colorblind: The Rise of Post-Racial Politics and the Retreat from Racial Equity* (San Francisco: City Lights Publishers, 2010).

Conclusion

1. Interview with Xu Bing, conducted by eChinaArt website, August 24, 2000, New York City. The interview was conducted in Chinese and translated into English. http://www.echinaart.com/interview/xu_bing/XiuBin_interview.htm/, accessed February 15, 2018.

2. World Health Organization factsheet: "Tobacco in China," http://www.wpro.who.int/china/mediacentre/factsheets/tobacco/en/.

3. Teemu Ruskola, "Ted Turner's Clitoridectomy, Etc.: Gendering and Queering the Corporate Person," Berkshire Conference on Women's History, Toronto, May 23, 2014.

4. Thurman W. Arnold, *The Folklore of Capitalism* (New Haven: Yale University Press, 1937), 50–51.

5. Philip J. Stern, "the Ideology of the Imperial Corporation: 'Informal' Empire Revisited," in "Chartering Capitalism: Organizing Markets, States, and Publics," Emily Erikson, ed., special issue, *Political Power and Social Theory* 29 (2015): 15–43.

6. For related discussions of this phenomenon, see Naomi R. Lamoreaux and William J. Novak, eds., *Corporations and American Democracy* (Cambridge: Harvard University Press, 2017), especially the introduction; Jefferson Cowie, *The Great Exception: The New Deal and the Limits of Politics* (Princeton: Princeton University Press, 2016).

7. Robert N. Proctor, *Golden Holocaust: Origins of the Cigarette Catastrophe and the Case for Abolition* (Berkeley: University of California Press, 2011), 541.

8. Quoted in Jessica Glenza, "Big Tobacco Still Sees Big Business in America's Poor," *Guardian*, July 13, 2017, https://www.theguardian.com/world/2017/jul/13/tobacco-industry-america-poor-west-virginia-north-carolina/.

9. "Smoking and Tobacco Use: West Virginia," Centers for Disease Control and Prevention. Data is for 2011. https://www.cdc.gov/tobacco/data_statistics/state_data/state_highlights/2012/states/west_virginia/index.htm/; World Health Organization factsheet: "Tobacco in China," http://www.wpro.who.int/china/mediacentre/factsheets/tobacco/en/.

10. Lauren Fedor and James Fontanella-Khan, "BAT Eyes Resurgent US Tobacco With Reynolds Deal," *Financial Times* (London), January 17, 2017; Jennifer Maloney and Saabira Chaudhuri, "Tobacco Comes Back From the Brink: Booming US Cigarette Makers Shrug Off

Government Regulation, Legal Settlements, and the Decline of Smokers By Boosting Prices," *Wall Street Journal Asia* (Hong Kong), April 25, 2017, A1.

11. "RAI Shareholders Approve Proposals in Connection With Proposed Acquisition by BAT" (press release), Reynolds American Website, July 19, 2017, http://www.reynoldsamerican.com/about-us/press-releases/Press-Release-Details-/2017/RAI-shareholders-approve-proposals-in-connection-with-proposed-acquisition-by-BAT/default.aspx/.

12. "Reynolds American to Cut US Workforce by 10 percent," *Triad Business Journal*, May 14, 2012, https://www.bizjournals.com/triad/news/2012/03/14/reynolds-american-to-cut-us-work-force.html/.

13. "FLOC Speaks Out Against Abuses in the BAT Supply Chain," AFL-CIO Farm Labor Organizing Committee Blog, April 2017, http://www.floc.com/wordpress/floc-speaks-out-against-abuses-in-bat-supply-chain/.

14. Kenneth Rapoza, "China-Like Wages Now Part of U.S. Employment," *Forbes*, August 4, 2017.

15. Abram Chayes, "The Modern Corporation and the Rule of Law," in Edward S. Mason, ed., *The Corporation in Modern Society* (Cambridge: Harvard University Press, 1959), 40. Some readers may raise the movement for corporate social responsibility as a mode for corporate reform. While corporate responsibility might be preferable to corporate irresponsibility, CSR is a self-regulating mechanism that often becomes confounded by its simultaneous public relations function; it is far from a democratic institutional structure.

INDEX

Page numbers in italics refer to figures.

AFL-CIO, 125, 267
AH Motley Tobacco Company, 89–91
Alexander III (Russian czar), 33
Allen, George, 91–92
Allen, John F., 19–21
Allen & Ginter Tobacco Company, 16, 19, 23, 25–27, 30, 32–33, 34, 40–42, 55, 97, 280n13; merger with Duke, 52–53, 60, 66–67, 288n39
Allison cigarette machine, 60, 66
American Club, 205, 231, 319n39
American Tobacco Company (ATC), 4–5, 7, 12–16, 30, 31, 36, 37, 38, 43, 90–91, 96, 307n40; African Americans, 181–84; British American Tobacco Company (BAT), 82–83, 86; chew and pipe tobacco industry, 125–26; fragmentation of, 126, 145, 155, 165–68, 227, 300n23; incorporation, 51–58, 66–78, 84, 122, 157, 285n1, 288n39, 291n104; Lucky Strike, 177–81, 197–202; overseas expansion, 81–83, 119, 123–24, 292n118;

property laws, 78–80; workers, 133, 136–37, 145–50, 181, 273n3
American Trading Company, 43, 283n92
Anargyros, Sotirios, 34, 37, 38
Anderson, Ivy, 219
anti-imperialism, 45, 48–49, 82, 92, 171–77, 248–59
Armstrong, Louis, 215
Arnold, Thurman, 262
Arrington, Kate, 93
Arthur, Chester A., 33
Astor House Hotel, 104–5, 204, 231
Atlanta Daily World, 202–3
Aunt Jemima's, 181
Australia, 4, 7, 12, 16, 30, 42, 65, 81
Austria, 158

baby boomers, x
Bailey, Pearl, 215
Baker, Josephine, 196
Baltimore Afro-American, 210
Barbee, Anne Mack, 137, 147

325

Barnett, Claude A., 181
Baron, Bernhard, 71–72
baseball, 157, 171, 177–83
Bassett, Arthur, 222, 224–28, 241, 247
Bedrossian Brothers, 21
Belgium, 16, 30
Belmany, Mrs. George, 151
Bernays, Edward, 181, 198–201, 310n100
Bismarck, Otto von, 33
Black, Robert, 147
Black Swan Records, 201
Blackwell Tobacco Company, 288n34
Bonsack, James, 59, 64. *See also* Bonsack Machine Company
Bonsack Machine Company, 4, 7, 59–60, 63–66, 72
Boswell, E. V., 149, 151
Boyd, James M., 66
Brady, Pat Foy, 179
Briggs, Howard, 179
bright leaf tobacco, xii, 260, 295n21; cultivation in China, 89–90, 92–101, 107, 119, 295nn22–23; foreign novelty, 24–26, 30, 38, 43–45; network, 1–5, 9–15, 83, 86–121, 133, 140, 145, 152–54, 189, 222, 234, 262–63; nicotine content, 281n35; origins of, xiii, 3, 10–13, 16, 20, 22–25, 31; overseas marketing, 17–18, 24–26, 30, 42–43, 51, 85, 154, 227. *See also* Ginter, Lewis
Britain, xi, xiii, 4, 12–15; British cigarette, 24–26, 30, 34, 39, 42, 49; and Egypt, 17, 31–32; Opium Wars, 14, 81, 278n30
British American Tobacco Company (C): and China, x, xii–xiii, 1–4, 7–9, 14–15, 22, 42–51, 55, 82–124, 133, 140–44, 155–90, 197, 204–67, 317n5; origins and development of, 4, 13, 54, 82, 84; Southern identity, 1–2, 101–8, 112–13, 234; Yongtaihe, 221–59, 266, 320n53, 320n71. *See also* corporate power: multinational
British (English) East India Company, 13, 264, 278n30
Buck, Pearl, 161

Bull Durham, 167, 180, 311n105
burley tobacco, 167–68, 185, 277n20
Burnaby, Frederick Gustavus, 283n84
business partnerships, 3, 70
Butler, William, 70

cabarets, 187–220, 229. *See also* jazz music and dance
Cairo. *See* Egypt; Egyptian tobacco
Caldwell, C. C., 148
Calloway, Cab, 202–3, 215–16
Canada, 12, 81
Canaday, Frank H., 116, 160–62, 173–75, 204–6, 221–59; photography of, 236–46
cancer. *See* smoking: health risks
Canidrome, 191, 196, 208, 210, 316n81
captains of industry, xi
Carolina Times, 215
Carr, C. Stuart, 106
Carreras Tobacco Company, 72
Carter, Benny, 314n57
Carter, Jack, 209–10
Castle, Irene, 195
Castle, Vernon, 195
Cayce, Milton, 66
Centennial Exposition (Phila.), 22–23
Cercle Sportif Français (Shanghai), 192
Chandler, Alfred D., Jr., 6, 55, 58, 262, 275n9
Chaney, Lib, 136
Chayes, Abram, x–xi, 152, 267
Cheatham, Patsy, 137
Chen, H. T., 241
Chen Xuyuan, 176
Chen Zifang, 97–99
chewing tobacco, 2, 4, 10, 21–24, 42, 70–71, 125–26, 166
child labor, 128, 146, 302n44
China, x, 16, 317n5; Beijing, 1–2, 173–75, 206, 231, 247–53; Boxer Rebellion, 82, 104; cigarette marketing, xii, 1, 3–4, 8, 44, 82–85, 154–86, 204–20, 236, 242–47; Confucian, 47–48, 107, 233, 247–48; courtesans, 45–48, 115–16, 284n105, 284n109; factory workers, 120–40;

farmers, 1, 96–102, 236, *237*; May 4th Movement, 251; May 30th Movement, 50, 143, 156–57, 171–77, 232, 248–59; rickshaws, 211–13, 284n109; servants, x, 5, 9, 14, 106–15, 250, 264, 316n89; treaties, 81, 92, 251, 278n32; US southerners, 86–88, 94–96, 99–119; Western products in, 18, 42–50, 194, 204–20, 223, 236–51, 266. *See also* British American Tobacco Company (BAT); Shanghai

China Times, 253

Chinese Exclusion Act. *See* US Chinese Exclusion Act(s)

cigar. *See* smoking: cigar

cigarette advertisements, 26, 28–37, 160–86, 227, 236, 242–47, 308n43, 313n23; images of, *29*, *32–33*, *37*, *44*, *161*, *163*, *165*, *169*, *171*, *246*

cigarette brands, xiii, 3, 42; Black Cat, 72; Camel, 3, 34, 153–88, 197, 202, 314n57; Chesterfields, 137, 156, 170, 197; Chinese, 42–43, 155–86, 244; early brands, 16–18, 30, 34–35, 37, 167–69; Golden Dragon, 165; Hatamen, 172–76; Kool (Raleigh), 197; Lucky Strike, 126, 128–29, 136–39, 147, 156, 170–71, 177–88, 197–203, 216, 314n57, 315n60; modernity, 3, 9, 155, 236–45; My Dear, 165; Old Gold, 156, 170, 197; as property, 79–80; Purple Mountain, 156, *158*; Richmond Gem, 16, 24–26, 28–30; Richmond Straight Cuts, 32, 34; rise of big brands, 155–56; Ruby Queen/Da Ying, 3, 7, 18, 42, 49–50, 117, 153–86, 220–21, 225–26, 241, 244; Vanity Fair, 225–26; Victory, 244

cigarette cards, 28–30, *29*, 309n83

cigarette machine(s), 5, 36, 59–65, 72, 86, 120–22, 128, 131–33, 224, 256; and James Duke, 6–9, 53–65, 83, 286n15, 287n25, 287n27. *See also* Bonsack Machine Company; cigarette making: mechanization

cigarette making: factories, 36, 43, 60–64, 120–53, 267; hand-rolled, 50–51, 58–65, 287n28, 288n34, 311n105; images of, *132*, *134*; mechanization, 4–9, 36, 50–51, 55, 58–66, 120, 286n15, 287n25, 287n27

cigarette terminology, 280n17

Citizens United, 265

Civil Rights Act (1964), 78, 128, 265

Civil War. *See* US Civil War

Clark, Ernest "Slick," 210

Clayton, Buck, 209–12, *213*, 316n89

Cobbs, Thomas, 225

Cochran, Sherman, 101

Columbia Records, 201

Communism, 143, 173, 177, 252, 266

Compradore Ou, 97–99

Conrady, C. L., 230–31

Consolidated Tobacco Company, 292n118

corporate power, ix–xiii, 273n2, 323n15; antimonopoly sentiment, 69–77, 265; corporate personhood, 54–55, 75–78, 80, 83–84, 273n2, 291n104; incorporation laws, 52–54, 55–79, 84, 291n104; multinational, x, 1–4, 14, 54, 82–85, 104–5, 123, 145, 251; race or gender, x, 2, 4–5, 10–12, 77–78, 88, 107–19, 112, 122–53, 182, 221–59, 262–65; railroads, 68–69, 77

Cosmopolitan Magazine, 206–7

Cotton Club, 196, 201–3

Coundouris, Nicholas, 21

Cox, Laura, 63

creative destruction. *See* Schumpeter, Joseph

Crow, Carl, 160

Cuba, 13, 65, 81

Cushing, Caleb, 81, 278n32

Da Ying. *See* cigarette brands: Ruby Queen/Da Ying

dance. *See* jazz music and dance

Daniel, Joseph, 222

Daniels, Jonathan, 137

Davis, Ruth, 216–18

Davis, William, 136, 182–83, *184*, 216–18

Delancy, Ruby, 120–21, 128–30, 136, 139
democracy, 124, 265, 267, 323n15
Democratic Party, 12
Department of Labor (US), 135
Dong-Feng Company, 208–11, 315n78, 316n81
Dubec cigarette, 23
Ducreux, Joseph, 28, 33
Duke, James B., 4, 11, 30–31, 34, 70, 157, 260–62, 290n81; American Tobacco Company, 16, 51–54, 58, 67–72, 80–83, 86–87, 127, 165, 167, 292n118; British American Tobacco Company, 54, 82–85, 124, 165; as cigar smoker, 41–42; corporate monopoly, 51–53, 68–72, 91; earliest cigarette production, 42–43, 58, 61–66, 286n15, 287n25; as entrepreneur, 5–8, 16, 52–53, 58–59, 64, 83, 262, 275n9; image of, 62; overseas expansion, 124–25, 261, 276n14; overseas marketing, 42, 81–83; personality of, 5, 262. *See also* cigarette machine(s): and James Duke
Duke, Washington, 30, 63, 103
Duke's Mixture, 167
Duke University (Trinity College), 2, 102–3
Durante, Nicandro, 266
Durham, Eddie, 215
Dutch East India Company, 13

Edwards, Paul K., 181
Egypt, 8, 17–18, 22, 43, 60, 63, 123, 282n70
Egyptian cotton, 31
Egyptian tobacco, 15, 17–19, 23–24, 38–39, 50, 70, 125, 282n55; and Egyptomania, 31–36, 39
Einstein, Benjamin, 74
Ellington, Duke, 196, 202–3, 212, 215, 216–19
Ellison, Ralph, 218
Emerson, Ralph Waldo, 247, 319n44
Emery, Charles, 59
Emery, William, 59
Emery cigarette machine, 66

entrepreneur myth, 5–9. *See also* Duke, James B.
environmental history, 275n6
Europe, James Reese, 195
Evans, E. Lewis, 148–49, 151
extraterritoriality, 14–15, 81, 264, 278n32

F. S. Kinney, 21, 23, 30, 52, 60–63, 70–71, 290n83
Fang Xiantang, 42
Farthing, Evelyn, 136
Fatima, 168
Fellowship of Reconciliation, 253
Finnane, Antonia, 46
Fiske, CE, 43
Flood, Henry D., 67
flue curing system, 1, 12, 97, 99
Forum, 198
Fourteenth Amendment, 3–4, 53, 76–78, 291n104
France, 14, 16, 21, 29, 158, 188
Fu Manchu, 104
Fuller, W. W., 72

gender and sexuality, x, 41, 47, 51, 233–39, 247, 263–64, 286n12. *See also* smoking: gender
gentlemen's clubs, 16, 24–30, 39–40, 193, 229, 231, 319nn38–39
Germany, 4, 14, 16, 30–31, 81, 251, 280n14
Gianaclis, Nestor, 17, 34, 36, 38
Ginter, Lewis, 276n13, 280n11, 290n81; bright leaf innovations, 15–17, 22–24; business background, 19–25, 84; cigarette machine, 59–64; and James Duke, 4, 8, 16, 43, 50–52, 55–59, 63, 66–70, 275n9, 288n39; image of, 56; monopoly, 68–71; overseas marketing, 3, 7, 16–17, 23–36, 39, 67, 81, 282n51, 288n39; personal life, 40–41, 55–57, 71, 84
Gladstone, William, 33
Gompers, Samuel, 125
Goodman, Benny, 202, 202, 314n57
Goodson, J. W., 102

Goodwin Tobacco Company, 52, 59, 63
Gordon, Charles George, 32
Great Britain. *See* Britain
Great Depression, 198, 214
Great Migration (African Americans), 214
Greece, 191
Greek cigarette producers, 21, 31, 35–36
Gregory, Hattie (Arrington), 88, 96–97, 105–18, 157, 212
Gregory, Jane, 107, 110–11, 114
Gregory, John, 110
Gregory, Richard Henry, 1, 88–105, 107, 113, 117–18, 157, 204, 255; images of, 93, 94
Grey, Stanley, 205
Griffin, Pat, 179
Grossman, Hyman, 61
Guam, 13

Hamilton House, 208–9
Hancock, Theodore, 74
Hardy, Marion, 203
Harris, Martha Gena, 146
Harris, William R., 124
Harvard University, 221, 232
Hawai'i, 13, 54, 81
Henderson, Fletcher, 314n57
Herrod, William, 151
Hewitt, A. M., 34
Hill, George, 146–47, 181, 198–99
Hill, Percival, 168
Hoggard, Sherman, 179
Holiday, Billie, 215
Hong'an Real Estate Corporation, 295n39
Huacheng Tobacco Company, 156, 165, 173
Huang Su'e, 112
Huang Yicong, 222, 224, 241
Huang Zhihao, 142
Hu Baoyu, 47
Hudson Bay Company, 13
Huff, Will, 149
Hughes, Langston, 211–12
human rights, 267
Hutchinson, James, 47, 102–5, 116, 160–62, 191, 197, 204, 232, 236

immigrant import/immigration, x, 24, 36, 39, 60–62, 70, 127, 188, 255, 280n14
Imperial Tobacco Company, 4, 13, 43, 82, 227
imperialism: assemblage theory, 275n6, 278n31, 311n2
India, 13, 16, 18, 31, 39, 42, 82, 106, 255–56
Italy, 158

Japan, xii, 4, 13–14, 18, 50, 81–82, 123–24, 158, 172, 188, 190–93, 251
jazz music and dance, x, xiii, 7, 9, 186–220, 314n57
Jeffress, A. G., 97, 102
Jeffress, Thomas F., 66
Jewish cigarette producers, 36, 61–63
Jian Zhaonan, 101
Jim Crow segregation, xii, 277n24; corporate racial understandings, 4–5, 10–12, 15, 88, 112, 118, 262–65; in factories, 122–40, 145–47, 151–53, 183, 189–90; isolation, 3, 214–19; jazz, 189–90, 214–19; laws, 78; translated in China, 87–88, 97, 100, 106–40, 210, 263. *See also* race; segregation
Johnson, Happy, 209–10
joint-stock companies, 13, 256
Jones, Esther, 149
Jones, Reginald, 210
Jones, Ruby, 135, 146
Joyner, James, 144, 206–8

Kinney Brothers Tobacco Company, 21, 23, 30, 52, 60–63, 70–71, 290n83
kinship networks, 2, 233–34, 256
Koo, V. K. Wellington, 249
Korea, 54, 81, 188, 191
Ku Klux Klan, 11, 149–50, 152, 304n99

Leary, Herbert "Rabbit," 178
Li Xinbao, 142
Liggett and Myers, 70, 90–91, 122, 137, 146, 156, 167–70, 290n81
Liu Lilin, 206–8

Locke, Alain, 183
Loews Inc., 201
London, 3, 7, 16–17, 20, 23–30, 32, 39–41, 65. *See also* Britain
Lone Jack Tobacco Company, 64–65
Lorillard, 122, 156, 166–67
Lunceford, Jimmie, 215

Ma Wenyuan, 142
MacKay Twins, 210–11
Mao Zedong, 42, 266, 303n77
Marsalis, Wynton, 217
McCoy, Frank, 73
Melachrinos, Miltiades, 17
Messenger, The, 196
Miller, Dora Scott, 146–47
missionaries, 45, 112, 163, 253
modernity. *See* China: Western products in; mythos of modernity; smoking: idea of modernity
Morris, William, 222
Motley, A. H. *See* AH Motley Tobacco Company
Mu Guilan, 142
Mu Shiying, 193
Murai Brothers Tobacco Company, 13–14, 43, 81–82, 123–24
Musical Digest, 200
mythos of modernity, 8–9, 54–55, 224, 232–33, 236–39, 257–59

Naegle, Walter, 286n12
Nance, Ray "Floorshow," 215
Nanyang Brothers Tobacco Company, 101, 131, 152, 156, 165, 317n5
national belonging, x, 18, 129–33, 188
National Broadcasting Company (NBC), 201–2
National Cigarette and Tobacco Company (NCTC), 71–74, 84, 291n86
National Labor Relations Act (Wagner Act) (1935), 145, 150–53
Neal, James, 150, 215, 217

New Deal, 265
New Jersey, 4, 52–53, 67–77
New York (state), 70–76
New York City: corporate headquarters, 3, 12, 19–21, 61–62, 67, 71, 92, 96; James Duke, 67, 71, 126; factories, 122–23, 138, 166; Ginter, 19–21, 61, 67; immigration/imports, 24, 35–36, 38, 42, 45, 61–62; radio shows, 201, 216; segregation, 122, 196
New York Syncopators, 209
New York Times, 38, 40–41
New York Tribune (magazine), 253
New Zealand, 7, 42, 81
North Carolina: bright leaf, xii, 10–11, 16, 20–22, 26, 85–87, 91–92, 96; business laws, 67–68, 79, 99; China connection, 1–2, 47, 83, 93–122, 125, 260–67; cigarette mechanization, 59, 62; factories, 123–28, 132, 134, 135, 141, 144–47, 153, 157, 265; jazz, 214–16; Lucky Strike baseball, 157, 171–77, 182. *See also* bright leaf tobacco

Olcott, William M. K., 76
Old Smoker character, 33
Ole Virginny Cigarette and Tobacco Stores, 25–26, 280n29
Omar, 168
opera, 196
orientalism, 31, 34–36, 48, 50, 54–55, 282n55
Ottoman Empire, 17, 21, 31–36, 39

Palmer, Dr. William P., 28
Paris, 23–24, 196
Paris Peace Conference (1919), 249, 251
Parker, Lee, 1, 8–11, 15, 47, 102–3, 105, 116
Parrish, Edward J., 81, 124, 300n11
Pater, Walter, 30
Payne, Oliver, 290n81
Peffer, Nathaniel, 253–55, 258
Penn, Charles, 126–27, 301n24

Perry, Hap, 179
Philippines, 13, 18, 42, 45, 49, 54, 81, 83, 190–91, 209, 255–56, 264
Pillsbury, 155
Polo, Marco, 104
Pope, John, 19, 40–41, 55–58, 66–67, 71, 84, 290n81; image of, 57
Porter, Thelma, 316n81, 316n89
postcolonial, 55
Pott, Francis Lister Hawks, 112
Powell, Radford, 149–51
Powhattan Club, 231, 319n38
Pratt, Bertie, 146
Prince Albert tobacco, 167–70, 180, 307n40
print culture, 9, 26, 181. *See also* cigarette advertisements
Proctor, Robert, 266
Prussian prince, 33
Puerto Rico, 13, 81

Qian Meifeng, 120–21, 129, 141

race: African Americans, 10–12, 78, 87, 128–29, 131–33, 145–53, 157, 177–86, 262–68, 315n60; China, 15, 48, 78, 107–24, 145; cigarette marketing, 18, 26, 157, 177–86, 188, 193–268; corporate privilege, 2, 4, 11–12, 87–88, 102, 145, 262; cross-racial encounters, 221–59, 264. *See also* Jim Crow segregation
radio shows, 196–220, 313n33, 314n57, 315n60
Reed, Alfred, 75
Reynolds, R. J., 11, 34, 122, 127, 146–47, 156, 166–70, 180, 300n23. *See also* Reynolds American
Reynolds American, 266–67. *See also* Reynolds, R. J.
Richards, John M., 24–25, 28–30
Riddick, Ivy, 1, 121, 141, 303n72
Rockefeller, John D., 38
Rolfe, B. A., 197, 199
Ross, E. A., 125

Russia, 18, 21, 188, 190–91, 280n14
Rustin, Bayard, 286n12

Samoa, 13, 54, 81
Saturday Evening Post, 169–70
Savoy, 196
Schooler, Samuel, 22
Schumpeter, Joseph, 6, 22, 58–59, 84, 262
segregation, x, xiii, 177–80; business culture, 115–17; in China, 107–17, 122, 136–37, 145, 152–53, 190, 228–29, 254–56; *Plessy v. Ferguson* (1898), 78. *See also* Jim Crow segregation
sex workers, x, 9, 205, 264, 284n105
Shafer, John C., 20, 28
Shanghai: anti-imperialism, 49, 92, 174–75; cigarette factories, 1–2, 43, 83, 86, 120–42, 172–73; cigarette marketing, 42–43, 44, 45–50, 160, 165; corporate headquarters, 14–15, 46, 140, 177, 224–30; culture shock, 2, 15, 46–47, 88, 104–5, 231, 235; entertainment, 17, 46–48, 186–216, 229, 313n33; gangs, 141; racism, 210–14, 221, 229, 231, 251–52; US Southern culture, 97, 106–19, 157, 231
sharecropping, 11, 69–70, 88, 98–100, 295n36
shareholder/stockholder, ix–x, 267
Sherman Anti-Trust Act, 70
Siam, 81
Siegel, David and J. M., 288n34
Sino-Japanese War, 14, 82, 124
Slade, Abisha, 12
slavery, 11–12, 26, 69–70, 102, 108, 114, 131
Smith, Fred Porter, 45
Smith, Irwin, 103, 113, 190, 204
Smith, Whitey, 209
Smoker's Guide, Philosopher and Friend, The, 39
smoking: cigar, 2, 23, 26–27, 41, 61, 63, 125, 205, 280n17; gender, 26–30, 36, 39–40, 47–48, 219–20, 266, 283n84; health risks, xi–xii, 68, 260–61, 265; idea of modernity, 8, 18, 45–46, 189, 194, 220,

smoking (*continued*)
 232–33, 236–47; leisure and ritual, xi–xii, 23, 25–29, 36, 47, 156, 245; photographs of, *242, 243, 245*; pipe, 2, 4, 10–11, 21–24, 26–27, 45–48, 70, 125, 167, 193, 301n105; urbanization, 2–3, 189; women, 219–20, 266, 313n23
Snow, Marion, 217
Snow, Valeida, 209
South (US). *See* US South
South Africa, 31, 42
South Carolina, 145
Spain, 158
Staley, S. James, 208–9
Standard Oil Company, 70
Stephens, John, 11
stock market, 3–4, 53, 78–80
Stockton, John, 73–74
strikes, 2, 143–45, 171–75, 252
Strouse, D. B., 65
Sun Yat-sen, 163
Switzerland, 30

Tang Foshu, 207
tariffs, 123–24
taxi dancers. *See* cabarets
Thomas, James A., 232, 300n18, 315n67, 318n20; ATC, 13, 66, 89–91; BAT-China, 2, 83, 88–97, 101–4, 122, 155–58, 197, 223–27, 253–59; Chinese labor, 125, 253–59; early life, 22–23, 89–90; image of, *90*; Southerner in China, 112–13
Thompson, John R., 28
Tian Junchuan, 99
Tilley, Nannie, 146
Tobacco, 39, 59
tobacco history, 6, 300n23
Tobacco Workers International Union (TWIU), 147–48, 153
Tonga, 81
transnational, xiii, 8, 18–19, 86, 105–18, 120–53, 266, 316n79. *See also* corporate power: multinational

Treaty of Wanghia, 278n32
Trinity College. *See* Duke University (Trinity College)
Turkish tobacco, 8, 17–18, 21–24, 31–34, 37, 38–39, 158, 166, 168, 185, 311n105

Umberto I (Italy), 33
Uncle Ben's, 155, 181
Union Commercial Tobacco Company, 158, 292n118
unions, ix, 125, 143–53, 265, 267, 302n58, 305n99
University of Chicago, 253
US Chinese Exclusion Act(s), 8, 17, 48–49, 61, 78, 92, 176, 231, 234, 250, 300n18; and *Yick Wo v. Hopkins* (1886), 78
US Civil War, 10–11, 19–22, 28, 31, 69, 87, 89
US Department of Labor, 135
US South, x, xii; cigarette factories, 120–27; connection to China, 86–88, 102–22, 154, 260–63; rural, 1–3, 10–11, 28; small towns, 127, 139, 190, 214–16. *See also* Jim Crow segregation; North Carolina; Virginia

Variety, 196
Victor Records, 201
Victorians, 35, 195
Virginia: bright leaf, xii, 1–3, 10–11, 15–16, 19–22, 34, 83–87, 96, 262; British colony, 26; business charters/laws, 52–59, 66–69, 84, 99; China, 85, 101, 105, 113, 231; cigarette mechanization, 59, 65; factories, 125, 265; "Virginia cigarette," 34, 38; and West Virginia, 266

W. D. & H. O. Wills, 30, 34, 43, 64, 96
W. Duke and Sons, 42, 55, 63–64, 288n34, 288n39
W. S. Kimball Company, 52
Wagner Act (National Labor Relations Act) (1935), 145, 150–53
Wake Forest College, 1–2

Wanghia, Treaty of, 278n32
Wang Yanzi, 97–99
War of 1898, 13, 81
Warnow, Mark, 314n57
Washington Naval Conference (1921–22), 251, 253
Waters, Ethel, 219
Watt, R. L., 125
Weatherford, Teddy, 209–12, 316n81
Weber, Max, 54–55, 77
West, Irene, 210
Western Union, 201
West Virginia, 266
Whiteman, Paul, 193
Wiebe, Robert H., 55
Wilde, Oscar, 27, 30, 39–41
Williams, Mary Lou, 314n57
Williamson, Alice, 148
Wilson, Woodrow, 223, 251–52
Wolseley, Garnet, 32
women: Chinese, 47, 206–8; and domesticity, 35, 107–15; in factories, 60–63, 120–42, 145–48, 301n39; idea of modernity and, 194, 219–20; rights of, 77–78. *See also* smoking: women
World War I, 17, 25, 122, 185, 190, 208, 227, 232, 251, 255, 311n105

World War II, xii, 6, 146, 258
World's Fair Columbian Exposition (Chicago, 1893), 35
Worsham, Barry, 179
Wright, Richard H., 42, 65, 275n9
Wu Tingsheng, 96, 155, 157, 222

Xu Bing, 260–61, 265–66

Yi, C. Y., 241
Yongtaihe. *See* British American Tobacco Company (BAT): Yongtaihe
Your Hit Parade, 200
Yu Huixian, 193
Yuan Yecun, 193

Zhang Jiatuo, 102
Zhang Tao, 45
Zhang Yongsheng, 140
Zhao Qizhang, 142–43
Zheng Bozhao: BAT, 15, 49, 96, 116–18, 157–77, 186, 220–41, 256–57; image of, *159*; marketing, 7–8, 17, 43, 45, 155–77, 186, 225
Zheng Guanzhu, 221–22, 224, 228–30, 233
Zubelda, 168

Printed and bound by CPI Group (UK) Ltd, Croydon, CR0 4YY
09/06/2025

14685686-0003